DATE DUE 7/17/94

AUG 1 1 1994	
JUN 1 0 1995	
JAN 2 1 1997	MAR 0 7 2006
JUN 2 4 1997	AUG 1 9 2009
JUL 1 4 1997	JAN 2 3 2012
JUL 2 8 1997	FEB 0 1 2012
FEB 1 7 1998	OCT 3 1 2012
MAY 2 6 1998	
FEB 1 4 2000	
SEP 1 3 2000	
NOV 0 1 2003	
FEB 1 5 2006	

The Library Store #47-0107

BASIC BLUEPRINT READING

JOHN A. NELSON

TAB BOOKS
Blue Ridge Summit, PA

FIRST EDITION
FIRST PRINTING

Library of Congress Cataloging-in-Publication Data

Nelson, John A., 1935–
Basic blueprint reading / by John A. Nelson.
p. cm/
ISBN 0-8306-3273-5 (pbk.) ISBN 0-8306-4273-0
1. Blueprints. I. Title.
t379.N36 1990
604.2'5--dc20 89-20287
CIP

Questions regarding the content of this book should be addressed to:

Reader Inquiry Branch
TAB BOOKS
Blue Ridge Summit, PA 17294-0214

Acquisitions Editor: Kim Tabor
Book Editor: Joanne M. Slike
Production: Katherine Brown

Contents

Acknowledgments v

Introduction vi

1 Title Block Study 1

Folding copies, 1 • Title block, 1 • Zoning, 9 • Microfilm arrowheads, 11

2 Rounding Off Decimals 13

Converting (fractions, decimals, metric), 13 • Reading the fractional-inch scale, 14 • Reading the decimal-inch scale, 16 • Reading the metric scale, 17 • Metric drawings, 17

3 Alphabet of Lines 21

Line thickness, 21 • Kinds of Lines, 21

4 One-View Drawings 31

Dimensioning systems, 31 • Kinds of dimensions, 31 • Dimensioning a diameter, 33 • Dimensioning a radius, 33 • Out-of-scale dimensions, 34 • Reference dimensions, 35 • Typical dimensions, 36 • One-view drawings, 37 • Dimensioning/Locating holes, 42

5 Multiview Drawings 47

Two-view drawings (front- and right-side view), 47 • Projections lines, 47 • Two-view drawings (front and top view), 52 • Hole call-offs, 53 • Three-view drawings, 58

6 Section Views 71

Cutting plane lines, 71 • Section lines, 71 • Kinds of section views, 72 • Conventional practices, 98

7 Auxiliary Views 111

Classification, 112 • Partial auxiliary view, 112

8 Threads and Fasteners 123

Classification, 123 • Tap and die, 123 • Right- and left-hand threads, 123 • Thread terms, 126 • Screw thread forms, 128 • Threads per inch, 129 • Pitch, 129 • Single and multiple threads, 130 • Screw representation, 130 • Thread relief or undercut, 132 • Chamfer, 134 • Thread call-offs, 134 • Screws and rivets, 134

9 Technical Information 141
Knurling, 141 • Surface finishes, 141 • Tabulated drawings, 144 • Dadum planes, 150

10 Metallurgy 155
Ferrous Metals, 155 • Nonferrous metals, 158 • Characteristics of metals and alloys, 160 • Heat treatment, 161 • Hardness testing, 162

11 Basic Welding 167
Welding methods, 167 • Basic welding symbol, 167 • Fillet weld, 168 • Process reference, 171 • Field welds, 173 • Welding joints, 174 • Types of groove weld joints, 176 • Contour symbol, 176 • Plug or slot weld, 176 • Flange weld, 179 • Multiple reference lines, 180 • Spot weld, 180 • Contour and finish symbols, 182 • Seam weld, 183

12 Geometric Tolerancing 189
Tolerancing, 189 • Kinds of fits, 191 • Geometric tolerancing symbols, 192 • Geometric characteristic symbols, 192 • When geometric dimensioning is required, 207

13 Kinds of Drawings 217
Design layout drawings, 217 • Assembly drawings, 217 • Subassembly drawings, 219 • Detail drawings, 220 • Purchased parts, 220 • Revision of drawings, 221 • Parts list, 222

A Worksheet Answers 241

B Conversion Chart 261

C Weights of Materials 263

Index 264

Acknowledgments

First off, I'd like to thank Walter A. Ryan, professor at New Hampshire Technical College, for his technical expertise and thorough review of the material in this book.

Thanks also to the editorial and production staff at TAB Books who put in their time, energy, and creativity. In particular: to Acquisitions Editor Kim Tabor, who signed the book and had confidence in it from the start, to Book Editor Joanne Slike, who designed and edited the book and saw the project through to its completion, Toya Warner, Page Make-up, who laid out the book and contributed to the overall design, and Proofreaders Joan Wieland and Linda King, who carefully checked and rechecked the galleys and mechanicals.

The efforts of everyone involved are greatly appreciated.

Introduction

In today's high-tech world, *everyone* should be able to read and understand the drawings used in industry. Industrial drawings are used in almost everyone's lives—for employment opportunities, for job promotion and satisfaction, for personal growth and knowledge. Learning about industrial drawings can even help you read and understand the instructions and plans when assembling a child's toy at Christmas.

Although this book is slanted somewhat toward the mechanical field of blueprint reading—those drawings used in industry—these "practices" from the mechanical field are also used in architectural drawings. Most people at some time or other build or remodel a home, so the ability to read and understand building drawings is very important.

This text can be used for both self-study and class use. The material is presented in a logical, step-by-step sequence. Material studied is directly followed by professional, industrial-type drawings that reinforce the material just presented.

Material in this text is broken down into the following areas:

1. **Objective of chapter**—The objective is a very simple and brief statement that explains what you will learn in the chapter. It is a good idea to know what you will be learning *before* you begin.
2. **Lecture pages**—These pages explain one or more drafting practices or methods.
3. **Worksheet pages**—These pages are professional, industrial-type drawings. Various questions will be asked about each drawing, using the material just covered in the lecture pages. This will give the reader the opportunity to apply and use the material just studied.

Answer keys for all worksheets can be found, in consecutive order, in Appendix A at the end of the book. It's best to check each answer as you proceed through the text.

Title Block Study

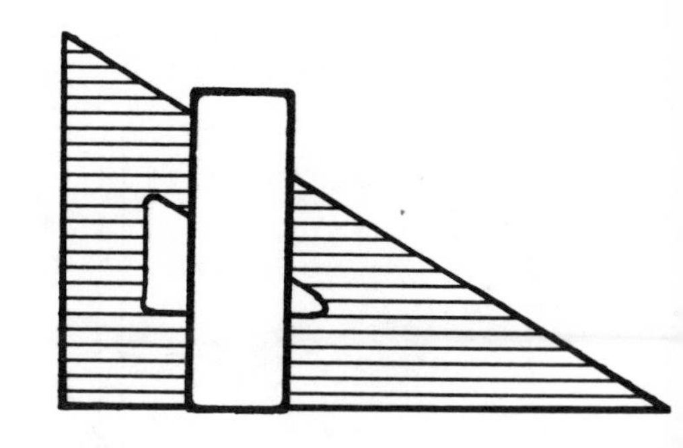

Objective: The reader will be able to identify the various paper sizes used in industry, be able to fold paper per the accepted standard, and know all material called off a standard title block.

Drawing paper, vellum, linen, or drawing plastic films are available in single sheets, pads, or rolls. Standard-size paper and borders are illustrated in FIG. 1-1. Drawings are "called off" according to their size. For example, a sheet of paper that measures $8^1/_2 \times 11$ inches is referred to as an "A" size, a sheet that measures 11×17 inches is called a "B" size, and so on. Drawing paper that comes in rolls is called *roll stock* and is designated by an "R" call-off.

FOLDING COPIES

Any standard-size drawing, regardless of its original overall size, is designed to be folded and filed into a standard file drawer. In order to fit and store neatly in a standard file drawer, *all* standard drawings fold down to an "A," $8^1/_2$-×-11-inch size. Thus, when a "B," "C," "D," or any other standard-size drawing is folded, it will be $8^1/_2 \times 11$ inches in overall size. FIGURE 1-2 illustrates how the paper is usually folded. Note the title block and drawing number is always left showing for easy identification, filing, and retrieving.

TITLE BLOCK

The title block provides supplementary information about the part, subassembly, or assembly. It provides drawing identification of the drawing and a means to file and retrieve the drawing when needed. The standard horizontal title block is located in the lower right-hand corner of the drawing (refer to FIG. 1-3).

Title blocks vary widely from company to company, but this example conforms to the latest ANSI Y14 standards (American National Standards Institute) and will be used throughout this book. Some companies use this title block, some use various elements of it, and some have not converted to the new standard at all. All blocks found in the title block example shown in FIG. 1-3 will be fully explained so that you will properly understand and interpret the given information.

As a rule, the same overall information is found in all title blocks regardless of their overall format; therefore, interpreting other title blocks should not be a problem. Referring to the circled numbers in FIG. 1-3, study each of the 18 items listed below. Be sure you understand what it is, why it is included, and how it should be used.

1. **Company name and address**—The company name and address, along with the company logo, is found in this block. If the company has several divi-

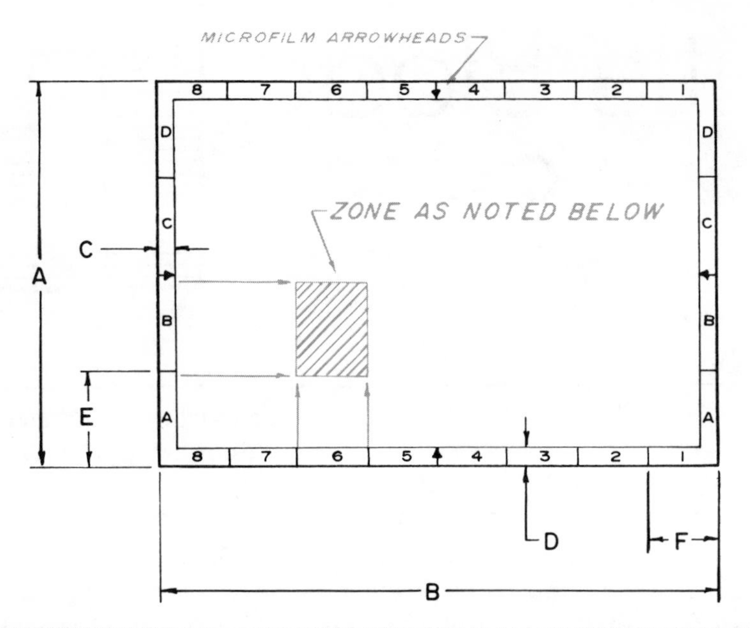

DRAWING SIZE	A	B	C	D	E	F
A HORIZONTAL	8.5	11.0	.25	.38	2 AT 4.25	2 AT 5.50
A4	210	297	10	10	4 AT 52.5	4 AT 74.3
A VERTICAL	11.0	8.5	.38	.25	2 AT 5.50	2 AT 4.25
A3	297	210	10	10	4 AT 74.3	4 AT 52.5
B "	11.0	17.0	.62	.38	4 AT 2.75	4 AT 4.25
A2	297	420	10	10	4 AT 74.3	6 AT 70
C "	17.0	22.0	.50	.75	4 AT 4.25	4 AT 5.50
A1	420	594	10	10	6 AT 70	8 AT 74.3
D "	22.0	34.0	1.00	.50	4 AT 5.50	8 AT 4.25
A0	594	841	20	20	8 AT 74.3	12 AT 70

NUMBERS IN GRAY ARE IN MILLIMETERS

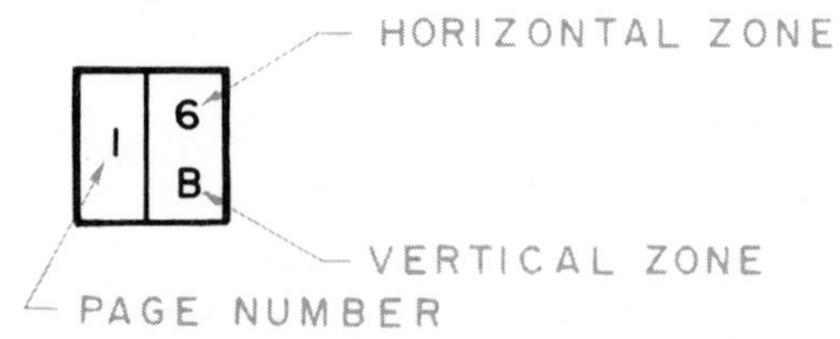

Fig. 1-1. Standard drawing paper, vellum, linen, and/or film sizes. Zoning is a specific area on the drawing.

sions, the division name is usually included also, along with any other identification information.

2. **Drawing number**—The drawing number is usually the same number as the part number. Some companies use a coded part or drawing number in order to indicate a particular product model, a particular kind of a part, or what kind of a drawing.

3. **Drawing title**—The drawing title usually is a very brief description of the part or tells what the part does. It begins with the name of the part or assembly followed by a descriptive modifier.

   Example:

   Bracket-Support—The name "bracket" tells that the part is a bracket of some kind. The modifier, "support," indicates that it is the support for the

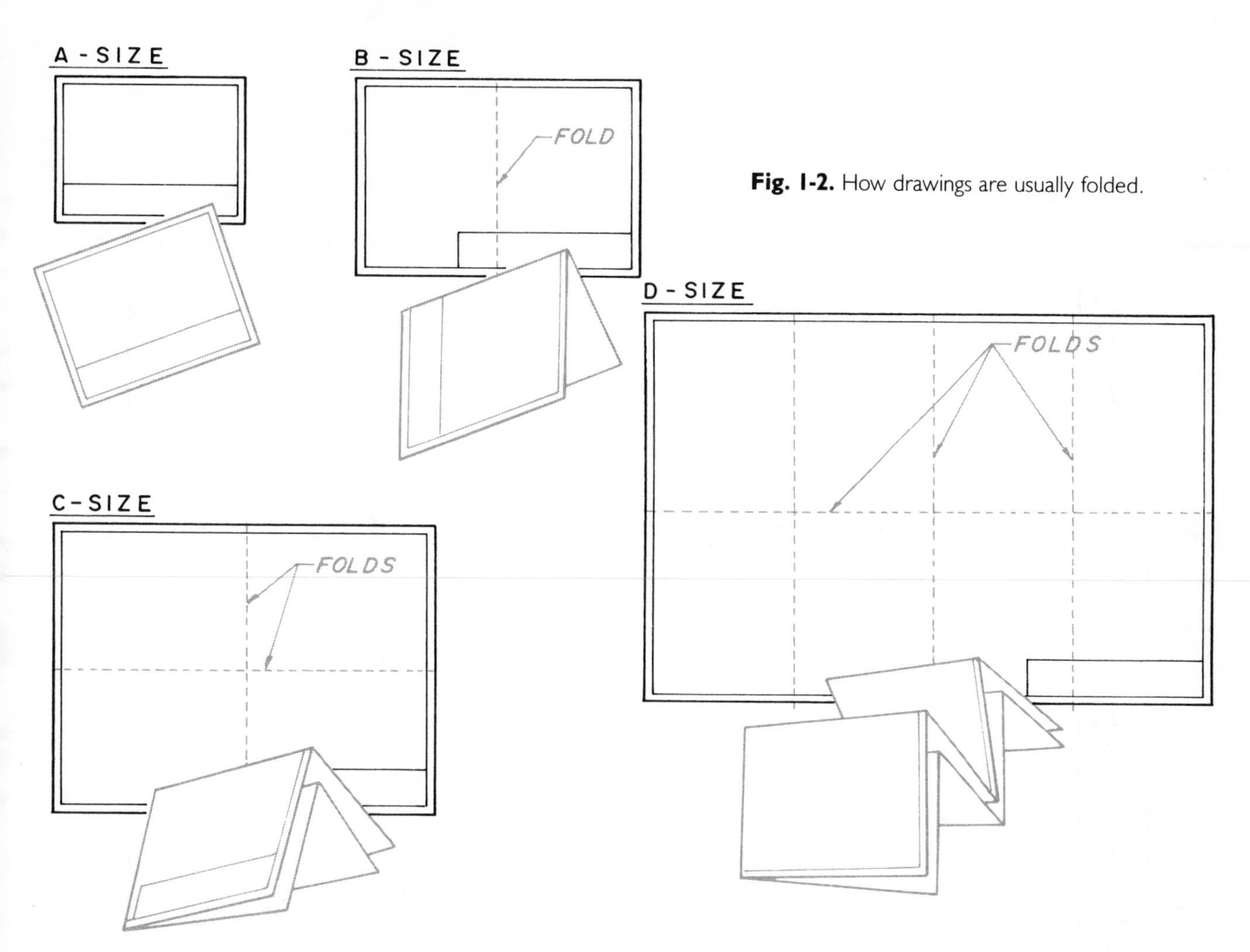

Fig. 1-2. How drawings are usually folded.

bracket. When reading or referring to this part's name, its modifier is read first, followed by its name; thus, this part is called off or referred to as a "support bracket."

4. **Drawing size**—The size of the paper is indicated by a large letter printed within the "size" block. Refer back to paper sizes (FIG. 1-1). A particular drawing is referred to by its physical *size*, followed by its drawing *number*.

   Example:

   A drawing with a number of A-80635

   The "A" indicates that drawing number 80635 is drawn on an "A" size paper.

5. **Sheet number block**—This block is used to indicate consecutive order, and the total number of sheets used for a part subassembly or assembly—all with the same drawing number.

   Example:

   PAGE 3 of 4—This is page number 3 and there are 4 total pages associated with this set of drawings with the same number. If only one page is used, it is called off as PAGE 1 OF 1.

6. **Scale**—The scale block indicates what scale is used, or proportion to the actual size the drawing is drawn to. Very large parts must be "scaled" down to fit on a sheet of paper; very small parts are drawn much larger than actual size in order to fully illustrate them.

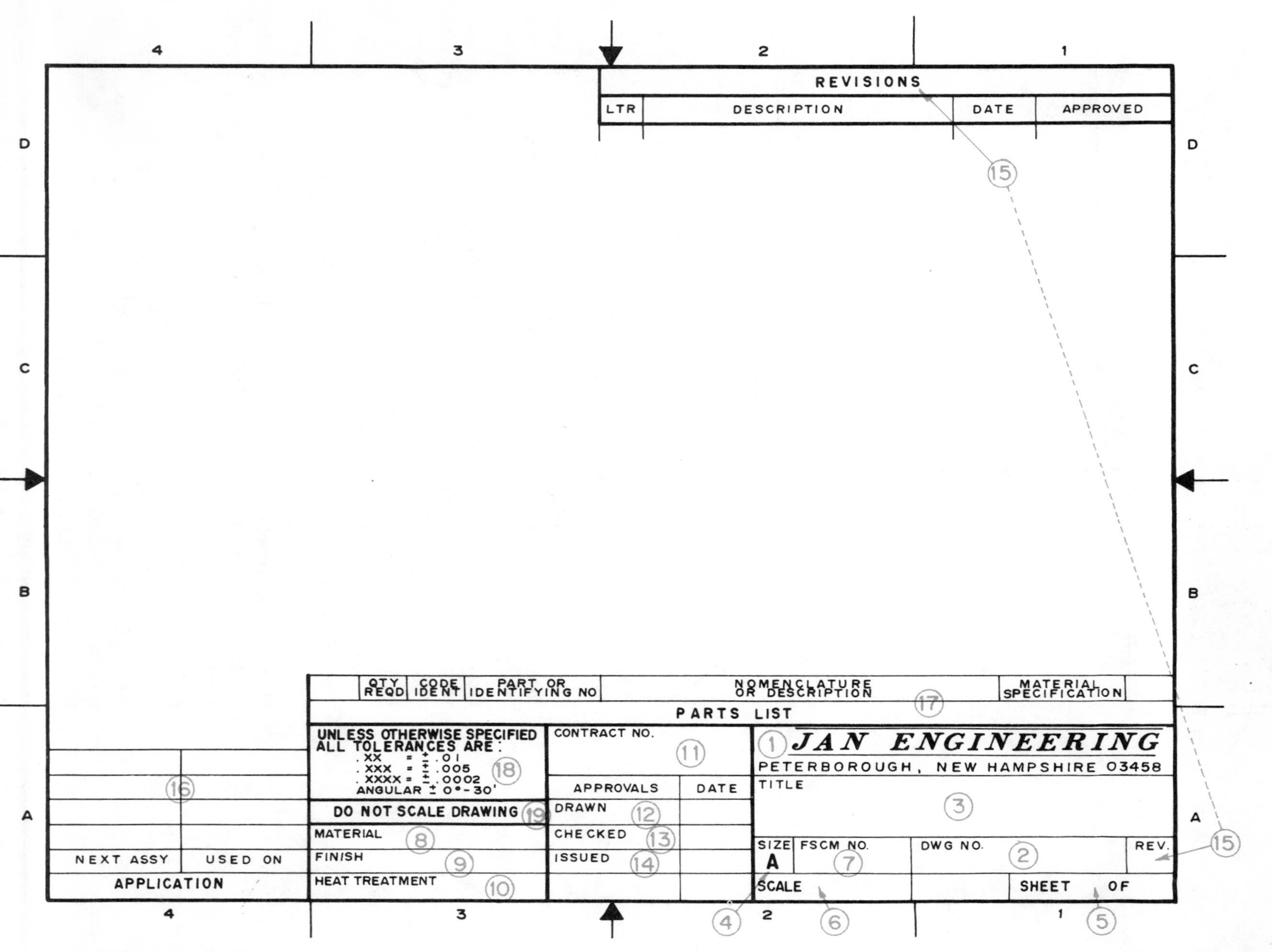

Fig. I-3. Standard title block.

Scales usually used:

Full-size	1/1	Double Size	2/1
Half-size	1/2	Four Times Size	4/1
Quarter-size	1/4	Ten Times Size	10/1
Tenth-size	1/10		

If more than one scale is used on a drawing, the scale of the major view is given in the title box. Views drawn at any other scale should have that scale printed directly below that view. Regardless of what size the part appears on the drawing, the *dimensions* indicate the exact size the part will be manufactured.

7. **FSCM block**—The FSCM means "Federal Supply Code for Manufacturers." It is a specific number assigned to a company that does government contract work.

8. **Material**—This notes the exact material that is to be used in manufacturing the part.

9. **Finish**—This block indicates any finish requirements, e.g., whether the part is to be primed, painted, or chrome-plated. If finish requirements are lengthy in detail, all specifications are given in the drawing itself and the word "noted" is placed in the finish block. If no finish is required, a dash line or the word "none" is entered in this space.

10. **Heat treatment**—If the part requires special heat treatment or hardness requirements, exact specifications are noted in this block. If no heat treatment is required, a dash or the word "none" is entered in this space.

11. **Contract number**—If this particular part is designed for a particular customer or job, its contract number is indicated. If the part is designed under no contract, a dash or the word "none" is entered in this space.

12. **Drawn**—This block indicates who actually made this drawing and the date that person *finished* the drawing.

13. **Checked**—This block indicates the person who checked the drawing for accuracy, completeness, clarity, and conformity with the latest ANSI drafting standards. It is this person who is responsible for any errors or omissions in the drawing. The date that the person finishes checking the drawing is recorded in the space provided.

14. **Issued**—The date the drawing is scheduled to be issued to the customer or the manufacturing plant is indicated in the space provided.

15. **Revisions**—The revision block usually appears in the upper right-hand corner of the drawing. Any and all revisions or changes must be carefully recorded within this area. It is extremely important to keep an accurate record of all changes made to the original drawing. This block briefly notes what change was made, the date of the change, and who approved the change. If the change is lengthy, the E.C.O. (Engineering Change Order) number is listed, as it goes into great detail about the change. The revision block is very important and must be kept up to date at all times.

    All revisions are noted in alphabetical order and noted in the block at the left side of the revision block. This *same* letter is added below in the "REV" block, down next to the drawing number (see FIG. 1-4). In most companies the last revision letter becomes part of the drawing number.

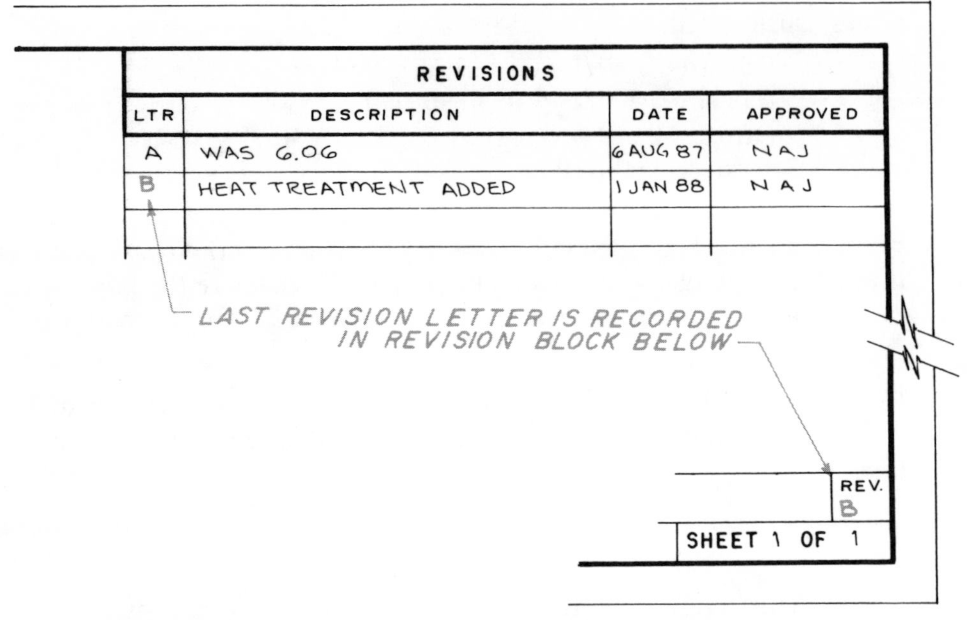

Fig. 1-4. The revision block is usually found in the upper-right-hand corner.

Example:

Part Number A-80635-B

The "A" indicates an A size paper, the assigned drawing-number is 806035, and the last revision done to the drawing was revision B.

On large drawings, the zone in which the change was made is also noted. (Zones will be covered later in this chapter.)

16. **Application block**—Sometimes referred to as the *usage block*, this block is used to note on which final assembly or assemblies, the part or subassembly shown on the drawing is used.

    Next assembly: Lists the drawing number this part will be used on.
    Used on: Notes the finished assembly drawing number.

17. **Parts list**—If a drawing is to be used for a Parts List or for listing the various parts in an assembly drawing, it is usually located directly above the title block. It lists, in tabular form, information on part numbers, titles of parts used, type of materials used on each part, code identification, and quantity of each part required to make up one complete assembly. This will be covered in more detail later in this workbook. The Parts List is sometimes referred to as materials list, bill of materials, or schedule of parts.

18. **Tolerance block**—As it is absolutely impossible to manufacture two things *exactly* alike, tolerances are "allowed," or given, on all dimensions listed on the drawing. Some dimensions are not as important as others, so larger tolerances are allowed. A hole that has a shaft through it turning at high speed would be an example of an important dimension and would require a closer tolerance than a hole that was put into a part for appearance only. The tolerance block lists the tolerances allowed for the various decimals. These tolerances vary from company to company and product to product, but for this workbook, the following tolerances will be used, unless otherwise specified:

    All two-place decimals (.XX) can vary plus or minus (±) .01 (one-hundredth of an inch) from the dimension size given on the drawing. All three-place decimals (.XXX) can vary plus or minus (±) .005 (five-thousandths of an inch) from the dimension size given on the drawing, and all four-placed decimals (.XXXX) can vary plus or minus (±) .0002 (two ten-thousandths of an inch). All angular dimensions can vary in actual size 30 minutes (1/2 of a degree) from the dimension given on the drawing.

Tolerance and Limits. To help you visualize and remember tolerance and limits, think of the usual road sign found along the interstate highways of the country (see FIG. 1-5). Listed is a 55 mile per hour *upper limit* and a 35 mile per hour *lower limit*. If you speed over the 55 m.p.h. upper limit, you could get a speeding ticket; if you poke along under the 35 m.p.h.

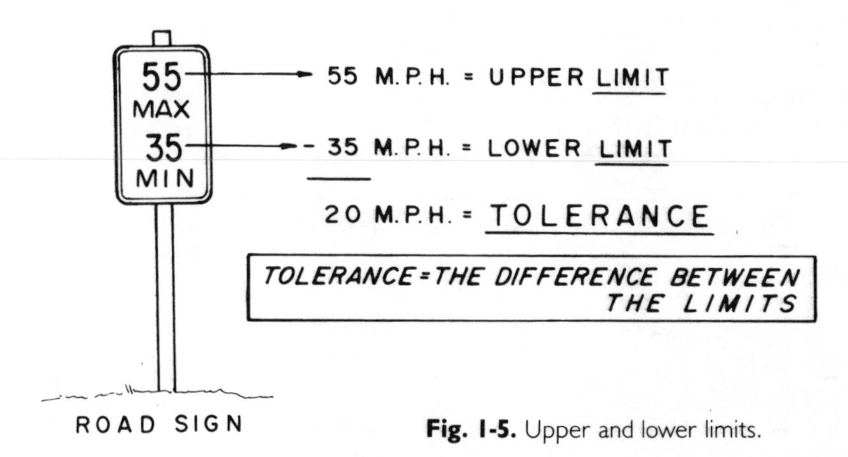

Fig. 1-5. Upper and lower limits.

lower limit, you could also get a ticket for going too slow. The *difference* between the upper limit and the lower limit is 20 miles per hour.

55 m.p.h.	(upper limit)
−35 m.p.h.	(lower limit)
20 m.p.h.	this is the *tolerance*

Applying this same idea to decimal sizes that are found on drawings (refer to FIG. 1-6). The *upper limit* is .505 and the *lower limit*is .500 thus the *tolerance* is .005.

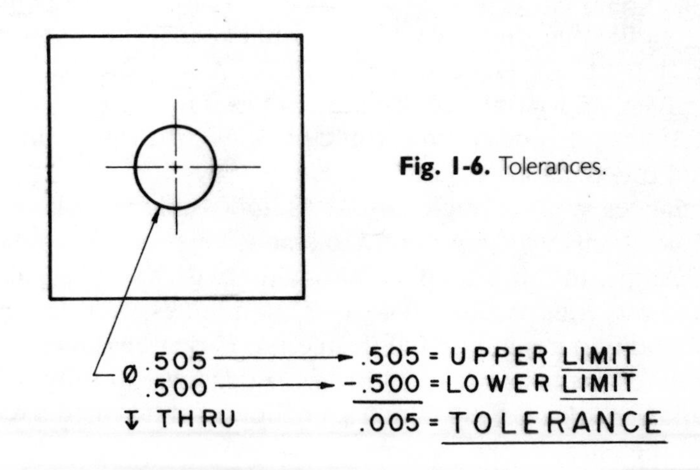

Fig. 1-6. Tolerances.

.505 maximum size hole allowed
−.500 minimum size hole allowed
.005 *Tolerance*

A decimal of .50, .500, and .5000 are mathematically equal, but in applying tolerances, they could vary a great deal (refer to FIG. 1-7). A two-placed dimension of .50 is given. All two-placed decimals can vary in size

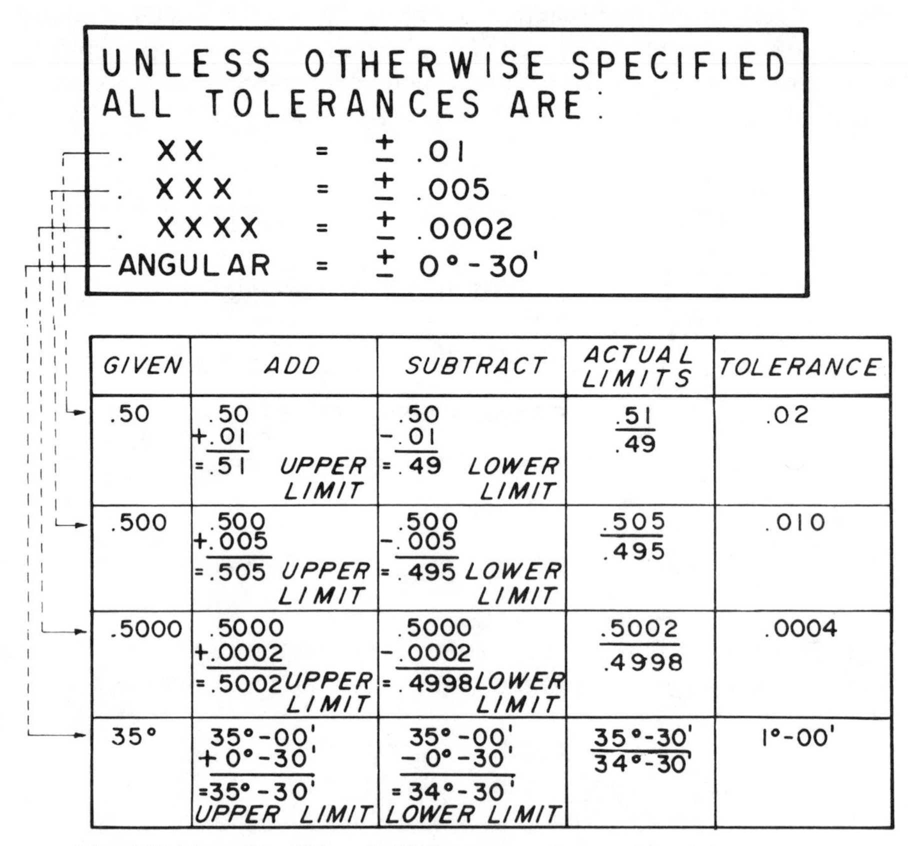
UNLESS OTHERWISE SPECIFIED
ALL TOLERANCES ARE:
.XX = ± .01
.XXX = ± .005
.XXXX = ± .0002
ANGULAR = ± 0°-30'

GIVEN	ADD	SUBTRACT	ACTUAL LIMITS	TOLERANCE
.50	.50 +.01 =.51 UPPER LIMIT	.50 −.01 =.49 LOWER LIMIT	.51 / .49	.02
.500	.500 +.005 =.505 UPPER LIMIT	.500 −.005 =.495 LOWER LIMIT	.505 / .495	.010
.5000	.5000 +.0002 =.5002 UPPER LIMIT	.5000 −.0002 =.4998 LOWER LIMIT	.5002 / .4998	.0004
35°	35°-00' +0°-30' =35°-30' UPPER LIMIT	35°-00' −0°-30' =34°-30' LOWER LIMIT	35°-30' / 34°-30'	1°-00'

Fig. 1-7. How .50, .500, and .5000 can vary when applying tolerances.

up to plus or minus (±) .01; thus, the lower limit would be .49, the upper limit would be .51, and the tolerance would be .02. A three-placed dimension of .500 is given. All three-placed decimals can vary in size up to plus or minus (±) .005; thus, the lower limit would be .495, the upper limit would be .505, and the tolerance would be .010. A four-placed dimension of .5000 is given. All four-placed decimals can vary in size up to plus or minus (±) .0002; thus, the lower limit would be .4998, the upper limit would be .5002 and the tolerance would be .0004.

Tolerances apply to angles also. It is impossible to make two surfaces at the same exact angle; therefore, a tolerance must also be applied to angular dimensions. In this example, unless otherwise specified, all angles are allowed to vary plus or minus 0 degrees, 30 minutes, or a 1/2 degree. (*Note:* There are 60 minutes in one full degree.) A 35-degree dimension is given. All angular dimensions can vary in size 0 degrees, 30 minutes, thus, the lower limit is 34 degrees, 30 minutes, the upper limit is 35 degrees, 30 minutes, and the *tolerance* is *1 degree*.

Fractions. The latest ANSI manual eliminates the use of fractions in machine drawings. Some companies still use fractions, but they are not considered vital dimensions, and they carry the same limits and tolerances as any two-placed decimals would have. In using the above example limits, all fractions are plus or minus (±) 1/64 inch (015).

WORKSHEET I-I

Instructions: Listed are 10 problems. Each has a given specified SIZE dimension in the left hand column. Using the GIVEN TOLERANCE, calculate the upper and lower allowable limits and place your answer in the block provided. Using the calculated limits, calculate the tolerance of each and place your answer in the block provided. Show all math work. (Answers in Appendix A.)

PROB.	DIMENSION	GIVEN TOLERANCE	UPPER LIMIT	LOWER LIMIT	TOLERANCE
1	.75	.XX ±.01			
2	.62	.XX ±.02			
3	.720	.XXX ±.00			
4	2.56	.XX ±.01			
5	7.0625	.XXXX ±.0002			
6	3.875	.XXX ±.010			
7	1.0032	.XXXX ±.0001			
8	.250-.251	(AS GIVEN)			
9	0.625	.XXX ±.010			
10	3/4	FRAC. ± 1/64			

19. **Do not scale drawing**—This note is usually found on all drawings. Although a drawing should be drawn to exact scale (proportion) and have all the required dimensions, a drawing *should not be scaled* in order to determine a dimension. A good drawing should be fully dimensioned, so the part can be fully made without any information missing or any calculations. A drawing should not be scaled because some drafters do not draw to exact size all the time, a dimensional change could have been made on the drawing without actually changing the figure, and most copied whiteprints actually stretch or shrink in the copying process which changes the size of the drawn object.

ZONING

Zoning is used on larger drawings to help locate a particular feature, dimension, detail, and so on. The zoning system is exactly like that used on highway road maps to locate towns, roads or points of interest. The zones are noted just outside the border (refer to FIG. 1-1). At the two side edges, the zone indicators are

WORKSHEET 1-2

Instructions: Listed are 6 problems. Answer each in the block provided to the right. Show all math work. (Answers in Appendix A.)

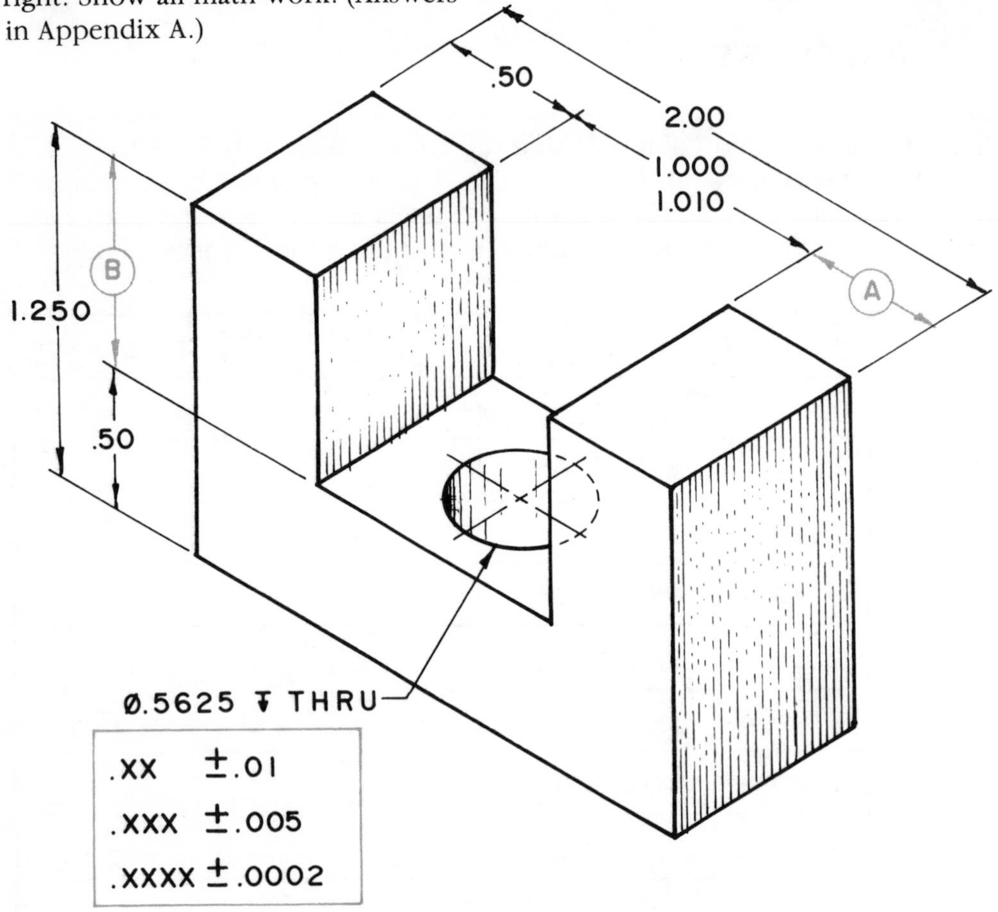

1	SMALLEST SIZE (A) COULD BE	
2	LARGEST SIZE (A) COULD BE	
3	SMALLEST SIZE (B) COULD BE	
4	LARGEST SIZE (B) COULD BE	
5	SMALLEST SIZE OF HOLE	
6	LARGEST SIZE OF HOLE	

listed alphabetically at even intervals from bottom to top. Along the top and bottom edges, the zone indicators are listed numerically at even intervals from right to left. (Although A-size paper usually does *not* have zoning, zoning indicators have been added to provide practice in zoning.

Zoning is usually illustrated by a $^{1}/_{2}$-inch square. The left side of the square indicates the page number of the zone. This is needed on multiple-page drawings. The right side of the zone square lists the horizontal number and vertical letter in order to pinpoint a specific area on the drawing. Refer again to FIG. 1-1. In this example the specific area called off is on page "one," and is located between the horizontal area "B" and the vertical area "6."

MICROFILM ARROWHEADS

The four large arrowheads just outside the borders are located at the center of each edge of the paper to line up the drawing when it is microfilmed.

WORKSHEET 1-3

Instructions: Using Drawing Number A080635 on p. 12, answer each question in the spaces provided. (Answers in Appendix A.)

1. What is the title of this drawing?

2. What is the full drawing number?

3. What is the contract number?

4. When was this drawing issued?

5. At what scale was this drawing made?

6. What size is this paper?

7. Up to this time, have any changes been made to this drawing?

8. What is the tolerance on .XXX (3 place) decimal dimensions?

9. Which size hole has the smallest tolerance?

10. What is the overall length of "A?"

11. What is the overall height of this object?

12. What is the overall thickness of this object?

13. What is this part made of?

14. This part is used on what model number?

15. List the upper and lower limits of the .75 diameter hole.

16. What dimensions falls inside the 1-B zone?

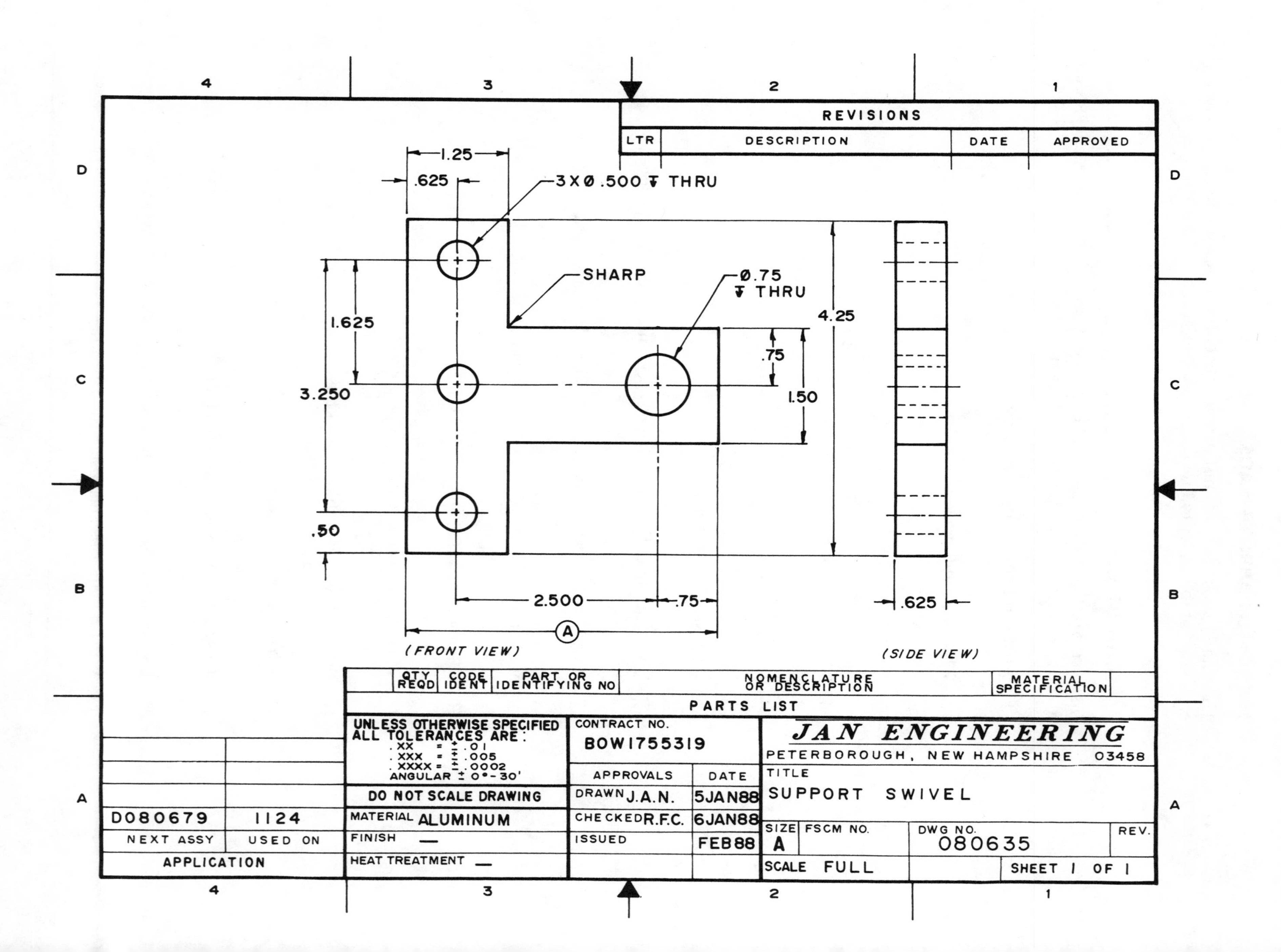
REVISIONS
LTR
DESCRIPTION
DATE
APPROVED
1.25
.625
3X Ø.500 ↧ THRU
SHARP
Ø.75
↧ THRU
4.25
1.625
.75
3.250
1.50
.50
2.500
.75
.625
A
(FRONT VIEW)
(SIDE VIEW)
QTY REQD
CODE IDENT
PART OR IDENTIFYING NO
NOMENCLATURE OR DESCRIPTION
MATERIAL SPECIFICATION
PARTS LIST
UNLESS OTHERWISE SPECIFIED ALL TOLERANCES ARE:
.XX = ± .01
.XXX = ± .005
.XXXX = ± .0002
ANGULAR ± 0°-30'
DO NOT SCALE DRAWING
MATERIAL ALUMINUM
FINISH —
HEAT TREATMENT —
CONTRACT NO.
BOW1755319
APPROVALS
DATE
DRAWN J.A.N. 5JAN88
CHECKED R.F.C. 6JAN88
ISSUED FEB88
JAN ENGINEERING
PETERBOROUGH, NEW HAMPSHIRE 03458
TITLE
SUPPORT SWIVEL
SIZE A
FSCM NO.
DWG NO. 080635
REV.
SCALE FULL
SHEET 1 OF 1
D080679
1124
NEXT ASSY
USED ON
APPLICATION

Rounding Off Decimals

Objective: The reader will be able to round off decimals, convert from the inch system fractions to decimals and to the metric system, and read the basic scales used in industry.

If you are mathematically rounding off a decimal, calculate the figure to one more decimal place than actually required and round off by one of the following methods:

Where the digit dropped is larger than 5, the last digit kept is increased by 1:

6.8216 Round off to 3 decimal places = 6.822
(the last digit, 6, is *larger* than 5)

Where the digit dropped is less than 5, the last digit that is kept is not changed:

5.2534 Round off to 3 decimal places = 5.253
(the last digit, 4, is *smaller* than 5)

Where the digit dropped *is* 5, the last digit that is kept is an even number:

1.765 Round off to 2 decimal places = 1.6
(the last digit kept, 6, is already even)

0.775 Rounded off to 2 decimal places = 0.78
(the last digit kept, 7, is odd and so is raised to the next highest even number)

CONVERTING (Fractions, Decimals, Metric)

Current ANSI standards suggest that fractional dimensions not be used on all new machine drawings. Some companies still use fraction, however, and most still have and use fractions on older drawings. Therefore, an understanding of fractional dimensions is very important, even today. For many years to come, the United States will be going through a transitional period from the inch system to a full metric system. Until that time, everyone will be converting from one system to the other. This can be done mathematically by using a conversion chart (see TABLE 2-1).

Mathematically, 1mm = 0.03937 inch and 1 inch = 25.4 mm.

Using the Conversion Chart

There are many kinds of conversion charts available today. FIGURE 2-1 is an example of a standard conversion chart used to convert from fractional inches to decimal inches to millimeters. If you know any one of these three, it is very simple to convert by using a conversion chart.

Table 2-1. Conversion Chart

INCH/METRIC–EQUIVALENTS					
	Decimal Equivalent			Decimal Equivalent	
Fraction	Customary (in.)	Metric (mm)	Fraction	Customary (in.)	Metric (mm)
$^1/_{64}$	.015625	0.3969	$^{33}/_{64}$	.515625	13.0969
$^1/_{32}$	.03125	0.7938	$^{17}/_{32}$	.53125	13.4938
$^3/_{64}$	.046875	1.1906	$^{35}/_{64}$	.546875	13.8906
$^1/_{16}$	.0625	1.5875	$^9/_{16}$	.5625	14.2875
$^5/_{64}$	.078125	1.9844	$^{37}/_{64}$	.578125	14.6844
$^3/_{32}$	.09375	2.3813	$^{19}/_{32}$	.59375	15.0813
$^7/_{64}$	.109375	2.7781	$^{39}/_{64}$	.609375	15.4781
$^1/_8$	.1250	3.1750	$^5/_8$	.6250	15.8750
$^9/_{64}$	.140625	3.5719	$^{41}/_{64}$	.640625	16.2719
$^5/_{32}$	.15625	3.9688	$^{21}/_{32}$	.65625	16.6688
$^{11}/_{64}$	.171875	4.3656	$^{43}/_{64}$	.671875	17.0656
$^3/_{16}$	.1875	4.7625	$^{11}/_{16}$	.6875	17.4625
$^{13}/_{64}$	.203125	5.1594	$^{45}/_{64}$	.703125	17.8594
$^7/_{32}$	.21875	5.5563	$^{23}/_{32}$	.71875	18.2563
$^{15}/_{64}$	.234375	5.9531	$^{47}/_{64}$	.734375	18.6531
$^1/_4$	.250	6.3500	$^3/_4$	.750	19.0500
$^{17}/_{64}$	.265625	6.7469	$^{49}/_{64}$	.765625	19.4469
$^9/_{32}$	.28125	7.1438	$^{25}/_{32}$	.78125	19.8438
$^{19}/_{64}$	.296875	7.5406	$^{51}/_{64}$	.796875	20.2406
$^5/_{16}$	.3125	7.9375	$^{13}/_{16}$	.8125	20.6375
$^{21}/_{64}$	.328125	8.3384	$^{53}/_{64}$	.828125	21.0344
$^{11}/_{32}$	.34375	8.7313	$^{27}/_{32}$	.84375	21.4313
$^{23}/_{64}$	.359375	9.1281	$^{55}/_{64}$	.859375	21.8281
$^3/_8$	.3750	9.5250	$^7/_8$	.8750	22.2250
$^{25}/_{64}$	.390625	9.9219	$^{57}/_{64}$	.890625	22.6219
$^{13}/_{32}$	.40625	10.3188	$^{29}/_{32}$	.90625	23.0188
$^{27}/_{64}$	.421875	10.7156	$^{59}/_{64}$	.921875	23.4156
$^7/_{16}$	.4375	11.1125	$^{15}/_{16}$	.9375	23.8125
$^{29}/_{64}$	.453125	11.5904	$^{61}/_{64}$	.953125	24.2094
$^{15}/_{32}$	.46875	11.9063	$^{31}/_{32}$	.96875	24.6063
$^{31}/_{64}$	.484375	12.3031	$^{63}/_{64}$	.984375	25.0031
$^1/_2$	.500	12.7000	1	1.000	25.4000

$$^3/_{64} \text{ inch} = .046875 \text{ inch} = 1.1906 \text{ mm}$$
$$.125 \text{ inch} = {}^1/_8 \text{ inch} = 3.1750 \text{ mm}$$
$$4.7625 \text{ mm} = {}^3/_{16} \text{ inch} = .1875 \text{ inch}$$

If an exact number is not listed on the chart, use the mathematical conversion factor to make the conversion.

Example:

What is the nearest fractional inch equivalent of 4.82 mm?

$$4.82 \text{ mm} \times 0.03937 \text{ inches/mm} = .18976 \text{ inches}$$

.18976 is between .1875 and .203125, but is closer to .1875.

Therefore, the nearest fractional equivalent is $^3/_{16}$.

READING THE FRACTIONAL-INCH SCALE

Although fractions are not usually used on drawings today, they are found on older drawings. The fractional-inch scale is often divided into 16 equal parts—

INCH/METRIC		
Fraction	Decimal Equivalent	
	Customary (in)	Metric (mm)
1/64	.015625	0.3969
1/32	.03125	0.7938
3/64	.046875	1.1906
1/16	.0625	1.5875
5/64	.078125	1.9844
3/32	.09375	2.3813
7/64	.109375	2.7781
1/8	.1250	3.1750
9/64	.140625	3.5719
5/32	.15625	3.9688
11/64	.171875	4.3656
3/16	.1875	4.7625
13/64	.203125	5.1594

Fig. 2-1. A standard conversion chart.

each part equals 1/16 of one inch (see FIG. 2-2). Three dimensions are illustrated: 3/4 inch, 1 3/8 inch, and 2 inches. (*Note:* The starting point on the scale is at the "0" and not the end of the scale.) To use the fractional scale:

Step 1 Place the scale along the distance to be measured with the "0" at one edge of surface.

Step 2 Read to the nearest 16th unit. Remember to include any full inches and to add this to the total.

Step 3 Convert the fractional answer into a decimal number.

In the bottom illustration nearest the scale, there are 12 divisions, or 12/16. 12/16 = 3/4 of an inch. Convert this into a decimal inch using the chart in FIG. 2-2. The correct answer is .75 of an inch.

In the center illustration (FIG. 2-2B), there is one full inch plus 6 divisions, or 1 and 6/16. 1 6/16 = 1 3/8 of an inch. Converted to a decimal, it equals .375.

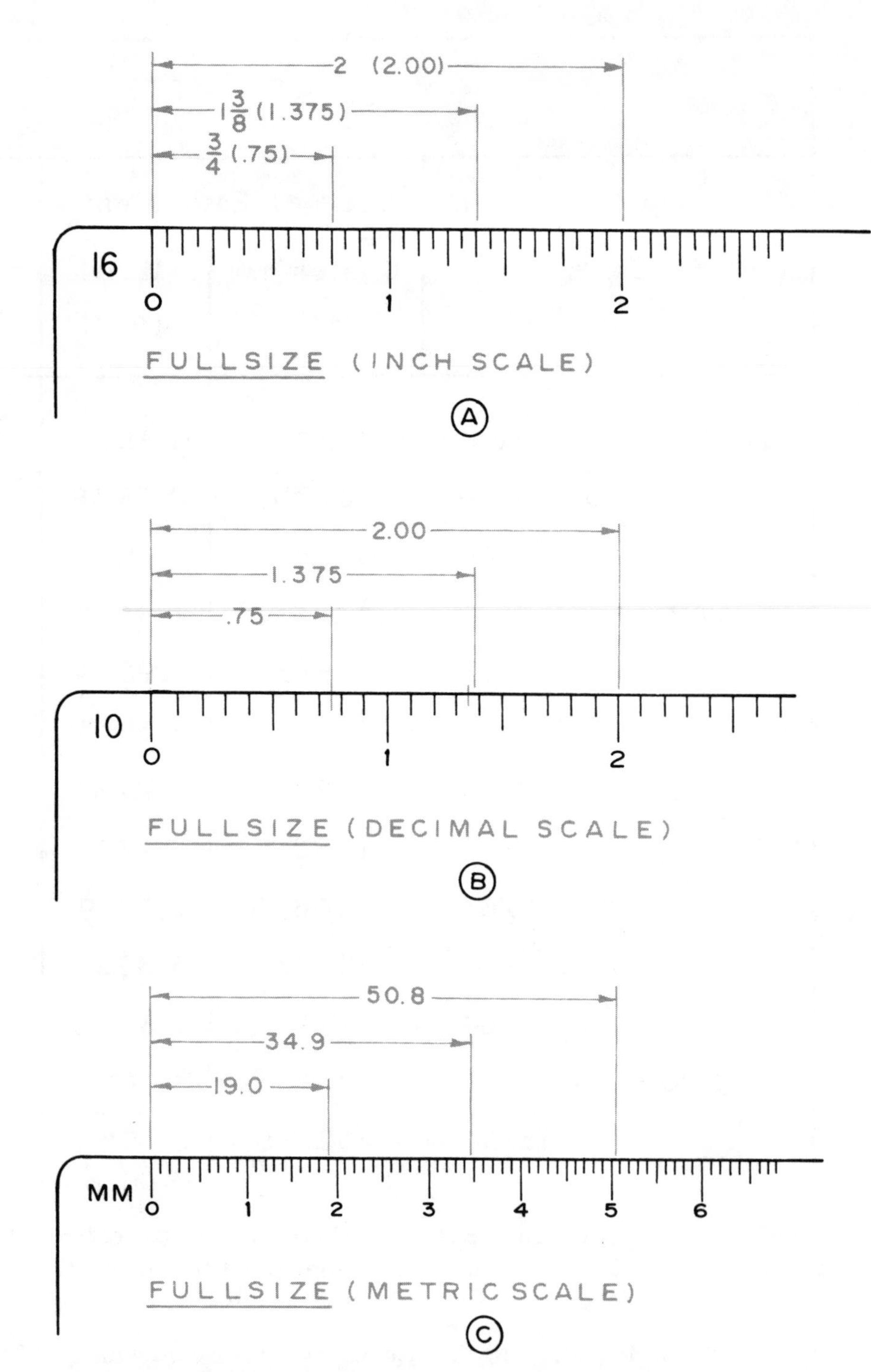

Fig. 2-2. (A) Reading a fractional-inch scale; (B) Reading a decimal-inch scale; (C) Reading a metric scale.

In the top illustration, there is an even 2 inches. To convert a full inch fraction to a decimal, the inch is called off as a two-placed decimal, or 2.00.

READING THE DECIMAL-INCH SCALE

The decimal-inch scale is often divided into 10 equal parts—each part equals .1 inch. See Fig. 2-2B. It is read directly to the nearest decimal. In the illustration nearest the scale, the measurement is centered between division 7 and division 8, and is read as .75.

In the center illustration there is 1 full inch, plus a little over 3 divisions (estimated as about 3/4 of a division), or .375. in the top illustration, there is an even two inches, which is expressed as 2.00.

READING THE METRIC SCALE

The metric scale is read like the decimal-inch scale except in millimeters. See FIG. 2-2C. Each division equals 1 millimeter. That means that the distance between the zero and the one on the scale is 10 millimeters. In the illustration nearest the scale, the measurement is 19 full dimensions and is read as 19 mm. In the center illustration there are 34 full divisions and most of the 35th. Therefore, the measurement is estimated to be 34.9 mm. In the top illustration, there are 50 full divisions and most of the 51st. Therefore, the measurement is estimated to be 50.8 mm.

METRIC DRAWINGS

Drawings using metric dimensions in place of the decimal inch units of measure are indicated by a rectangular block, located directly above the title block, with the words *metric.*

WORKSHEET 2-1

Instructions: Listed are 10 problems. Convert each "given" dimension to its equivalent fractional, decimal, or metric dimension as required. Use the conversion chart in TABLE 2-1. List each converted number in the spaces provided. (Answers in Appendix A.)

PROB.	INCH FRACTION	INCH DECIMAL	METRIC (mm)
1	1/2		
2		.062	
3			17.46
4	2 15/16		
5		7.44	
6			50.80
7		.6875	
8			36.51
9			114.30
10		10.88	

WORKSHEET 2-2

Instructions: Listed are 12 problems. Using the three scales provided, list the measurements indicated by the black triangles. Convert each measured dimension to its equivalent fractional, decimal, or metric dimension as required. (Answers in Appendix A.)

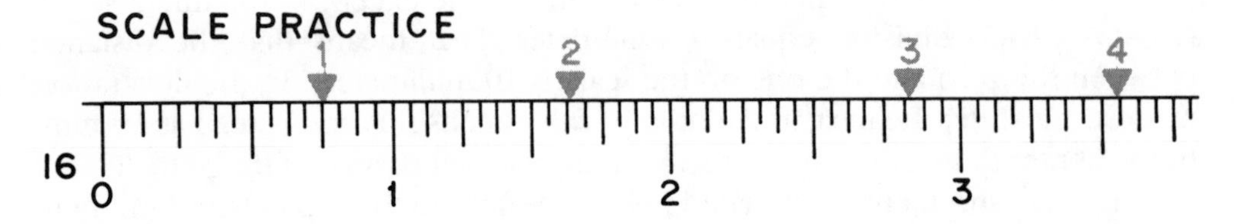

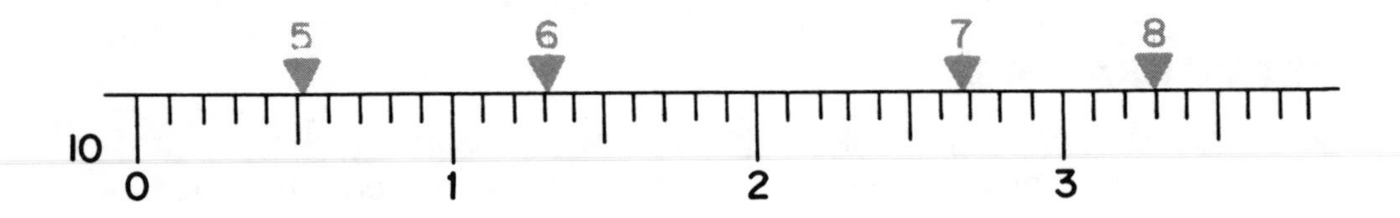

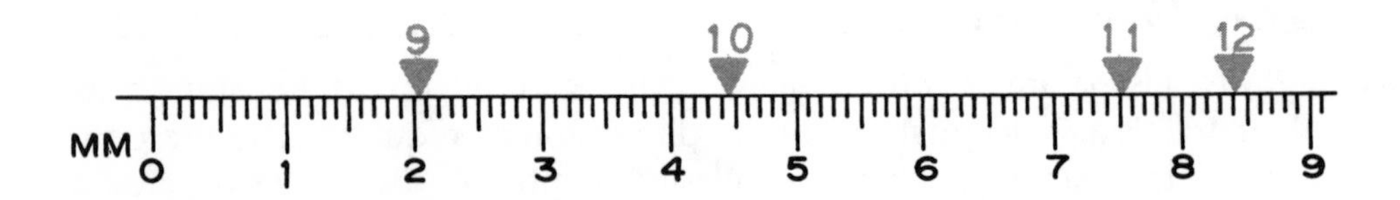

PROB	FRACTION	DECIMAL	METRIC
1			
2			
3			
4			
5			
6			
7			
8			
9			
10			
11			
12			

WORKSHEET 2-3

Instructions: Using Drawing Number A795513 on p. 20, answer each question in the spaces provided. (Answers in Appendix A.)

1. What is the maximum length (3.250) the collar could be and still be within limits?

2. What is the next assembly where this part is used?

3. What is dimension "A"?

4. What dimension falls within the 3-D zone?

5. What is the FSCM number and what does it mean?

6. When was this drawing released?

7. What is the largest overall diameter this part can be made and still be within tolerance?

8. Indicate the tolerance of the 2.00 diameter dimension.

9. List the upper and lower limits of the 35-degree dimension.

10. To what scale was the collar drawn to?

11. What is the metric equivalent of the 2.50 diameter?

12. What is this part made of?

13. Who made this drawing and when was it started?

14. How many changes have been made to this drawing?

15. What is the lower limit of the .750 wide groove?

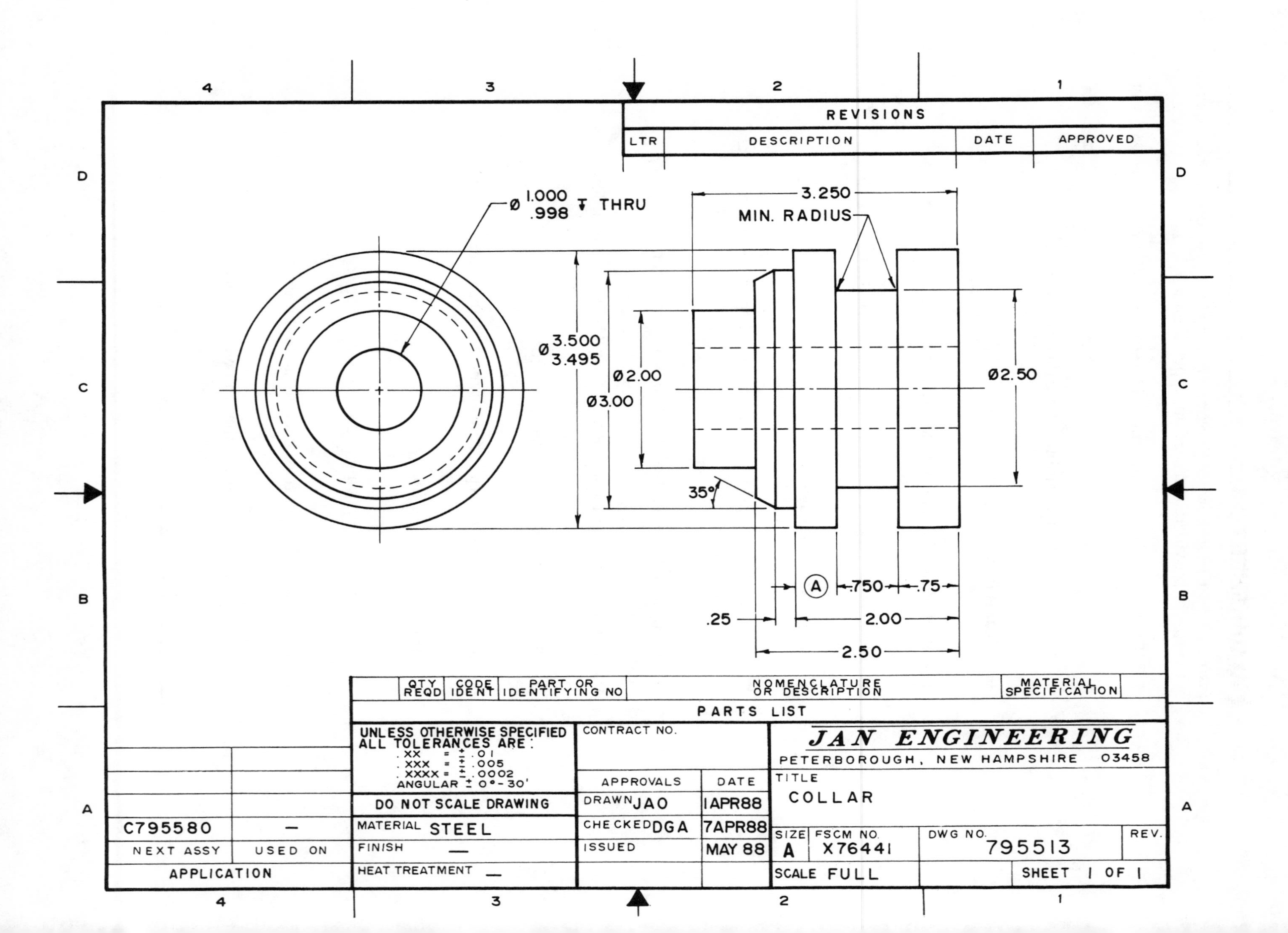
REVISIONS
LTR
DESCRIPTION
DATE
APPROVED
Ø 1.000
.998 THRU
3.250
MIN. RADIUS
Ø 3.500
3.495
Ø2.00
Ø3.00
Ø2.50
35°
A
.750
.75
.25
2.00
2.50
QTY REQD
CODE IDENT
PART OR IDENTIFYING NO
NOMENCLATURE OR DESCRIPTION
MATERIAL SPECIFICATION
PARTS LIST
UNLESS OTHERWISE SPECIFIED ALL TOLERANCES ARE:
.XX = ± .01
.XXX = ± .005
.XXXX = ± .0002
ANGULAR ± 0°-30'
DO NOT SCALE DRAWING
MATERIAL STEEL
FINISH —
HEAT TREATMENT —
CONTRACT NO.
APPROVALS
DATE
DRAWN JAO
1APR88
CHECKED DGA
7APR88
ISSUED
MAY 88
JAN ENGINEERING
PETERBOROUGH, NEW HAMPSHIRE 03458
TITLE
COLLAR
SIZE A
FSCM NO. X76441
DWG NO. 795513
REV.
SCALE FULL
SHEET 1 OF 1
C795580
—
NEXT ASSY
USED ON
APPLICATION

Alphabet of Lines

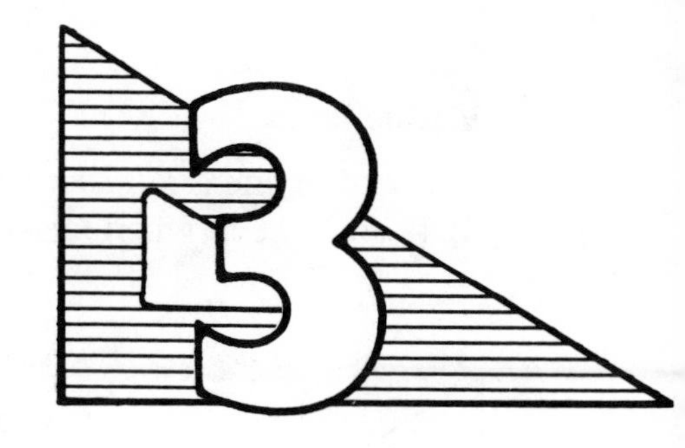

Objective: The reader will be able to identify all lines used in industry and know what each kind of line means.

A machine drawing is made up of many lines. Each line has a particular meaning and represents something: a visible surface, a hidden surface, the center of a hole, an extension of a surface, or a line with a dimension noted on it. The different kinds of lines are referred to as the *alphabet of lines.*

LINE THICKNESS

In order to help make drawings easier to read and understand, each kind of line is drawn with a different thickness or with a code of some kind. There are two thicknesses of lines used today: thick and thin (see FIG. 3-1). In the years before 1979, three line thicknesses were used, so any drawings before that date would have used a medium-thick line for a hidden line.

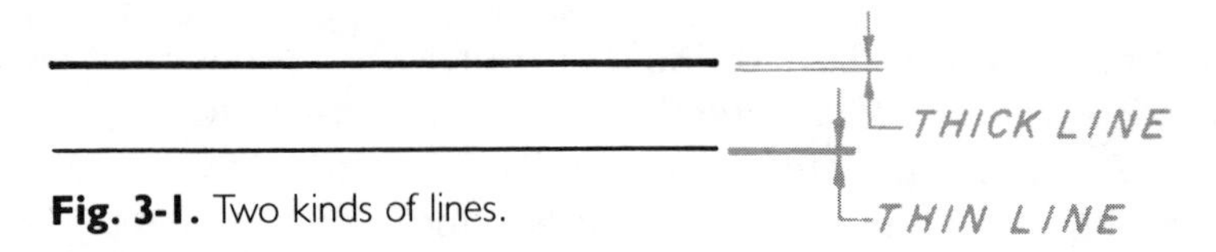

Fig. 3-1. Two kinds of lines.

KINDS OF LINES

The following 11 kinds of lines make up the alphabet of lines:

Thick lines are used for:

- object (or visible) lines
- cutting plane lines
- break lines

Thin lines are used for:

- hidden lines
- centerlines
- extension and leader lines
- dimension lines
- section lining lines
- phantom lines
- break lines
  (Z and S style)
- projection lines
  (very light and not seen)

Visible lines

The visible line is a *thick* continuous line that represents all edges and surfaces that can be seen when looking directly at the object (see FIG. 3-2).

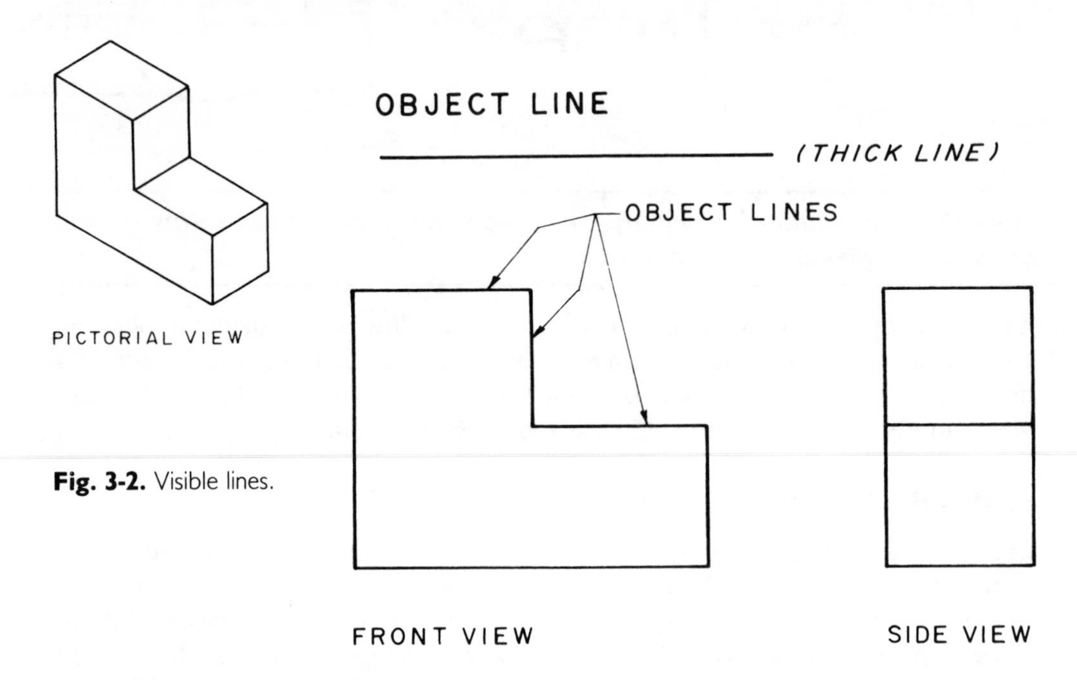

Fig. 3-2. Visible lines.

Hidden Lines

The hidden line is a *thin* dashed line that represents all edges or surfaces that cannot be seen when looking directly at the object, (see FIG. 3-3). In this example, the bottom of the "U" cannot be seen in the right-side view. There is a surface there and must be shown, thus a hidden line is drawn. (*Note.* On pre-1979 drawings the hidden line would have been drawn as a medium-thick line.)

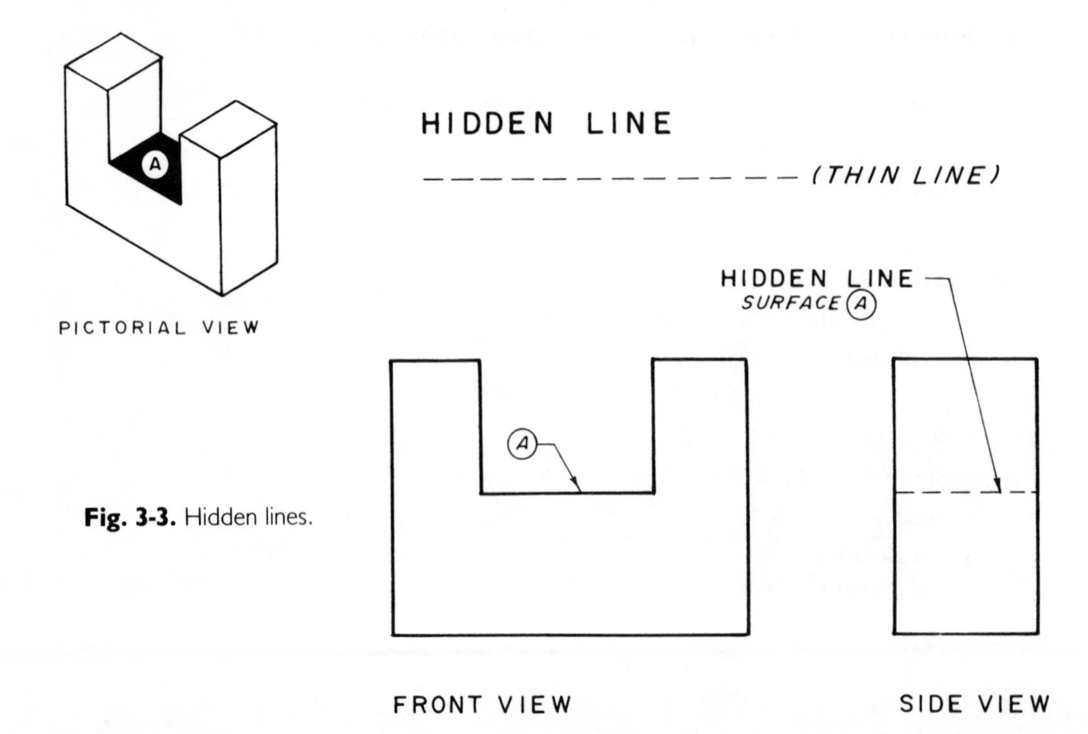

Fig. 3-3. Hidden lines.

Centerlines

The centerline is a *thin* line consisting of a series of long and short dashes. They locate and represent the center or axis of a particular circle or arc feature on the object (see FIG. 3-4). In this example, the centerline locates and represents the center of a .25-diameter hole. Note how the centerline is also drawn in the side view, representing the hole center of the hole. The two hidden lines represent the top and bottom surfaces of the hole.

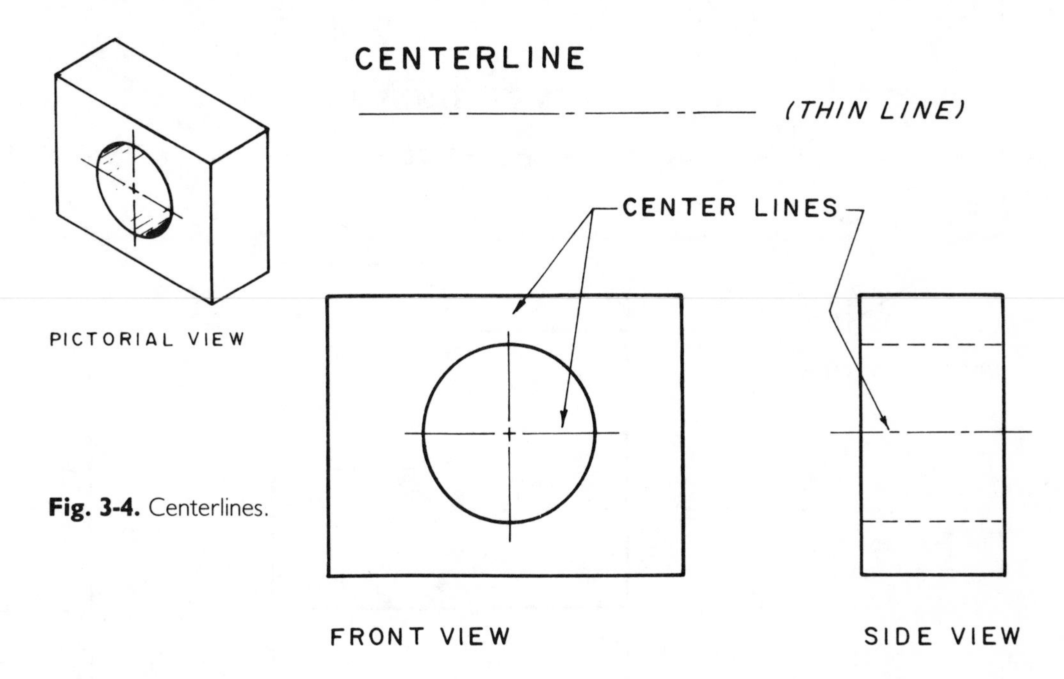

Fig. 3-4. Centerlines.

Extension Lines

The extension line, as its name implies, extends from the object. They are *thin* lines that are used in conjunction with dimension lines, (see FIG. 3-5). Note that a slight gap is left between the object line and the extension line.

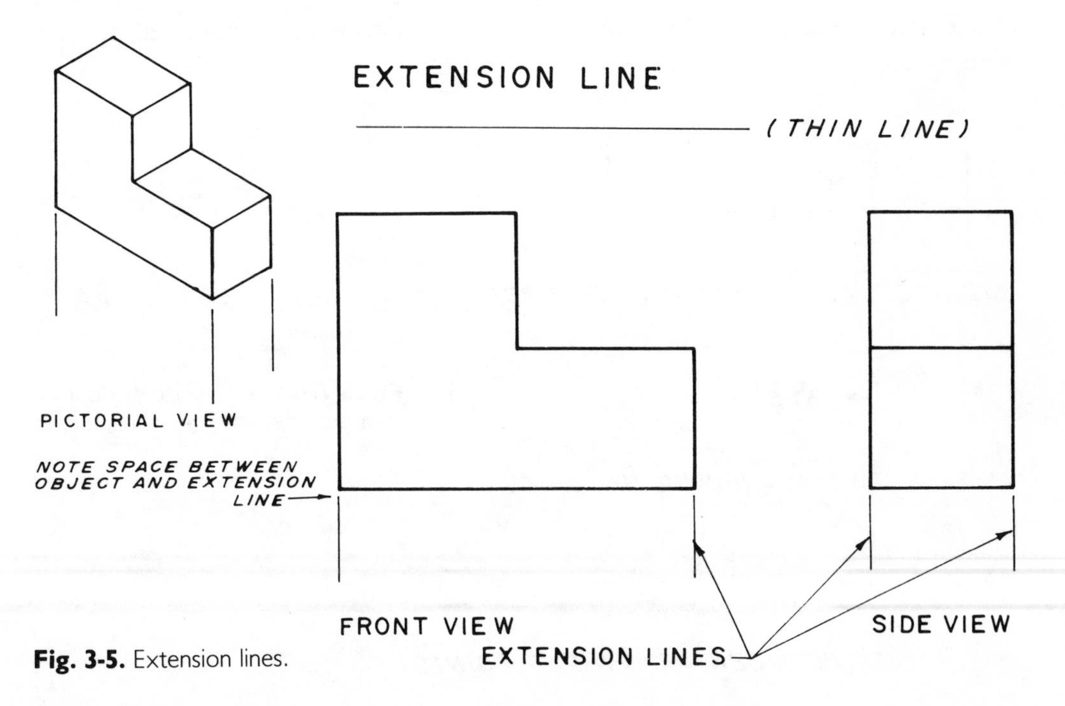

Fig. 3-5. Extension lines.

Dimension Lines

Dimension lines are *thin* lines that have the dimensions located on or near them. A dimension line is easily identified because it also has arrowheads located at its ends (see FIG. 3-6). Dimension lines are used in conjunction with extension lines, and the arrowheads always end at and touch the extension line.

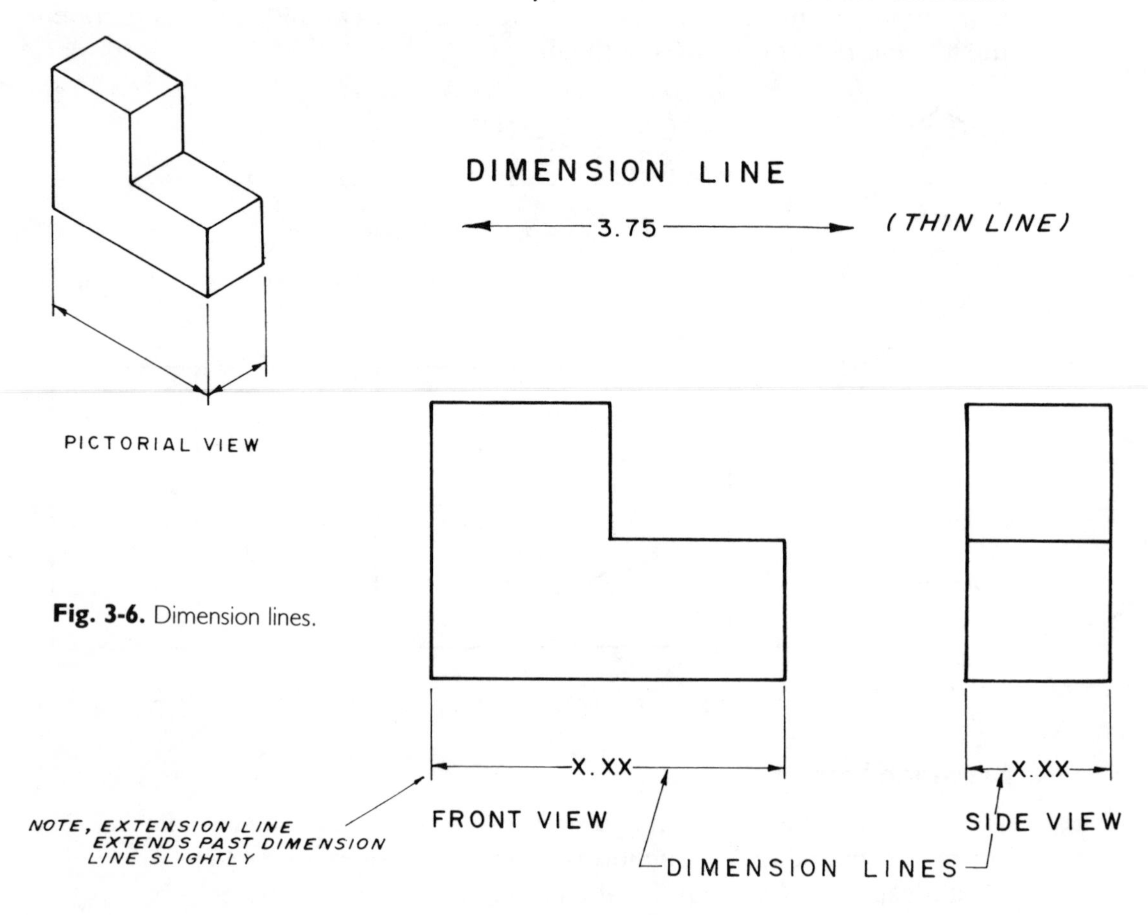

Fig. 3-6. Dimension lines.

(*Note:* The actual size or distance is measured *between* the arrowhead points. Refer to FIG. 3-7.)

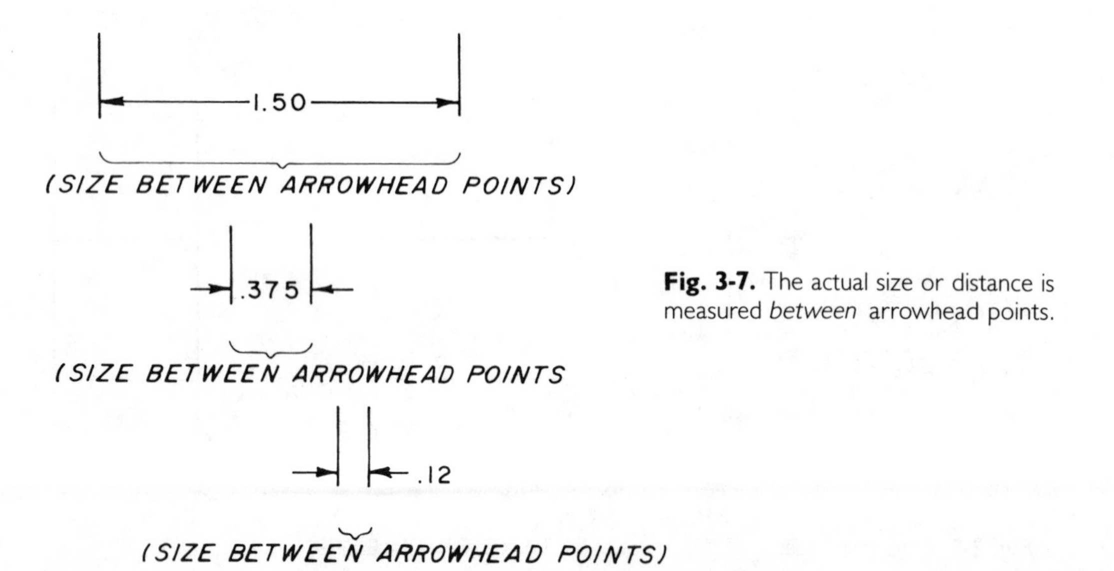

Fig. 3-7. The actual size or distance is measured *between* arrowhead points.

Leader Lines

Leader lines are *thin* lines that extend out and away from the object, in order to add a dimension or label (see FIG. 3-8). One end always ends with an arrowhead against or near the feature being called off.

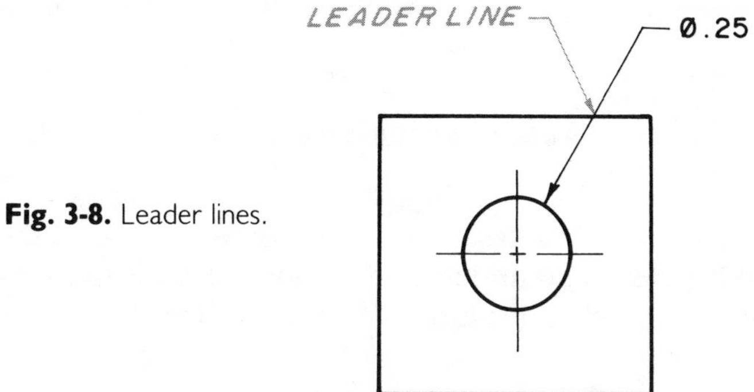

Fig. 3-8. Leader lines.

Phantom Lines

A phantom line is a *thin* line made up of a series of long and short dashes (see FIG. 3-9). They are used for three purposes:

1. To indicate alternate positions or travel of the part
2. To illustrate adjacent parts to the object
3. To simplify a drawing that has repeated features of details

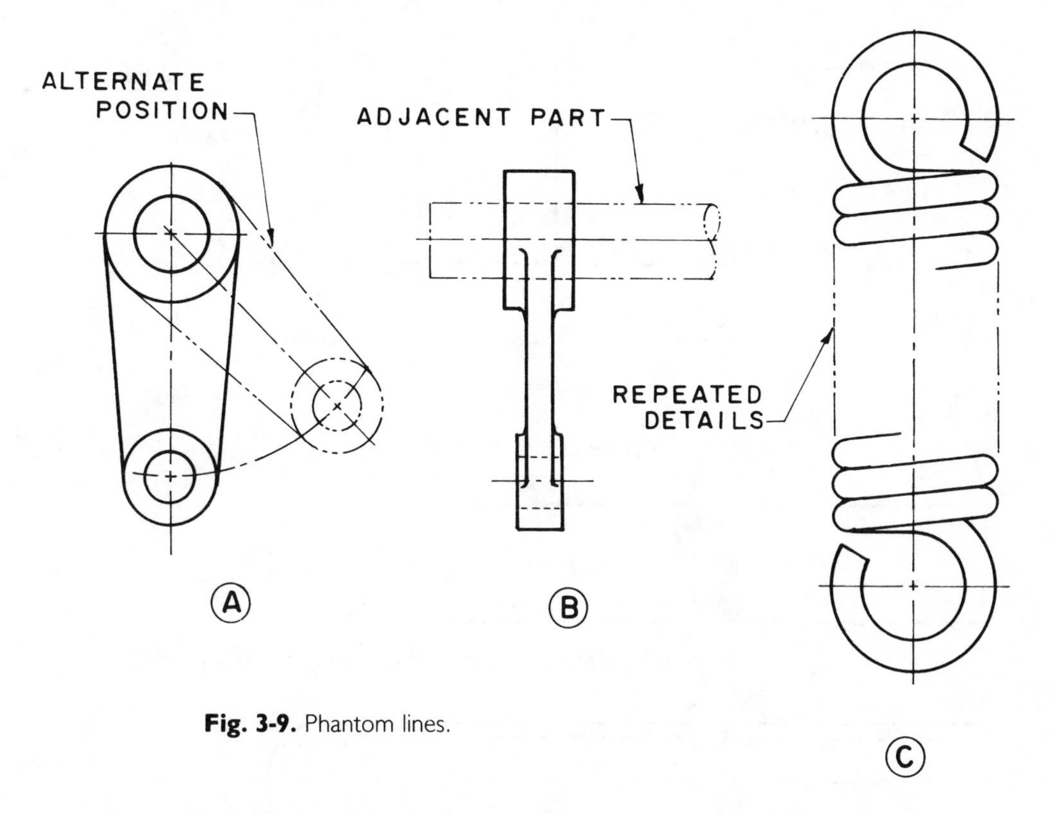

Fig. 3-9. Phantom lines.

Break Lines

Extra-long objects that have no unusual features along their length can be drawn at a larger scale for clarity by using break lines. The break line indicates that material has been removed between the break lines. The other views, dimensions, and all other features are drawn full size. Refer to FIG. 3-10. The

Fig. 3-10. Break lines.

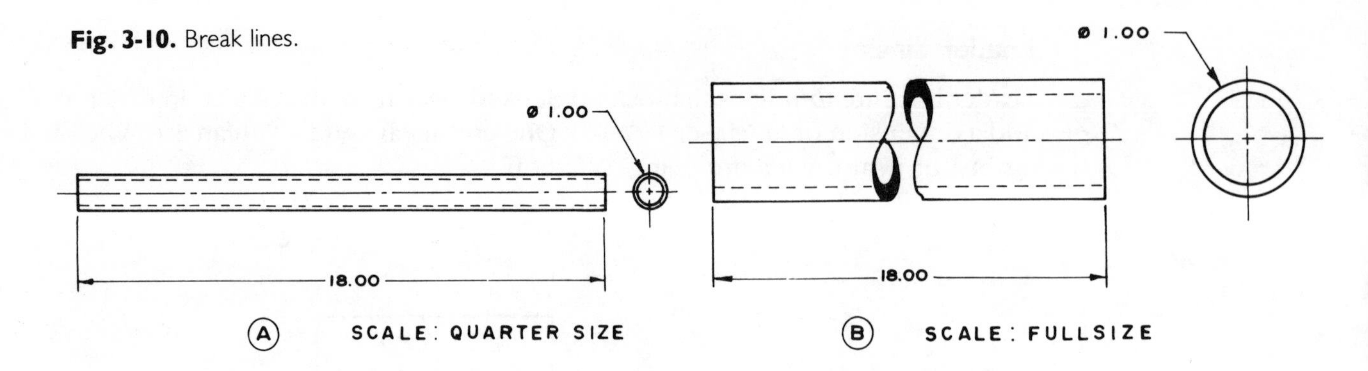

illustration at the top was drawn *quarter size* in order to fit it on the paper. Note how it is difficult to read and understand. Figure 3-10B, is the *same* object but was drawn full size by using break lines and is much easier to read and understand.

There are three kinds of break lines (see FIG. 3-11):

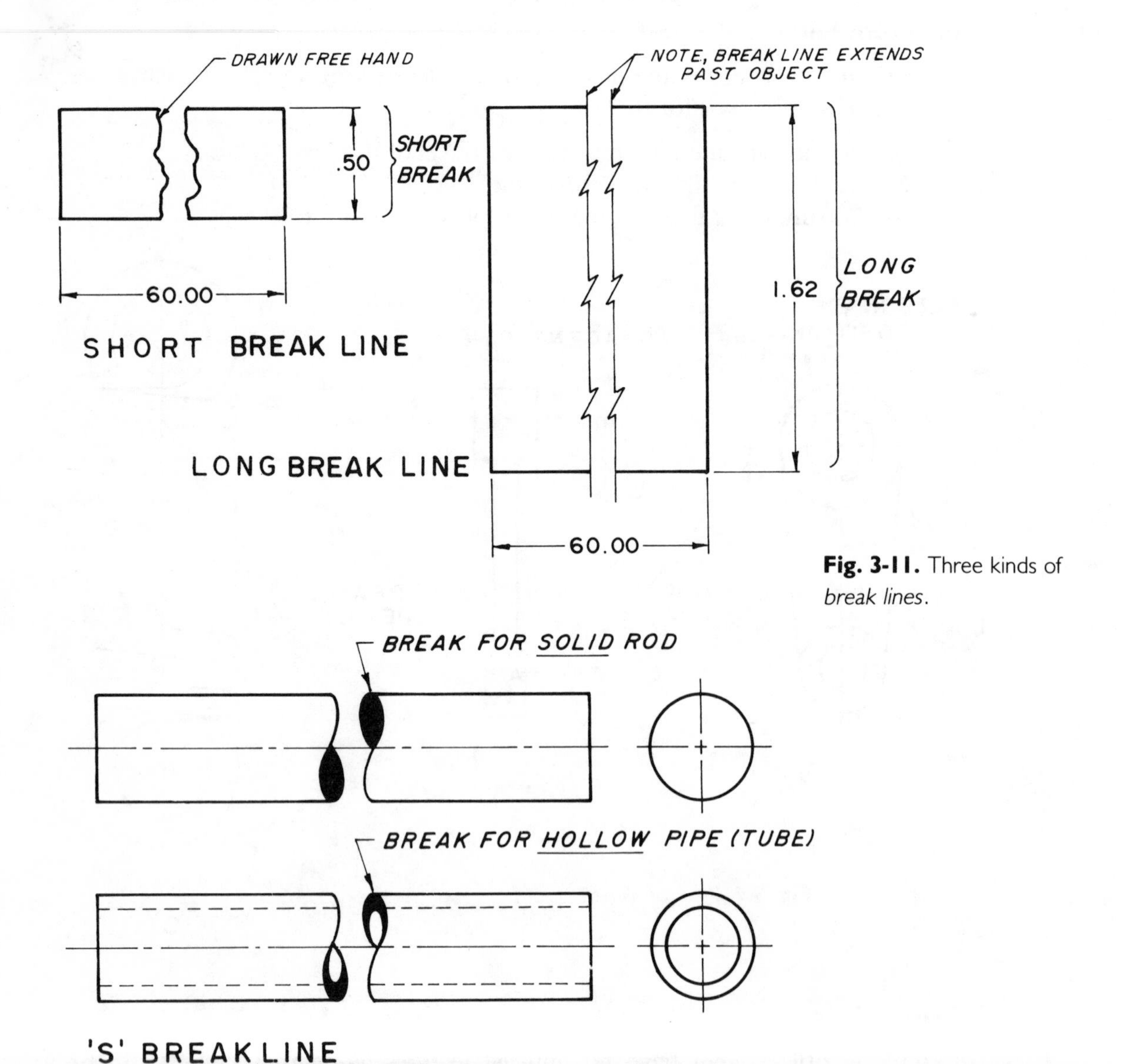

Fig. 3-11. Three kinds of *break lines*.

1. The short break line is a *thick* line drawn freehand to the same thickness as the object line.
2. The long break line is a *thin* line with "Z's" located at various intervals. (*Note.* This kind of break line extends beyond the object slightly.)
3. The "S" break line uses *thick* lines the same thickness as an object line and is used for round objects only. (The top illustration is solid like a rod, the bottom illustration is a hollow tube.

Other Kinds of Lines

There are three kinds of lines, each of which will be fully illustrated in following chapters: *cutting plane lines* (thick), *section lining* (thin), and *projection lines* (thin and very light).

WORKSHEET 3-1 (Part One)

Instructions: Using Drawing Number A40937143 on p. 29, list the 10 kinds of lines noted by the letters A through K. Indicate also the pattern or style of each line. (see example, line one, for letter A.) (Answers in Appendix A.)

PROB.	LET	KIND OF LINE	HOW LINE IS IDENTIFIED
EXAMPLE	A	*HIDDEN LINE*	*THIN DASHED LINE*
1	B		
2	C		
3	D		
4	E		
5	F		
6	G		
7	H		
8	I		
9	J		
10	K		

WORKSHEET 3-1 (Part Two)

Instructions: Using Drawing Number A40937143 on p. 29, answer each question in the spaces provided. (Answers in Appendix A.)

1. What is the drawing number?

2. What scale was used in drawing this drawing?

3. How many thicknesses of lines were used on this drawing?

4. What are the limits of the 3.500 diameter hole?

5. What is the finish of the roller guide?

6. What is the smallest size the 2.000 dimension could be and still be within tolerance?

7. What size is this "A"-size drawing?

8. What is the part made of?

9. What is the smallest size the "L" could be and still pass inspection? (use the smallest 2.000 limit, minus a maximum .75 limit)

10. What dimension falls within the 3-B zone?

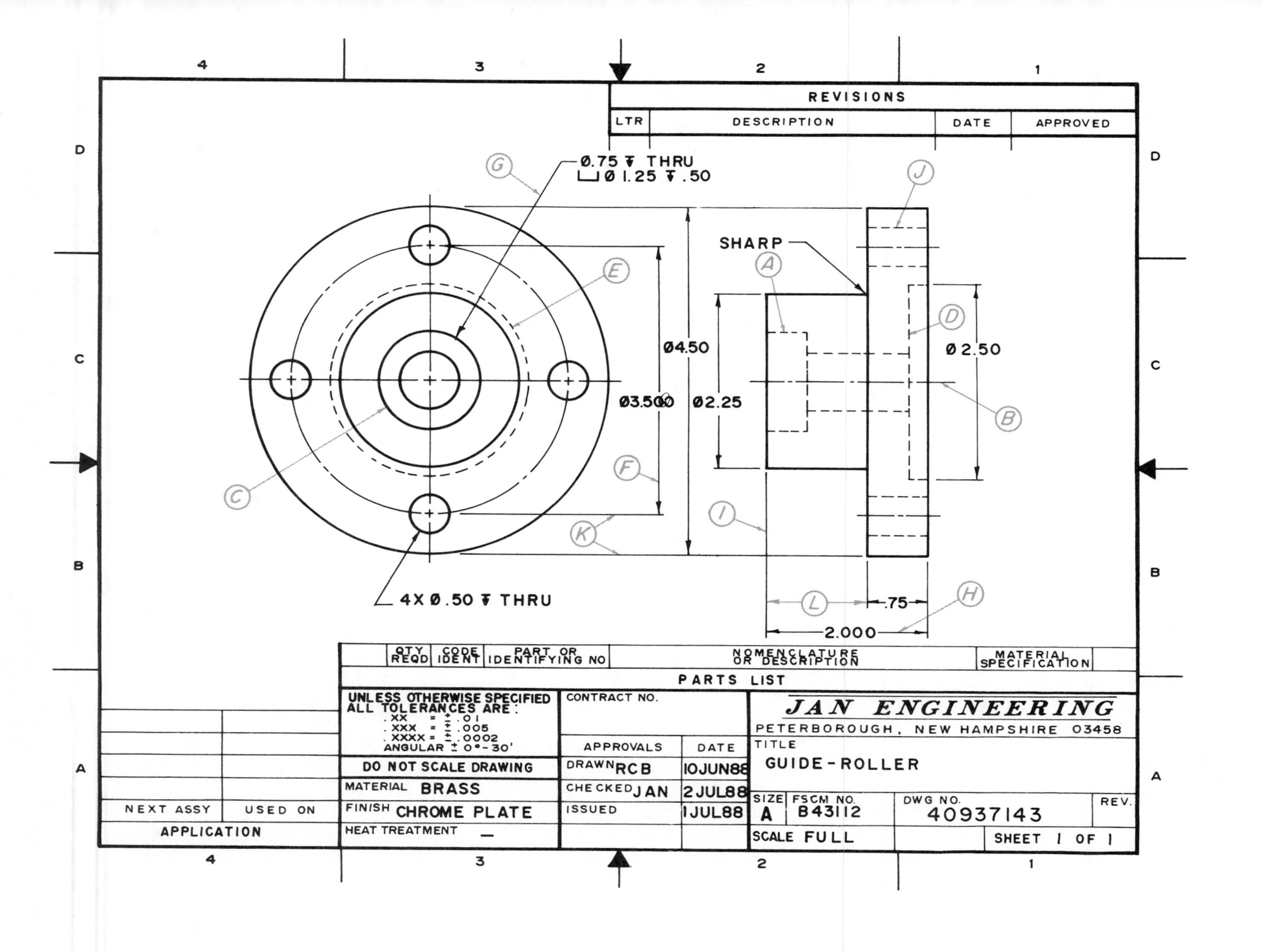
REVISIONS
LTR
DESCRIPTION
DATE
APPROVED
Ø.75 ↧ THRU
⌴Ø1.25 ↧ .50
SHARP
Ø4.50
Ø3.500
Ø2.25
Ø2.50
4X Ø.50 ↧ THRU
.75
2.000
A
B
C
D
E
F
G
H
I
J
K
L
QTY REQD
CODE IDENT
PART OR IDENTIFYING NO
NOMENCLATURE OR DESCRIPTION
MATERIAL SPECIFICATION
PARTS LIST
UNLESS OTHERWISE SPECIFIED ALL TOLERANCES ARE:
.XX = ±.01
.XXX = ±.005
.XXXX = ±.0002
ANGULAR ± 0°-30'
DO NOT SCALE DRAWING
MATERIAL BRASS
FINISH CHROME PLATE
HEAT TREATMENT —
CONTRACT NO.
APPROVALS
DATE
DRAWN RCB 10JUN88
CHECKED JAN 2JUL88
ISSUED 1JUL88
JAN ENGINEERING
PETERBOROUGH, NEW HAMPSHIRE 03458
TITLE
GUIDE-ROLLER
SIZE A
FSCM NO. B43112
DWG NO. 40937143
REV.
SCALE FULL
SHEET 1 OF 1
NEXT ASSY
USED ON
APPLICATION

One-View Drawings

Objective: The reader will be able to read and understand a one-view drawing and know the basic standard dimensioning practices used today in industry.

The function of a machine drawing is to *graphically* illustrate an object with enough dimensions, call-offs, and details so that it can be manufactured without question, any place in the world, exactly as it was intended to be. In order to do this, a drawing *must* be drawn in accordance to accepted and recognized standard methods. There are one-view drawings, two-view drawings three-view drawings, and sometimes even four or more views of the same object used. Each of those views is a particular kind of view, and each must be positioned in accordance to the recognized standard. In this chapter, simple one-view drawings will be presented.

Before studying a simple one-view drawing, a few basic dimensioning practices should be explained. These and all dimensioning practices apply to all views, regardless of how many views are used.

DIMENSIONING SYSTEMS

There are two methods used to place dimensions on a drawing, the *aligned system* and the *unidirectional system*. The aligned system is not used on machine drawings of today. It is, however, still used on architectural drawings. The unidirectional system is used on all machine drawings of today. It is easier to read and use.

In the *aligned* system all dimensions are placed in line, or parallel to the dimension line (see FIG. 4-1). All dimensions are read from the *bottom* of the paper and from the *right side* of the paper.

In the *unidirectional* dimensioning system *all* dimensions are placed *one* direction on the paper, regardless of what direction the dimension line is placed (see FIG. 4-2). All dimensions are placed so that they can be read from the *bottom* of the paper.

KINDS OF DIMENSIONS

A drawing must represent the object *exactly* as it is to be manufactured and must convey each and every detail *fully dimensioned.* There are two kinds of dimensions: size dimensions and location dimensions. As each name implies, *size dimensions* indicate sizes; *location dimensions* indicate locations (see FIG. 4-3). An object is dimensioned by a series of many dimensions, each noting a particular size or location. Note that the actual size or location is indicated by the distance *between the tips* of the arrowheads (see FIG. 4-4).

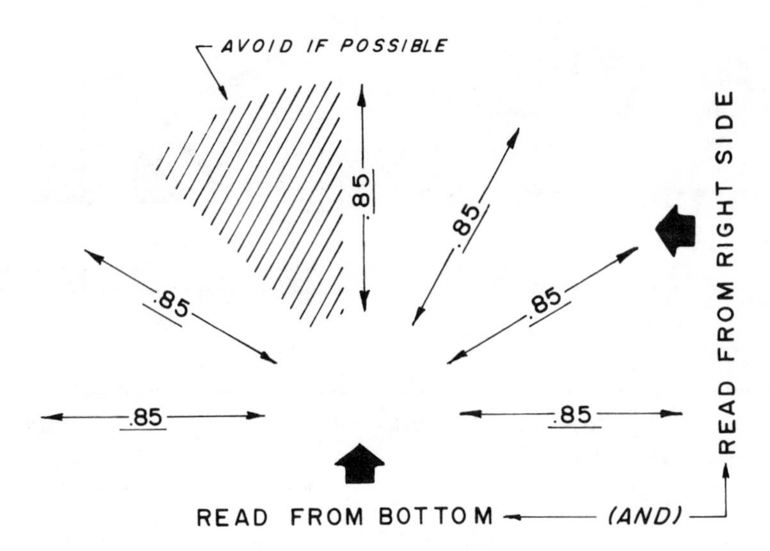

Fig. 4-1. Aligned dimensioning system.

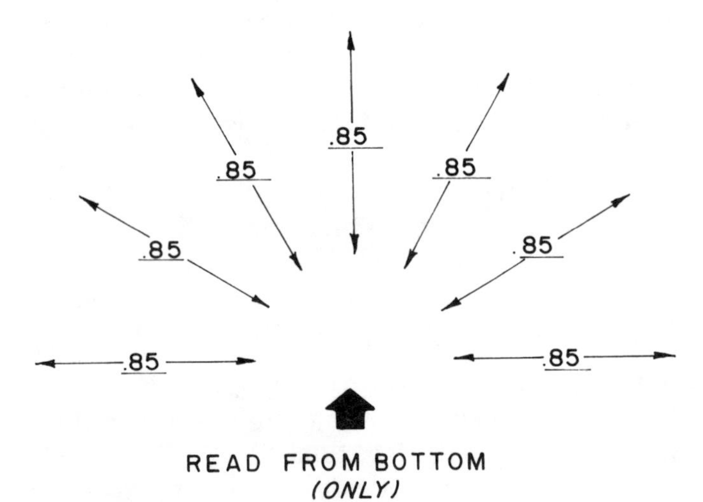

Fig. 4-2. Unidirectional dimensioning system.

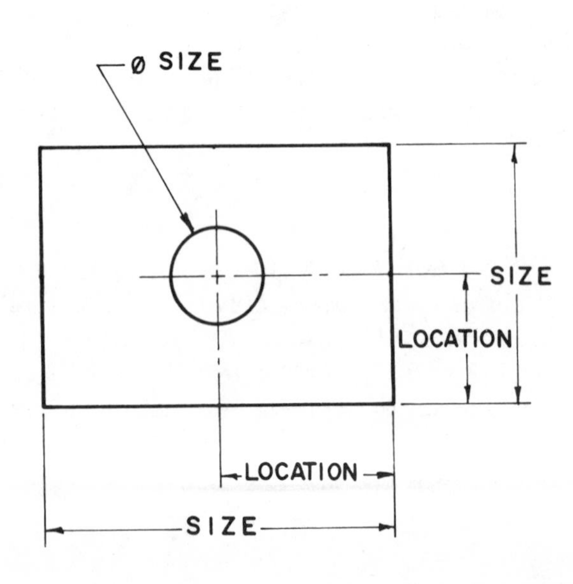

Fig. 4-3. Size and location dimensions.

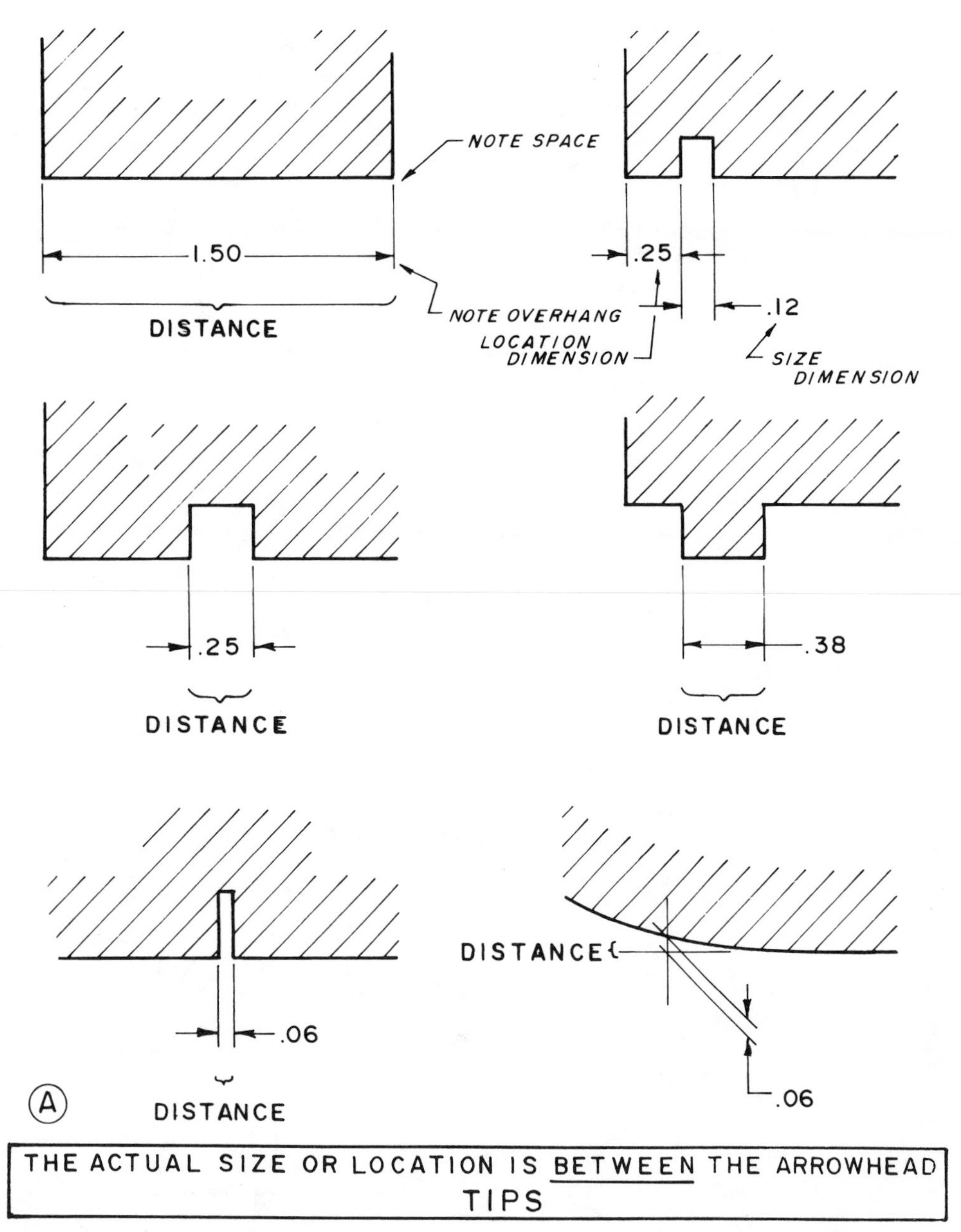

Fig. 4-4. The actual size or distance is located *between* the arrowhead points. (Continued on p. 34.)

DIMENSIONING A DIAMETER

A diameter is the distance from one side of a circle to the opposite side of the circle. To find a diameter, if only the radius is given, you must multiply the radius by two. All diameters, whether they are holes or a round area on an object, are dimensioned as illustrated in FIG. 4-5. Note the new symbol used to indicate the word "diameter." This new symbol is placed before the actual diameter size. Older drawings place the diameter size first, followed by the word "DIA," (for example, .500 DIA).

DIMENSIONING A RADIUS

A radius is half the distance across a diameter. To find a radius, if only the diameter is given, you must divide the diameter size by two. A circular arc is usually

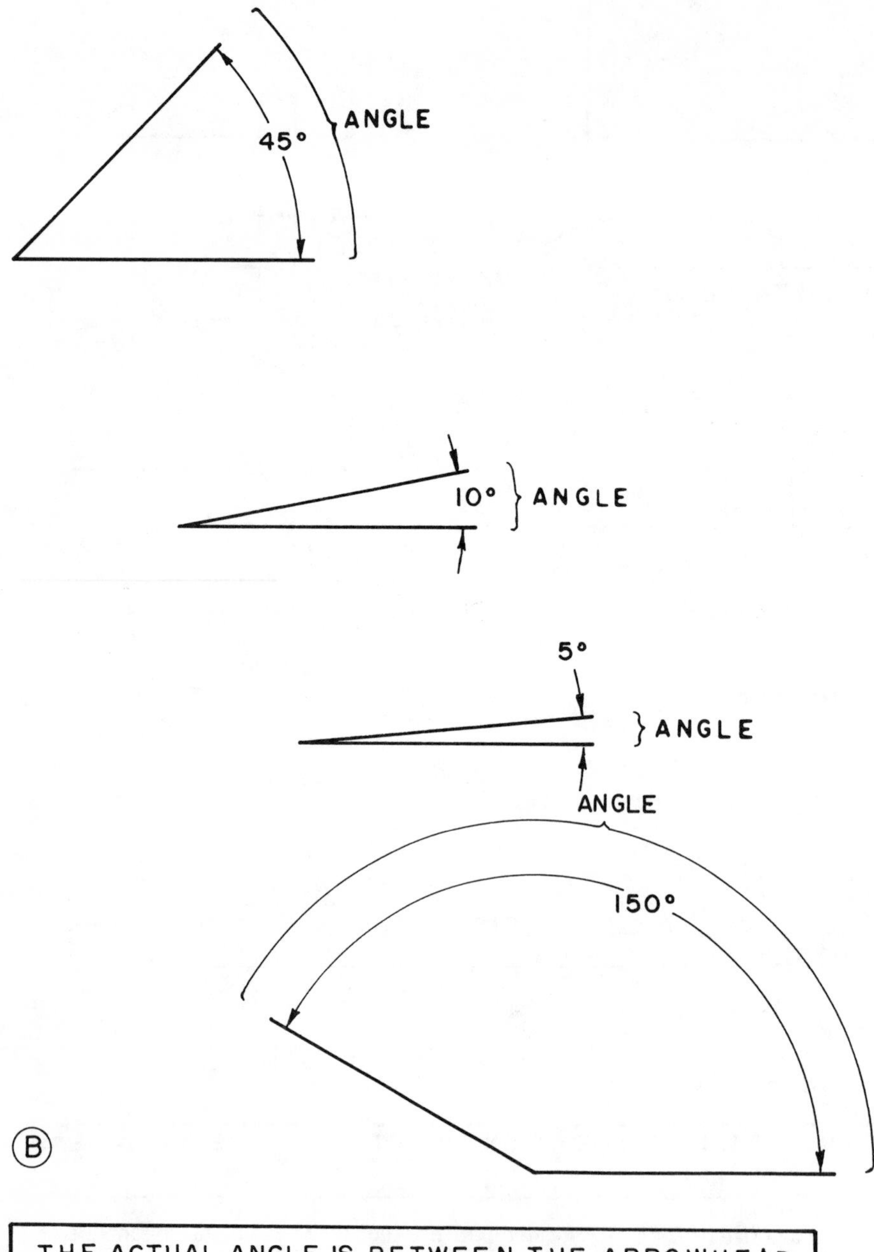

Fig. 4-4. Continued.

dimensioned by indicating its radius. The "R" placed before the dimension indicates that actual size of the arc, (see FIG. 4-6). The actual distance is the distance from the swing point to the *tip* of the arrowhead. In the event the swing point falls off the paper, the drafter will draw a swing point on the paper but *not* in the actual, correct location. This kind of dimension uses a broken dimension line (refer again to FIG. 4-6).

OUT-OF-SCALE DIMENSIONS

If a drawing is drawn slightly out of scale or if a change had been made without actually changing the object, dimensions will also be out of scale. Any out-of-

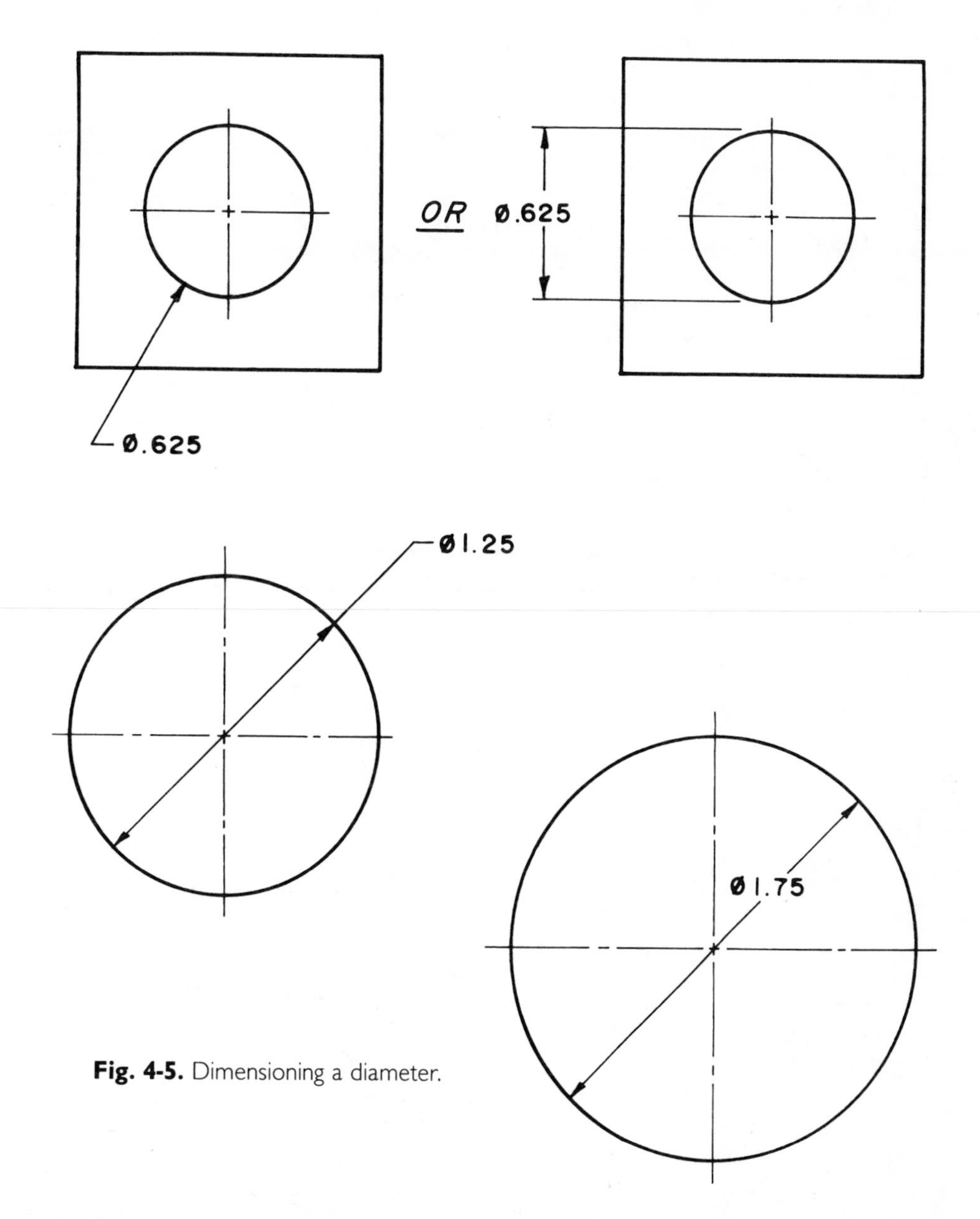

Fig. 4-5. Dimensioning a diameter.

scale dimensions are indicated by a *thick solid* line under the particular dimension (see FIG. 4-7). On older drawings, this was indicated by a thin, wavy line placed under the dimension. Usually, if the drawing had been changed, the original dimension size should be indicated in the *change block*, along with the date of change.

REFERENCE DIMENSIONS

As its name implies, a reference dimension is a dimension used for *reference only*. Under no circumstances should this dimension be used to manufacture or inspect the part. All reference dimensions are noted by a dimension inside a pair of parentheses (see FIG. 4-8). Older drawings simply added the dimension followed by the letters "REF" (for example, 1.25 REF). In FIG. 4-8, the slot size of 1.000(three decimal places) is more important than the .75 (two decimal places) location dimension. The (.75) reference dimension is there only to give reference to the fact that .75 is reference size of the right side. A correctly dimensioned drawing must *never* have a string of dimensions that total up to the outer, overall dimension; therefore, a reference dimension is used.

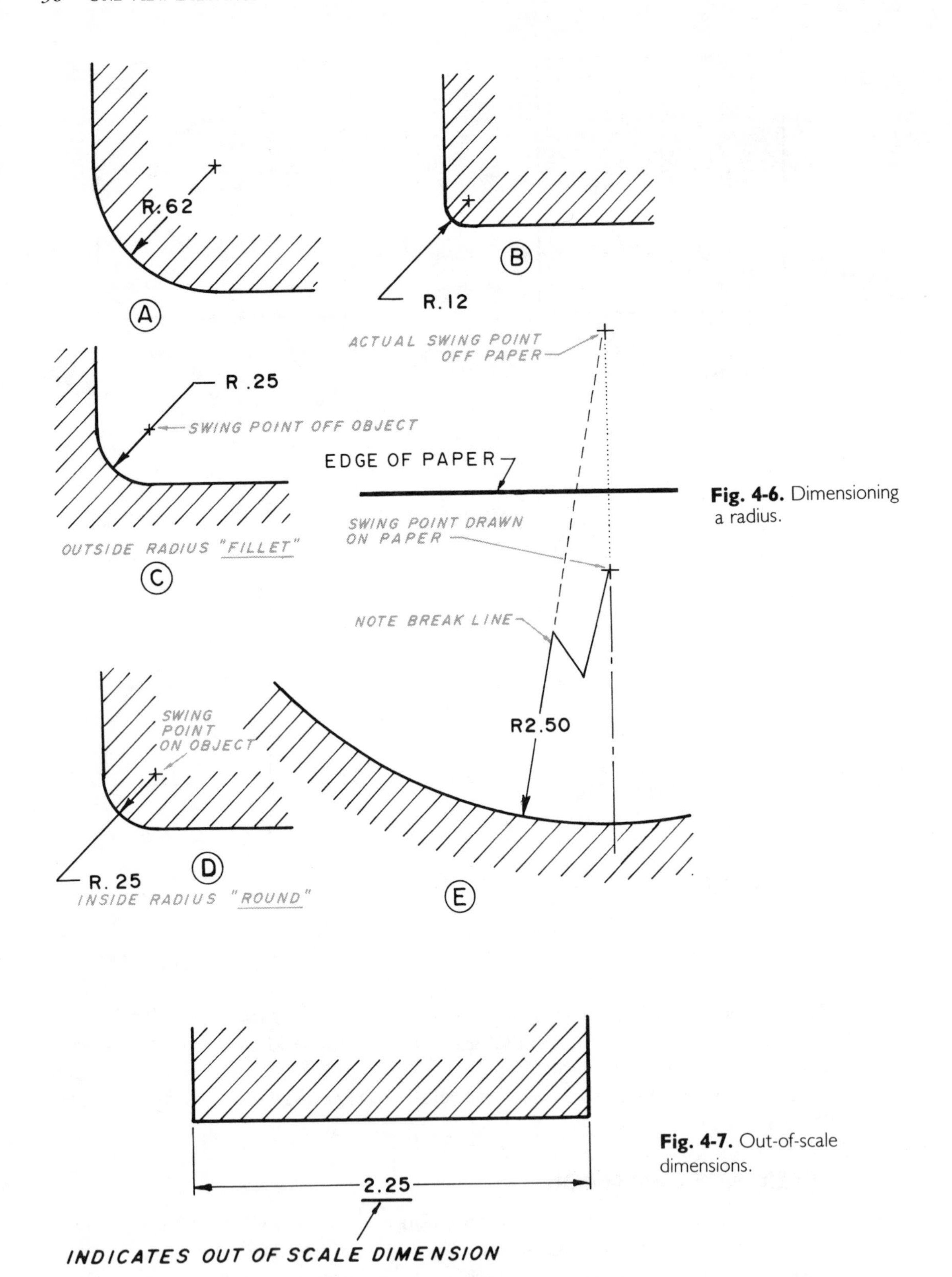

Fig. 4-6. Dimensioning a radius.

Fig. 4-7. Out-of-scale dimensions.

TYPICAL DIMENSIONS

A dimension followed by the letters TYP means "typical," that is, all such features are considered the exact same size or dimension.

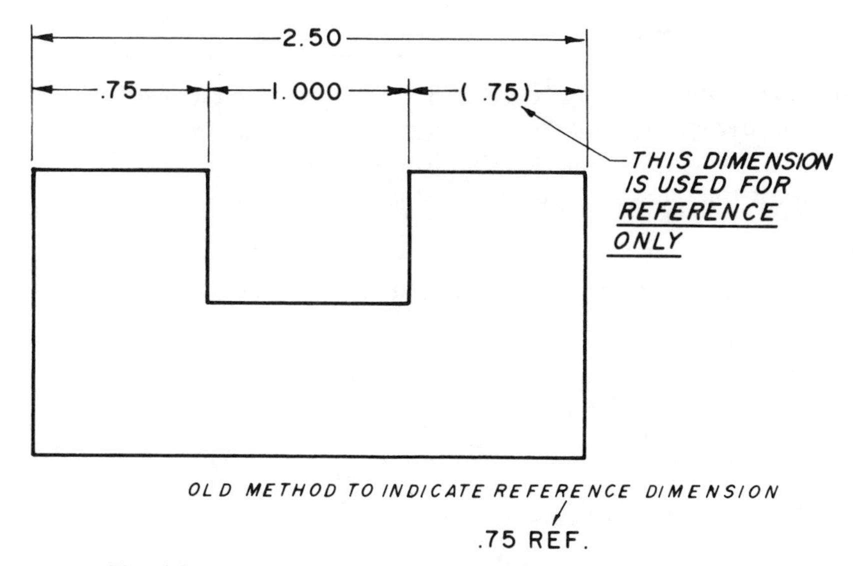

Fig. 4-8. Reference dimensions are used for *reference only.*

ONE-VIEW DRAWING

FIGURE 4-9 shows a simple thin gasket. If you held the gasket in your hand, it would look like FIG. 4-9A. The *line of sight* is the direction from which you are actually viewing the object. The drafter would draw the gasket as a one-view drawing as illustrated in FIG. 4-9B. Note that on this one-view drawing and other such simple objects, depth is not shown in this view; its thickness is noted in the title block under "MATERIAL." In single one-view drawings, the one view is always considered the *front view.*

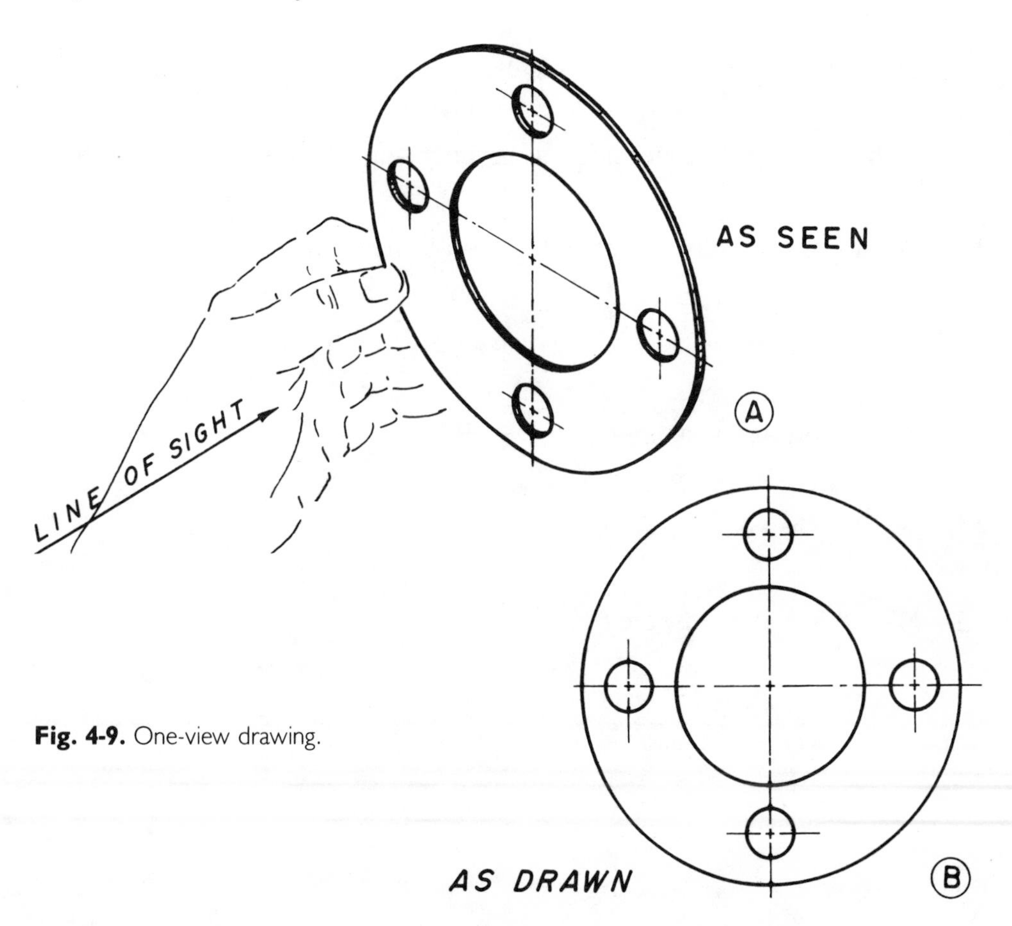

Fig. 4-9. One-view drawing.

WORKSHEET 4-1

Instructions: Using Drawing Number A113791 on p.39, answer the following questions in the spaces provided. (Answers in Appendix A.)

1. What is the overall length of the shim, dimension A?
2. Assuming the centerline indicates the center of the shim, what is dimension B?
3. What is dimension C in mm?
4. Calculate radius D.
5. What is the overall height of the shim, dimension E?
6. Calculate dimension F.
7. What kind of a line is G?
8. What kind of a line is H?
9. What is the thickness of the shim?
10. If the four .62 diameter holes are evenly spaced, how far apart are they?
11. What is the maximum size of the 1.00 diameter hole?
12. What was the size of the 1.00 diameter hole when the part was issued?
13. What kind of line is line I?
14. What is the contract number.
15. What kind of line is line J?

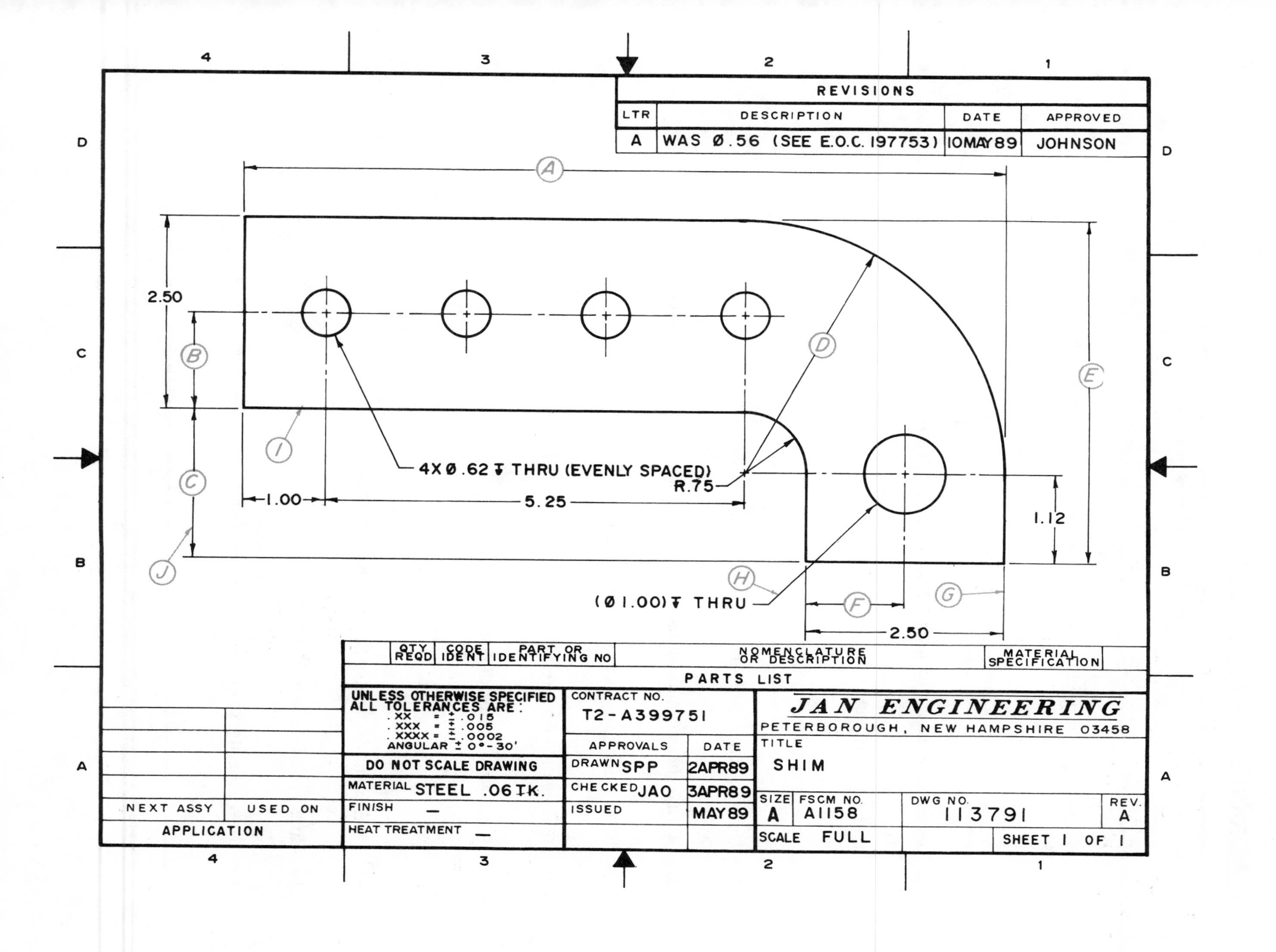
REVISIONS
LTR
DESCRIPTION
DATE
APPROVED
A
WAS Ø .56 (SEE E.O.C. 197753)
10MAY89
JOHNSON
A
B
C
D
E
F
G
H
I
J
2.50
4X Ø .62 ↧ THRU (EVENLY SPACED)
R.75
1.00
5.25
1.12
(Ø 1.00) ↧ THRU
2.50
QTY REQD
CODE IDENT
PART OR IDENTIFYING NO
NOMENCLATURE OR DESCRIPTION
MATERIAL SPECIFICATION
PARTS LIST
UNLESS OTHERWISE SPECIFIED ALL TOLERANCES ARE:
.XX = ± .015
.XXX = ± .005
.XXXX = ± .0002
ANGULAR ± 0°-30'
DO NOT SCALE DRAWING
MATERIAL STEEL .06 TK.
FINISH —
HEAT TREATMENT —
CONTRACT NO.
T2-A399751
APPROVALS
DATE
DRAWN SPP
2APR89
CHECKED JAO
3APR89
ISSUED
MAY 89
JAN ENGINEERING
PETERBOROUGH, NEW HAMPSHIRE 03458
TITLE
SHIM
SIZE
A
FSCM NO.
A1158
DWG NO.
113791
REV.
A
SCALE FULL
SHEET 1 OF 1
NEXT ASSY
USED ON
APPLICATION

WORKSHEET 4-2

Instructions: Using Drawing Number A872751-21-A, on p. 41, answer the following questions in the spaces provided. (Answers in Appendix A.)

1. What is the overall width of the link, see dimension A?

2. Assuming the centerline indicates the center of the link, what is dimension B?

3. What is the minimum limit dimension C could possibly be?

4. What is the maximum limit dimension D could possibly be?

5. Line E is what kind of a line?

6. How thick is this part?

7. On what day, month and year was this part checked?

8. What is the maximum distance the two .62 diameter holes could be apart?

9. What kind of an extension line is line F?

10. What is the tolerance for the 1.50 diameter hole?

11. What was the 1.62 dimension when the part was drawn?

12. What scale is used to draw this part?

13. The radius of the 1.50 diameter hole is what?

14. Indicate the contract number.

15. How far is the swing point of the 4.62 radius from the edge of the link?

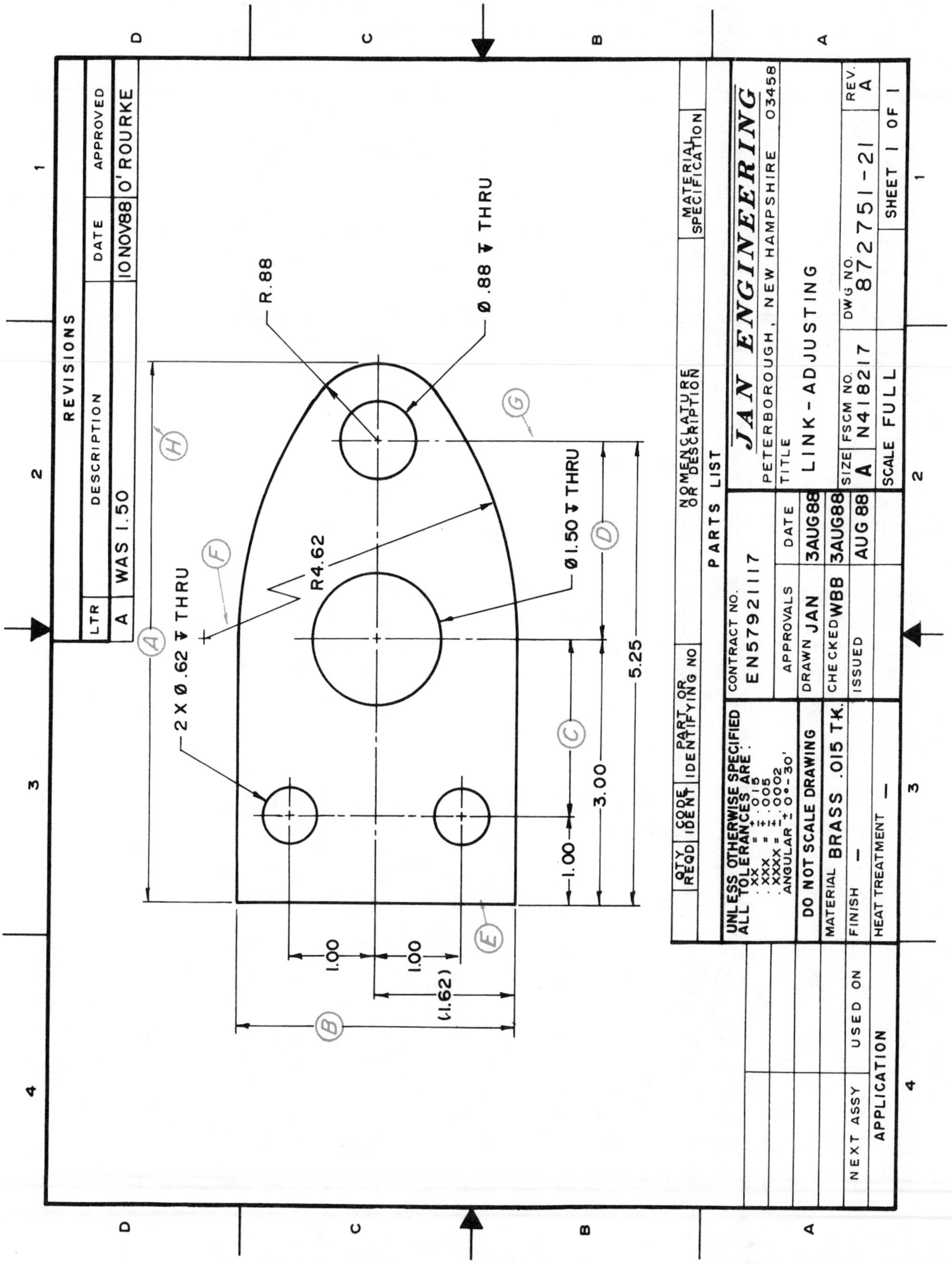

REVISIONS			
LTR	DESCRIPTION	DATE	APPROVED
A	WAS 1.50	10NOV88	O'ROURKE

2 X Ø .62 ↧ THRU

R.88

R4.62

Ø .88 ↧ THRU

Ø 1.50 ↧ THRU

1.00

1.00

(1.62)

1.00

3.00

5.25

A B C D E F G H

QTY REQD	CODE IDENT	PART OR IDENTIFYING NO	NOMENCLATURE OR DESCRIPTION	MATERIAL SPECIFICATION

PARTS LIST

UNLESS OTHERWISE SPECIFIED ALL TOLERANCES ARE:
.XX = ± .015
.XXX = ± .005
.XXXX = ± .0002
ANGULAR ± 0°-30'

DO NOT SCALE DRAWING

MATERIAL BRASS .015 TK.

FINISH —

HEAT TREATMENT —

CONTRACT NO. EN57921117

APPROVALS		DATE
DRAWN	JAN	3AUG88
CHECKED	WBB	3AUG88
ISSUED		AUG 88

JAN ENGINEERING
PETERBOROUGH, NEW HAMPSHIRE 03458

TITLE LINK-ADJUSTING

SIZE	FSCM NO.	DWG NO.	REV.
A	N418217	872751-21	A

SCALE FULL

SHEET 1 OF 1

NEXT ASSY	USED ON
APPLICATION	

DIMENSIONING/LOCATING HOLES

There are many methods used to locate holes. If holes are to be located on a circle, the circle is called the *bolt circle.* The bolt circle is a diameter and is usually indicated by the initials B.C., e.g., 2.25 B.C. (see FIG. 4-10).

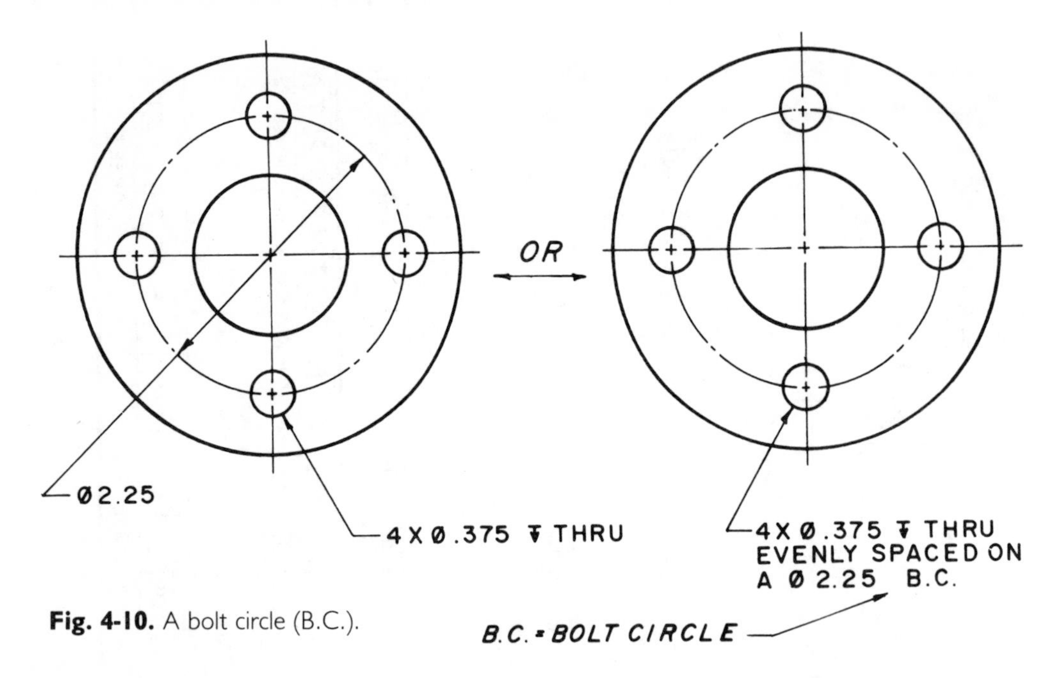

Fig. 4-10. A bolt circle (B.C.).

There are 360 degrees in a circle and holes are usually called off by the number of degrees they are apart (see FIG. 4-11). If the holes are to be equally spaced and the number of degrees is not given, you must divide the number of holes into 360 degrees. For example: there are six, .375-diameter holes, equally spaced on

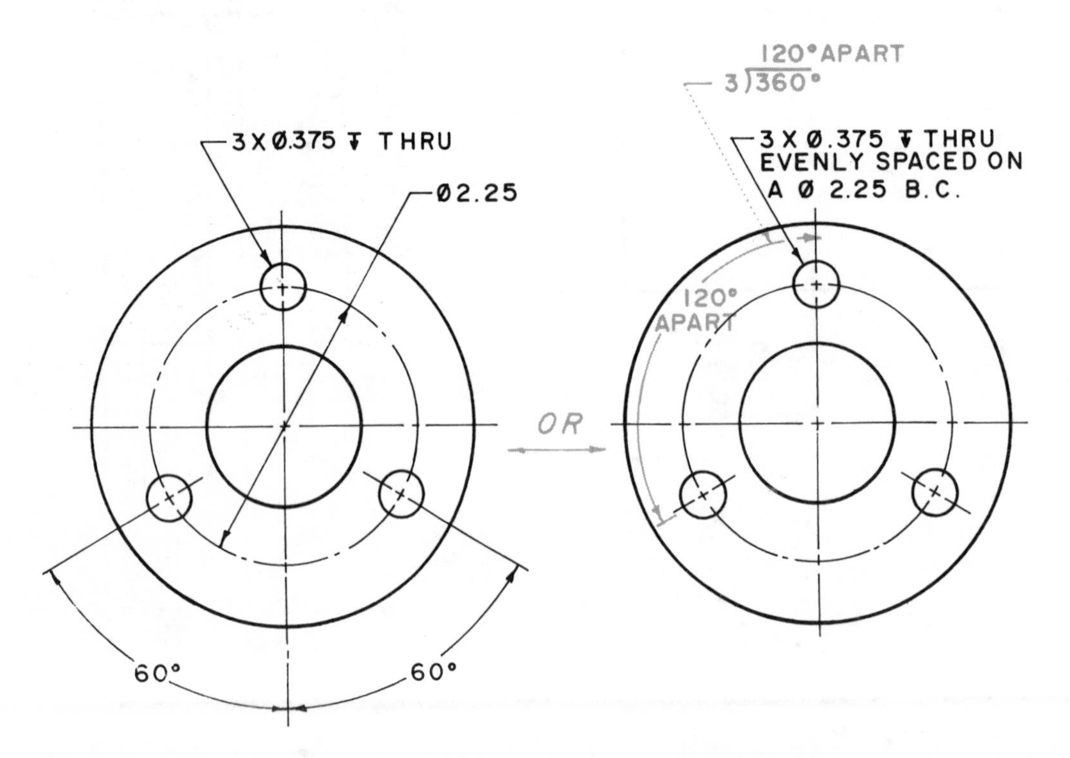

Fig. 4-11. Holes on a bolt circle are usually located by the number of degrees they are apart.

a 2.25 B.C. The holes are 60 degrees apart. (Six divided into 360 degrees.) If the holes are not located on the usual horizontal or vertical centerlines, one hole must be located, and the other holes must be dimensioned in relation to that hole (see FIG. 4-12). Note in hole calloffs, the number place *before* the diameter indicates how many holes are required. For example, 5 × DIAMETER .25 means there are *five*, .25-diameter holes required (see FIG. 4-13).

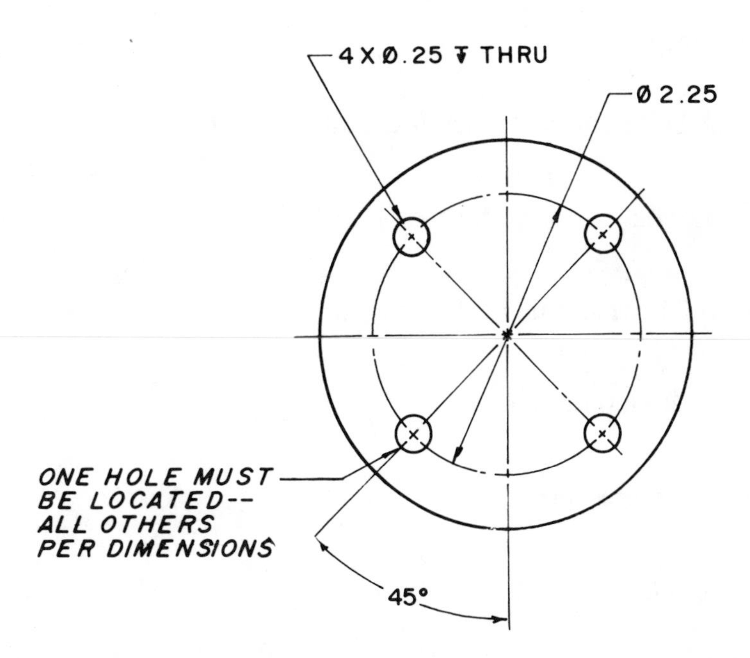

Fig. 4-12. If holes are not located on the usual horizontal or vertical centerlines, one hole must be located, and the other holes are dimensioned in relation to that hole.

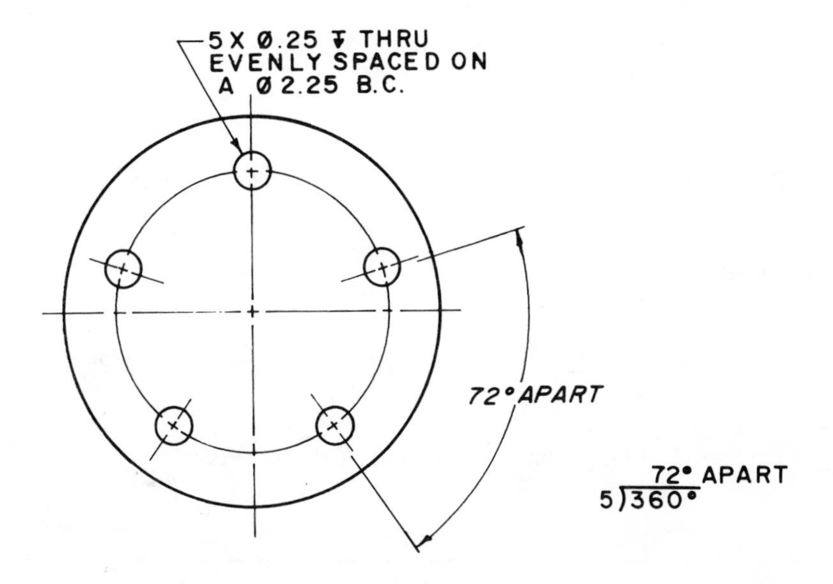

Fig. 4-13. The number before the diameter shows the number of holes required.

WORKSHEET 4-3

Instructions: Using Drawing Number A541223 on p. 45, answer the following questions in the spaces provided. (Answers in Appendix A.)

1. What is angle A in degrees?

2. What is the diameter at B?

3. What is the largest possible measurement for angle C?

4. What kind of dimension is dimension D?

5. Dimension E is what kind of dimension?

6. Line F is what kind of line?

7. What does G represent?

8. What is the fractional equivalent of the 0.50 holes?

9. What is the gasket made of?

10. What kind of a line is H?

11. Line I is what kind of line?

12. What is the I.D. (inside diameter) of the part?

13. How thick is the gasket?

14. How far apart, in degrees, is the hole "b" from hole "d"?

15. What is the O.D. (outside diameter) of the gasket?

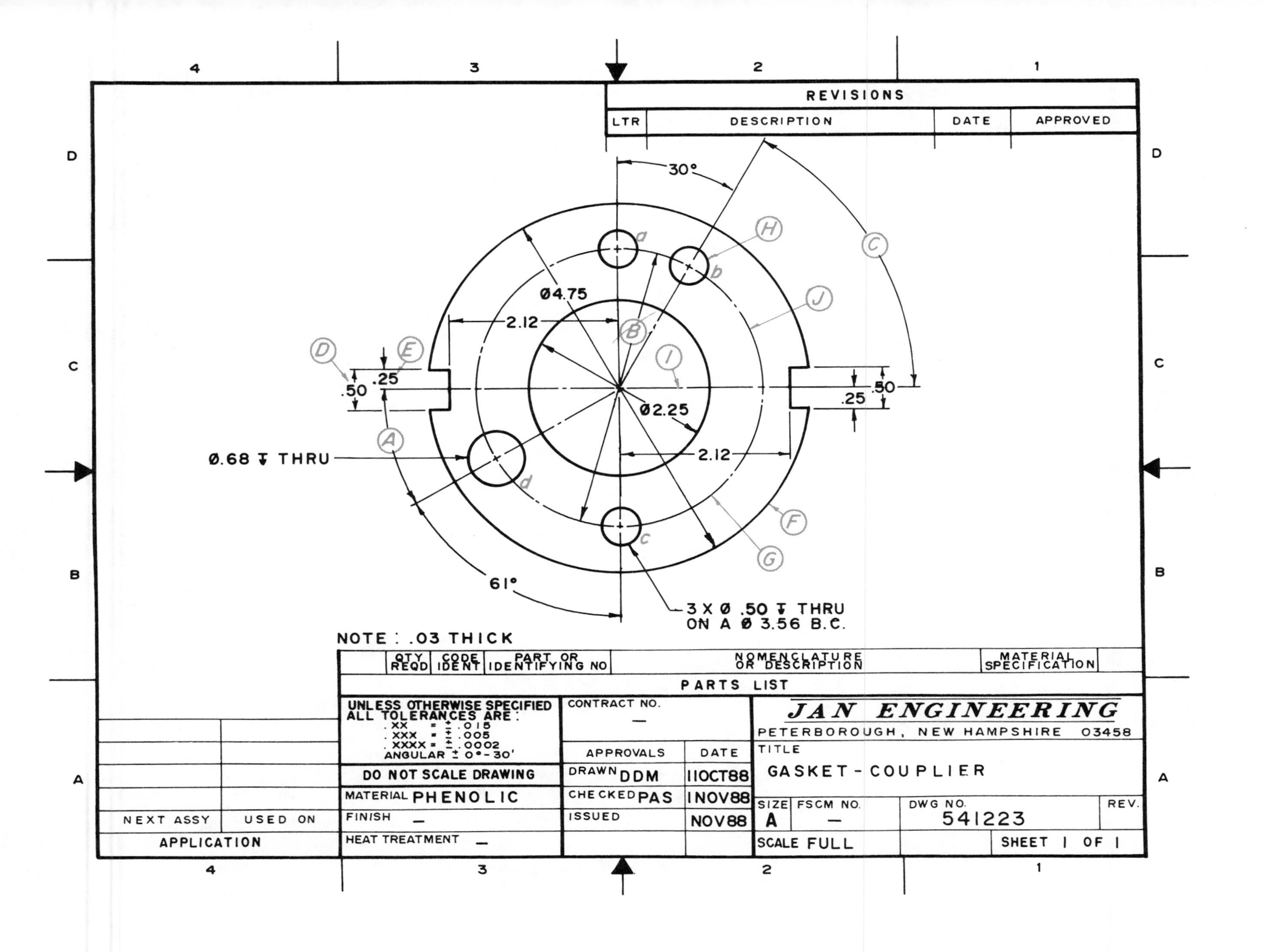
REVISIONS
LTR
DESCRIPTION
DATE
APPROVED
30°
Ø4.75
2.12
.25
.50
Ø2.25
.25
.50
Ø.68 ↧ THRU
2.12
61°
3 X Ø .50 ↧ THRU
ON A Ø 3.56 B.C.
NOTE : .03 THICK
QTY REQD
CODE IDENT
PART OR IDENTIFYING NO
NOMENCLATURE OR DESCRIPTION
MATERIAL SPECIFICATION
PARTS LIST
UNLESS OTHERWISE SPECIFIED ALL TOLERANCES ARE :
.XX = ± .015
.XXX = ± .005
.XXXX = ± .0002
ANGULAR ± 0°-30'
DO NOT SCALE DRAWING
MATERIAL PHENOLIC
FINISH —
HEAT TREATMENT —
CONTRACT NO. —
APPROVALS
DATE
DRAWN DDM 11OCT88
CHECKED PAS 1NOV88
ISSUED NOV88
JAN ENGINEERING
PETERBOROUGH, NEW HAMPSHIRE 03458
TITLE
GASKET - COUPLIER
SIZE A
FSCM NO. —
DWG NO. 541223
REV.
SCALE FULL
SHEET 1 OF 1
NEXT ASSY
USED ON
APPLICATION

Multiview Drawings

Objective: The reader will be able to read and understand a two-view drawing and know the basic standard dimensioning practice used today in industry.

Up to this point, all drawings have been of flat objects, requiring only one view. In order to fully describe the objects in this chapter, more than one view must be used.

TWO-VIEW DRAWINGS (Front- and Right-Side View)

The most important view is always the *front view*—it is the view with the most details. All other views, regardless of which views are used and regardless of how many views are used, are projected from the front view. The following example uses the front view and, projected directly to the right of the front view, the right side view.

Notice the pictorial drawing of a short piece of pipe (FIG. 5-1). Imagine holding this short piece of pipe in your hand. Look directly at the end of the pipe—this will be called the *front view*. Slowly rotate the pipe so that the right side is seen, this is the *right view*. The right-side view is located directly to the right of the front view, and although it is actually round in shape, it appears as a rectangle in the right-side view. The two hidden lines in the right-side view represent the inner walls of the pipe.

PROJECTION LINES

Projection lines are actually lines that are *not* usually seen on the drawing copy; they are the light lines the drafter used in laying out and drawing the views on the paper. FIGURE 5-2 illustrates the front view and side view of the pipe—the usually omitted projection lines have been added. Note how the projection lines project directly from one view to the other view. All features are also projected from one view to the next view. In this example, the hidden lines for the hole are projected into the right-side view and indicated by hidden lines.

In order to accurately interpret drawings, you must know and understand how projection lines are made and used. A more complicated object with many more features would have more projection lines, but the same application applies (see FIG. 5-3). In this example, surfaces A, B, and C and hole D in the front view project over into the right-side view. Study the right-side view and find surfaces A, B, C, and D in the right-side view.

Fig. 5-1. Two-view drawing (front view, and right-side view).

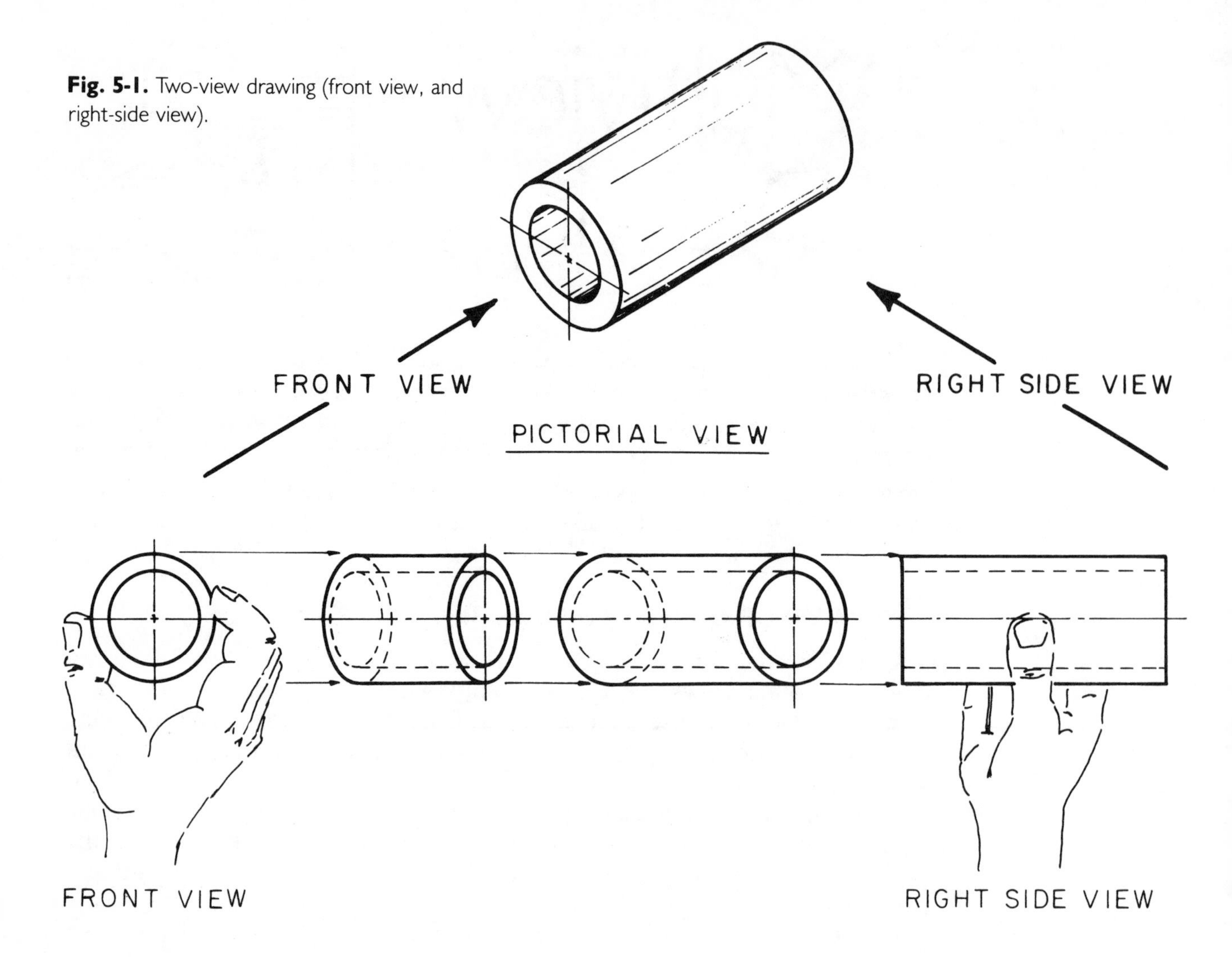

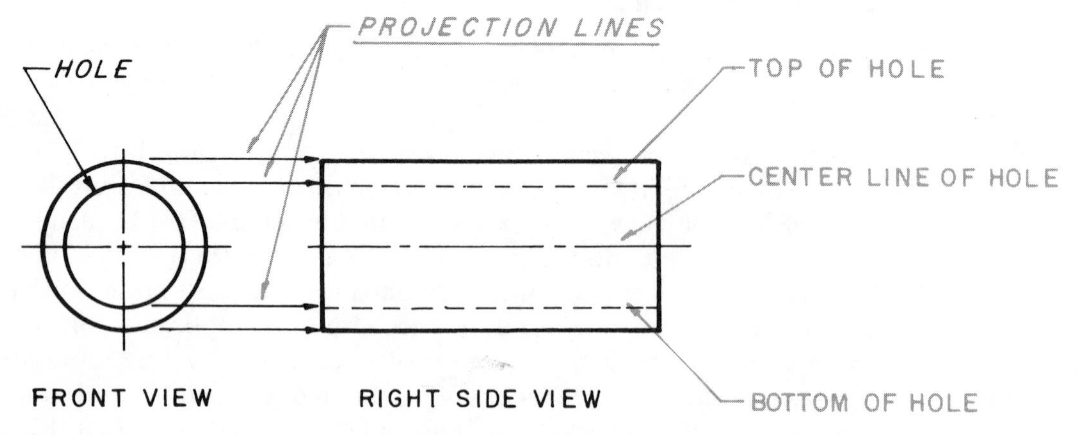

Fig. 5-2. Projection lines.

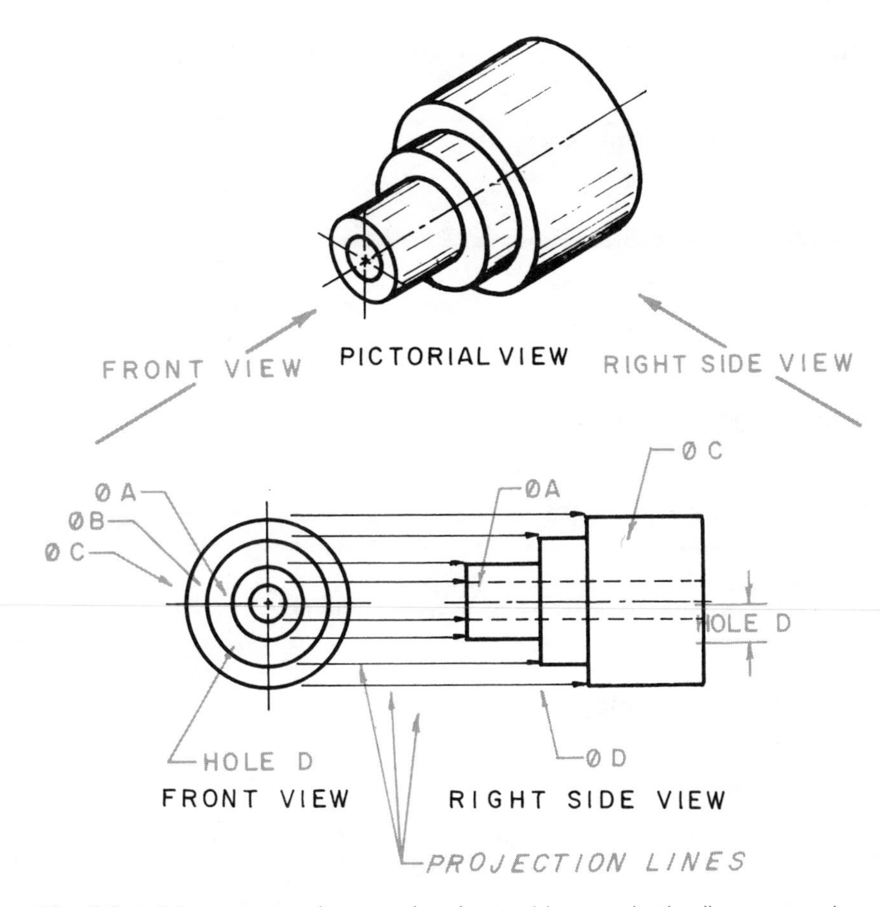

Fig. 5-3. It is important to know and understand how projection lines are made and used.

WORKSHEET 5-1

Instructions: Using Drawing Number A067835 on p. 51, answer the following questions in the spaces provided. To help in answering the questions, add light projection lines from the front view to the right-side view. (Answers in Appendix A.)

1. What kind of line is line F?

2. What is the size of diameter A?

3. What is the size of diameter B?

4. What is the size of diameter C?

5. What is the size of diameter E?

6. What kind of line is line T?

7. What is the distance between surface U and surface I?

8. Surface V in the front view is what surface in the right-side view?

9. What is the size of diameter M?

10. What kind of line is line Q?

11. Dimension R is how deep?

12. Calculate dimension H.

13. What kind of line is line S?

14. What is the distance from surface V to surface U?

15. What kind of line is line K?

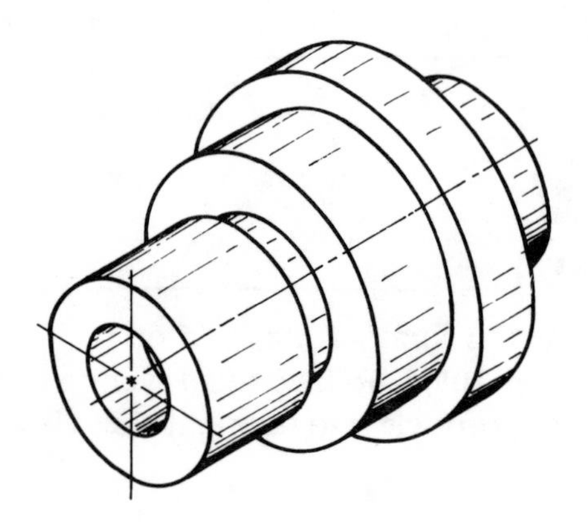

PICTORIAL VIEW

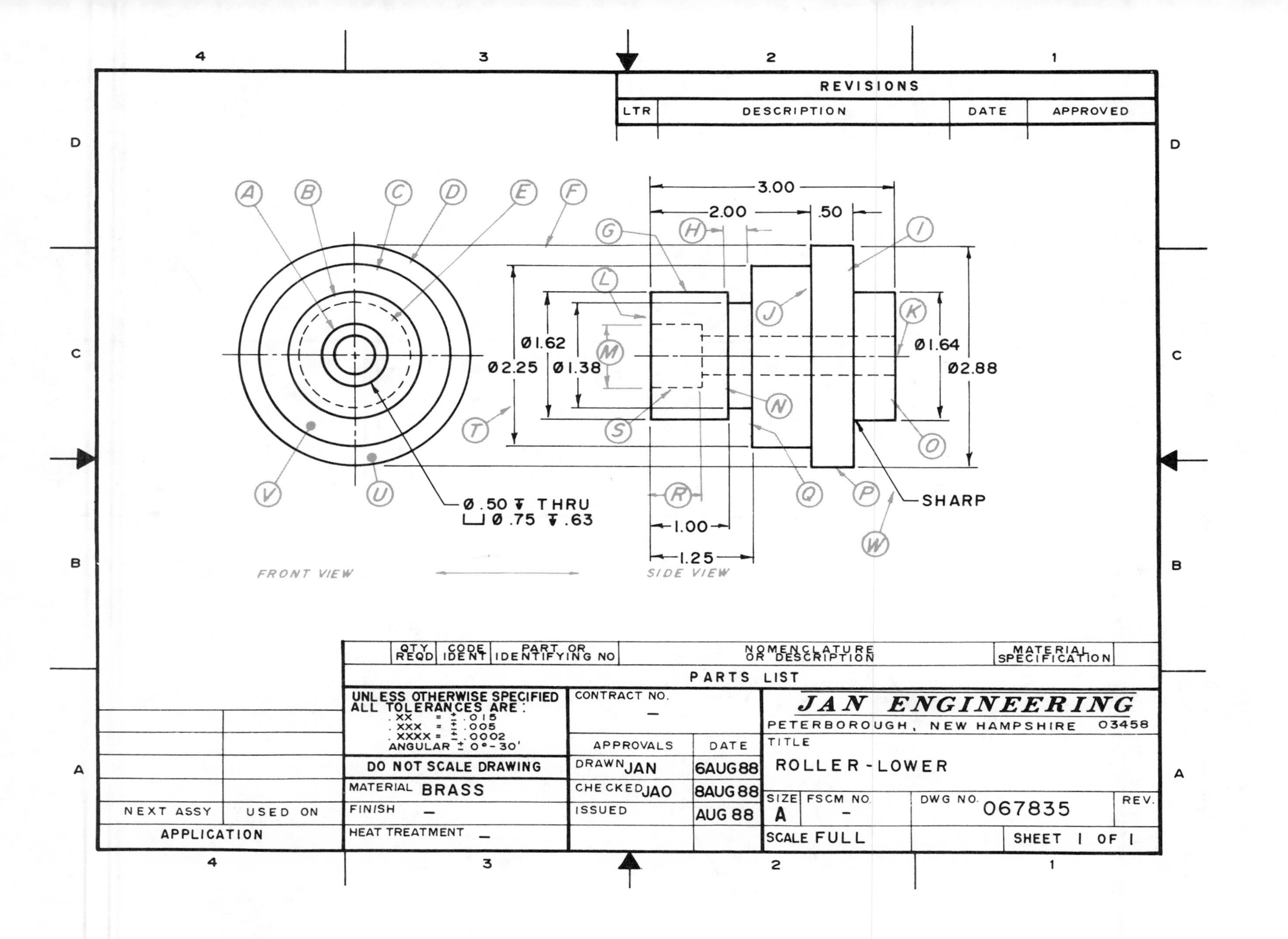
REVISIONS
LTR
DESCRIPTION
DATE
APPROVED
A
B
C
D
E
F
G
H
I
J
K
L
M
N
O
P
Q
R
S
T
U
V
W
3.00
2.00
.50
Ø1.62
Ø2.25
Ø1.38
Ø1.64
Ø2.88
Ø.50 ↧ THRU
⌴ Ø.75 ↧ .63
SHARP
1.00
1.25
FRONT VIEW
SIDE VIEW
QTY REQD
CODE IDENT
PART OR IDENTIFYING NO
NOMENCLATURE OR DESCRIPTION
MATERIAL SPECIFICATION
PARTS LIST
UNLESS OTHERWISE SPECIFIED ALL TOLERANCES ARE:
.XX = ± .015
.XXX = ± .005
.XXXX = ± .0002
ANGULAR ± 0°-30'
DO NOT SCALE DRAWING
MATERIAL BRASS
FINISH –
HEAT TREATMENT –
NEXT ASSY
USED ON
APPLICATION
CONTRACT NO. –
APPROVALS
DATE
DRAWN JAN
6AUG88
CHECKED JAO
8AUG88
ISSUED
AUG 88
JAN ENGINEERING
PETERBOROUGH, NEW HAMPSHIRE 03458
TITLE
ROLLER-LOWER
SIZE A
FSCM NO. –
DWG NO. 067835
REV.
SCALE FULL
SHEET 1 OF 1

TWO-VIEW DRAWINGS (Front View and Top View)

Another type of two-view drawing shows the front view and the top view. For instance, FIG. 5-4 shows a pictoral drawing of a simple block with a rectangular hole cut out. Try to imagine holding this simple block in your hand looking *directly* at one side, in this example it will be the *front view*. Then slowly rotate the block so the *top view* is seen. Notice the *top view* is located directly above the *front view*.

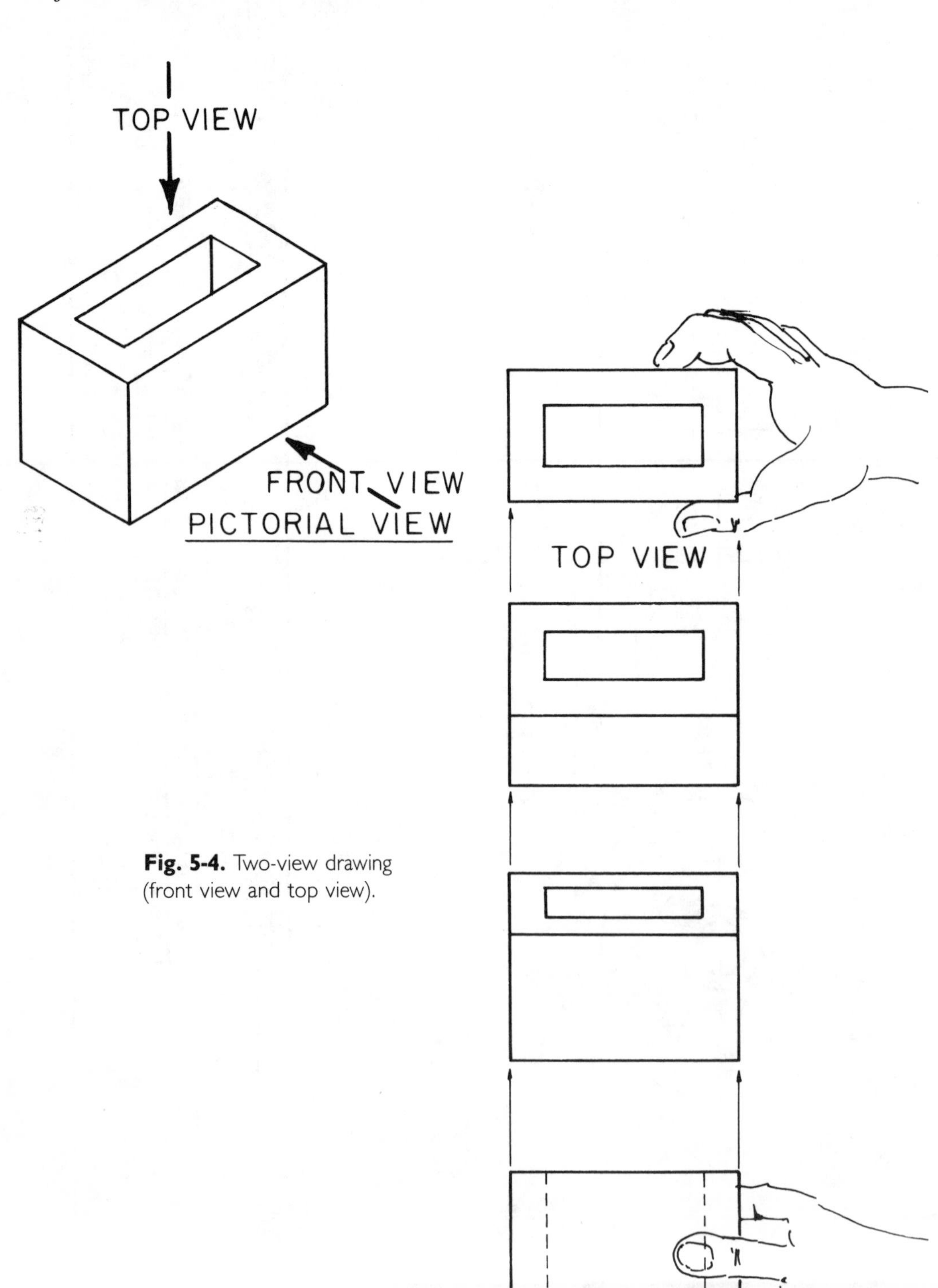

Fig. 5-4. Two-view drawing (front view and top view).

HOLE CALL-OFFS

A hole is usually made by a simple drilling operation. There are two major kinds of holes: thru holes and blind holes. A *thru hole* is, as its name implies, a hole drilled completely through the object. A *blind hole* is a hole that is drilled partly through the object (see FIG. 5-5). This is an example of a front view and a top view of two holes—a thru hole and a blind hole. There are two symbols used in hole call-offs—one for diameter and one for depth (refer back to FIG. 5-5). Using these two symbols, the drafter calls off exactly what is required. In the example to the left, a hole with a .50 diameter is drilled through. To the right, a blind hole with a .50 diameter is drilled to a depth of .88 deep. Note the .88 is the *full* depth at which the .50 diameter is maintained. The cone shape at the bottom of the hole represents the tapered end of the drill used to drill the hole. It is not included in the depth of the hole.

Another kind of hole is a *countersunk hole* (FIG. 5-6). FIGURE 5-6 also illustrates a front view and a top view of a countersunk hole. The symbol for a countersunk hole is a "V." Most all holes, regardless of what kind of hole, must be drilled first; therefore, the drilling information is listed first. In this example, a .50-diameter hole is drilled through the part first, as noted and as drawn. The countersinking operation is then called off. In this example, the "V" symbol indicates that the hole is to be countersunk until the outside diameter measures 1.13, at the *top* surface. The 82-degree call-off is the standard angle of most countersunk holes. Unless otherwise noted, all holes are countersunk to 82 degrees.

FIGURE 5-7 is an example of a *counterbored hole*. The counterbored hole symbol is illustrated by a rectangular "U." In this example, a simple .50-diameter hole is drilled through the part as noted. A 1.13-diameter counterbored is now used to a depth of .25. Study the front and top views of this example. Note how the counterbored hole is illustrated. The hidden lines illustrate the depth of the counterbore.

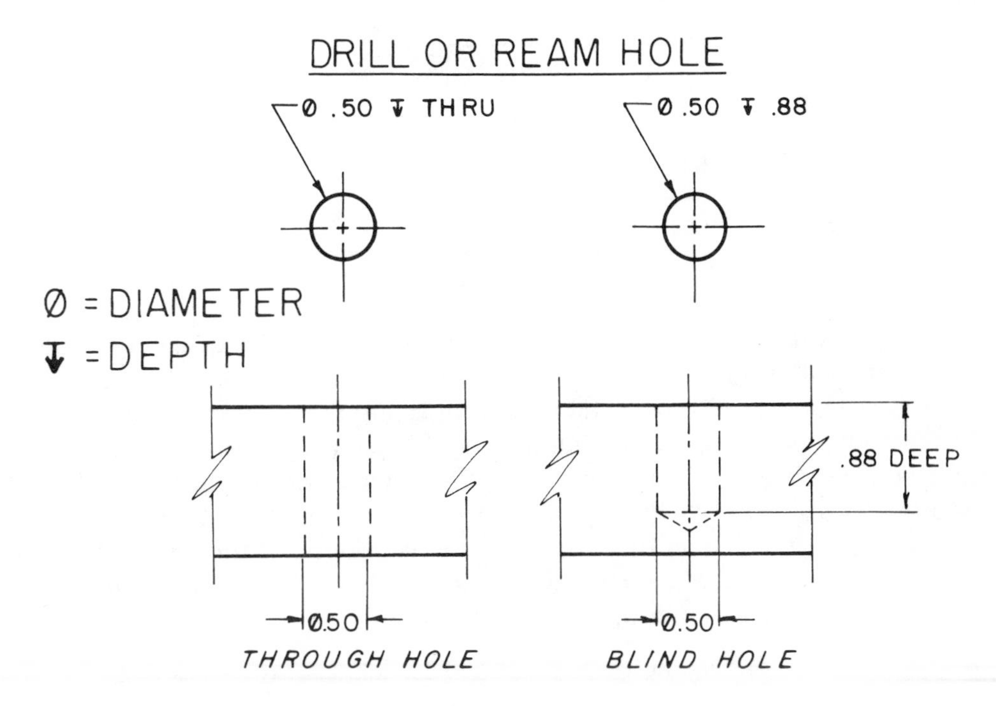

Fig. 5-5. Thru hole and blind hole specifications.

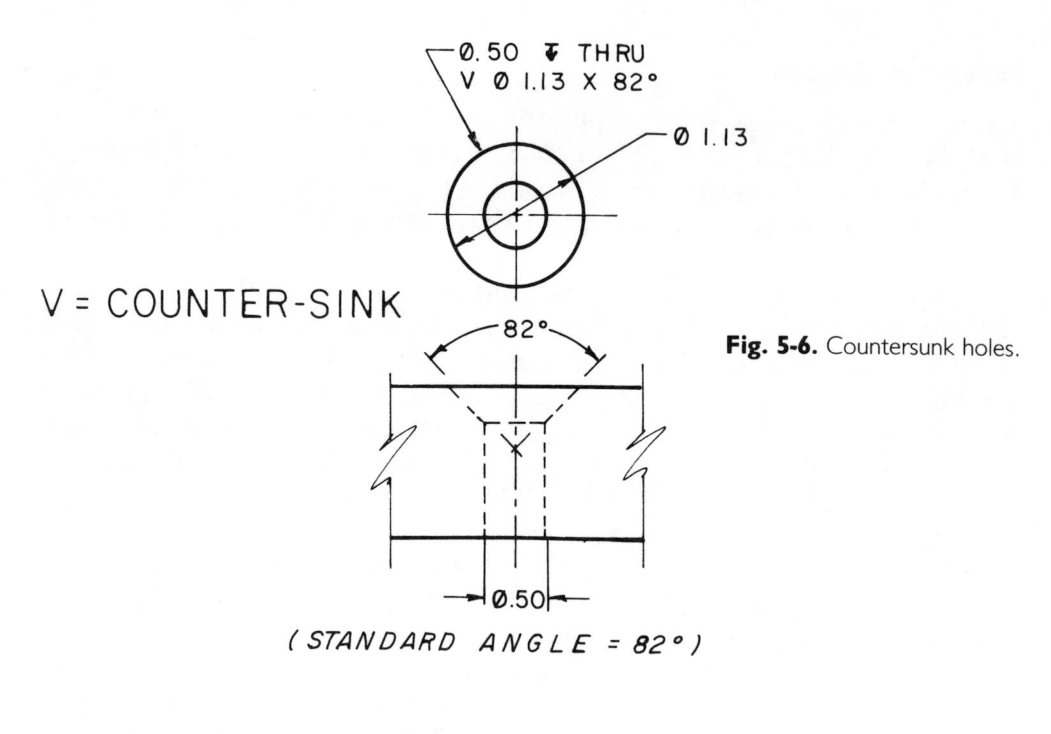

Fig. 5-6. Countersunk holes.

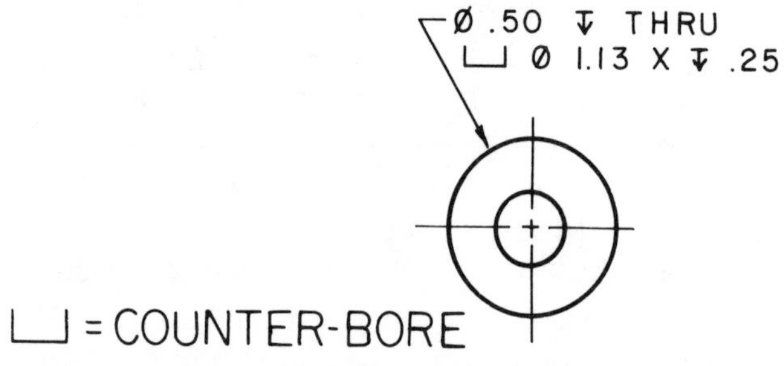

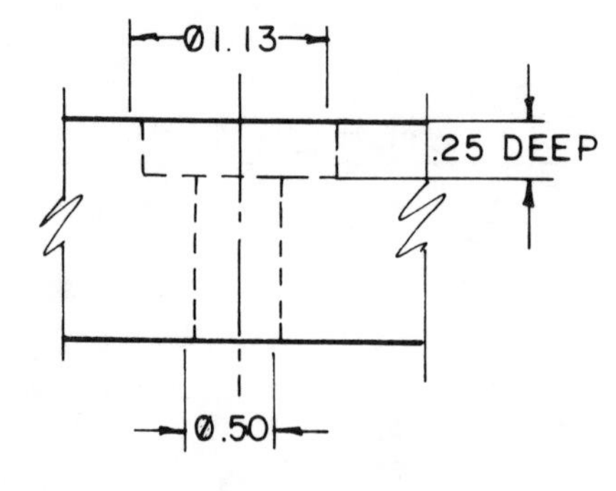

Fig. 5-7. Counterbore holes.

A hole, very similar to a counterbored hole, but not as deep, is the *spot-faced hole*. The spotface does not have a symbol. Instead, it uses the initials "S.F." (see FIG. 5-8). Note how the hidden lines indicate the depth of the spot-face. As with any hole, the drilling information is called off first, followed by the next operation. In this example, a .50 diameter is drilled through, followed by the 1.63-diameter spotface operation. Notice, the depth is *not* called off; in practice, the craftsperson determines the actual depth of the spotface at the time of manufacture. Generally, a spotface is made to a *minimum* depth required to produce a smooth *flat* surface.

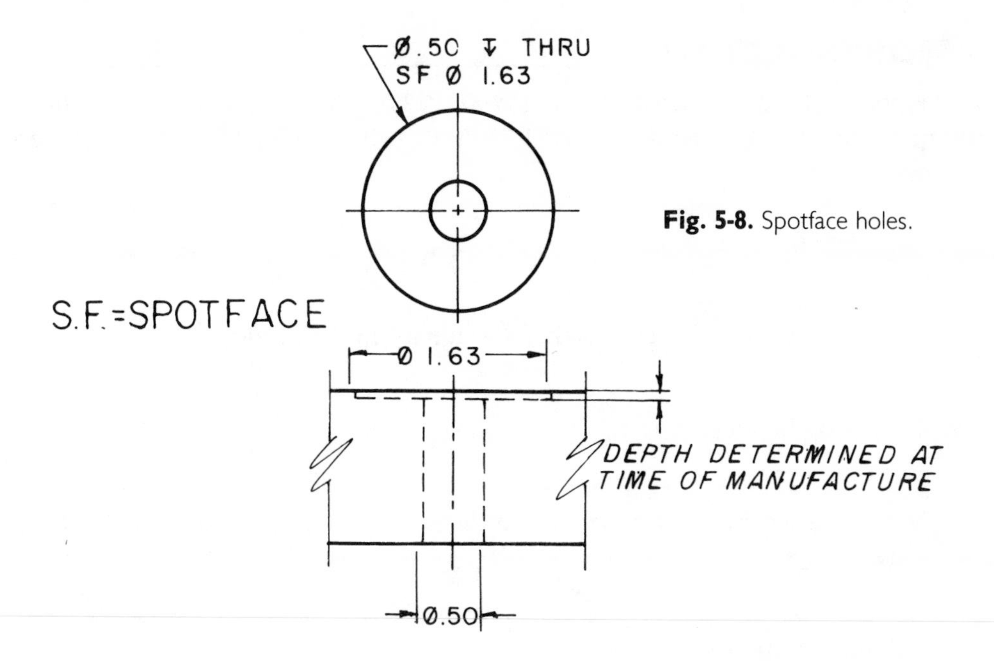

Fig. 5-8. Spotface holes.

Another type of a hole is the *slot* (FIG. 5-9). The drafter uses various call-offs in dimensioning a slot. In these three examples, the same size slot is called off three different ways. Notice that the actual radius is *not* indicated; the radius would be *half* the width of the slot, or .50. Each of the three examples call off a slot that is 1.00 wide × 3.00 long with a .50 radius at each end.

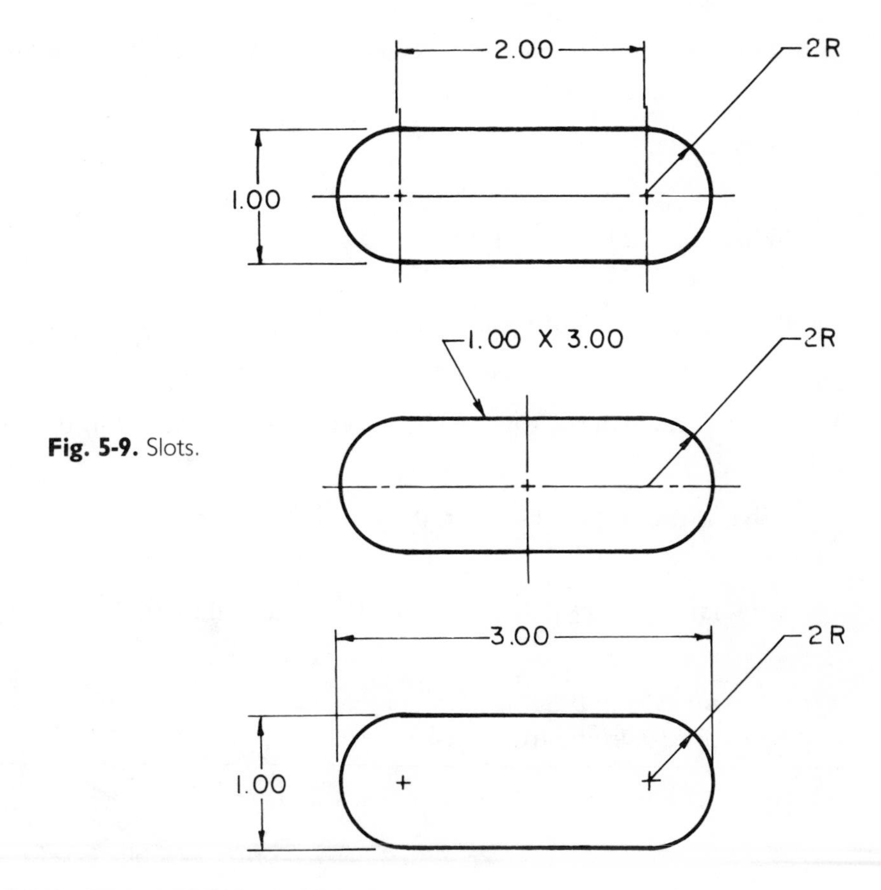

Fig. 5-9. Slots.

WORKSHEET 5-2

Instructions: Using Drawing Number A39812279 on p. 57, answer the following questions in the spaces provided. Note this drawing is in metric. (Answers in Appendix A.)

1. Calculate dimension E.

2. Surface J in the top view is what surface in the front view?

3. How long is dimension B?

4. What surface in the top view is indicated by hidden line U in the front view?

5. Calculate dimension W.

6. What surface in the top view is indicated by surface X in the front view?

7. How far is surface I in the top view from surface N in the top view?

8. What surface in the top view is indicated by surface R in the front view?

9. What is the size of diameter H?

10. What kind of line is line T?

11. What surface in the front view is indicated by surface G in the top view?

12. What surface in the top view is indicated by surface Q in the front view?

13. What is the size of diameter M?

14. How far is surface C from surface D in decimal inches?

15. What surface in the top view is indicated by surface Z in the front view?

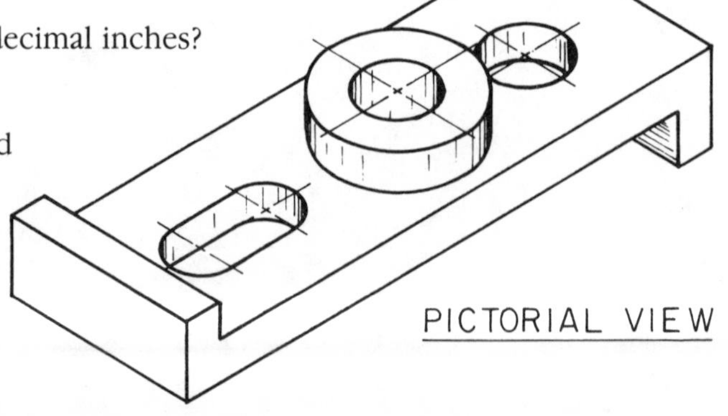

PICTORIAL VIEW

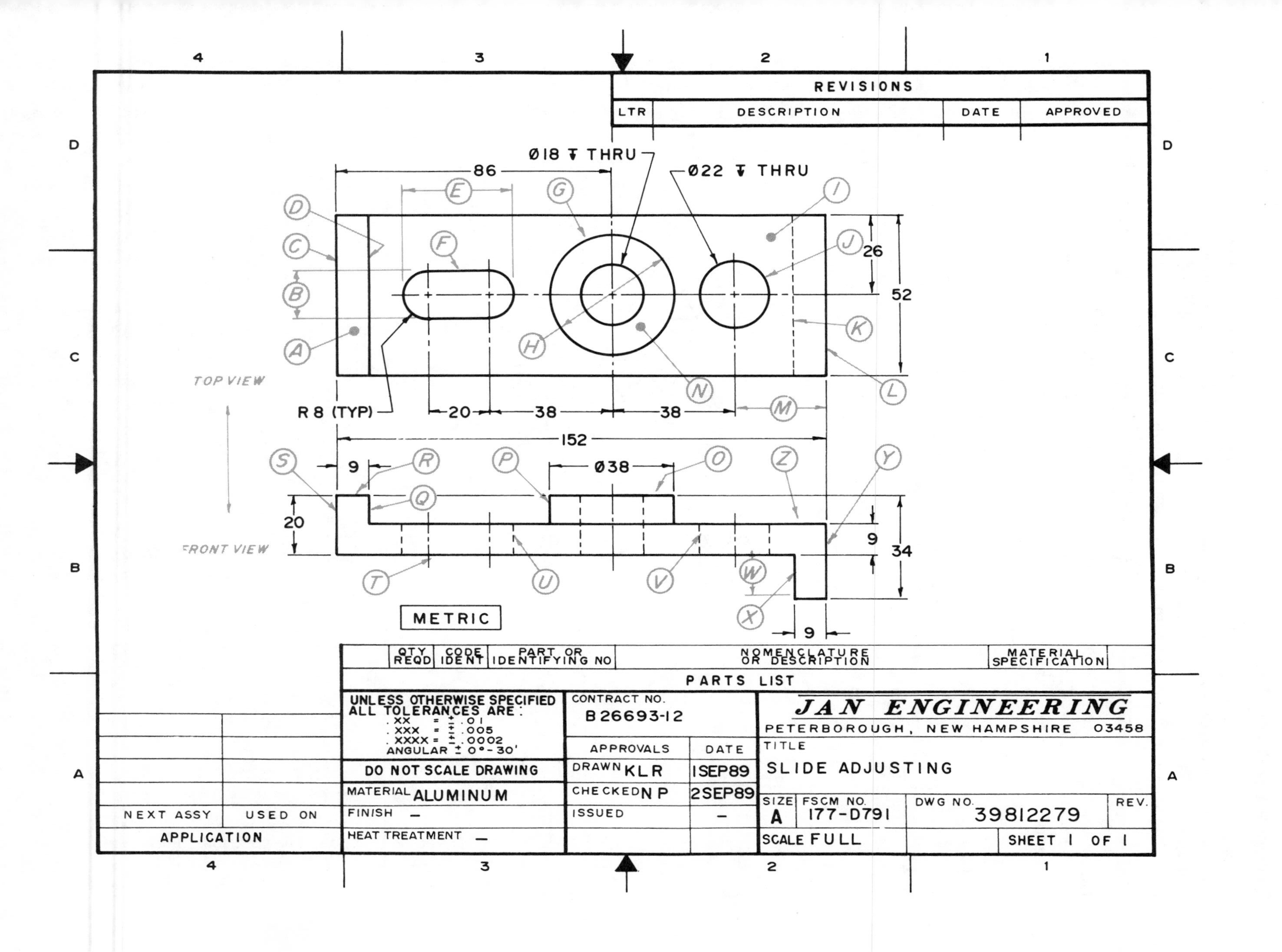
REVISIONS
LTR
DESCRIPTION
DATE
APPROVED
Ø18 ↧ THRU
Ø22 ↧ THRU
86
26
52
R 8 (TYP)
20
38
38
152
9
Ø38
20
9
34
9
TOP VIEW
FRONT VIEW
METRIC
QTY REQD
CODE IDENT
PART OR IDENTIFYING NO
NOMENCLATURE OR DESCRIPTION
MATERIAL SPECIFICATION
PARTS LIST
UNLESS OTHERWISE SPECIFIED ALL TOLERANCES ARE:
.XX = ±.01
.XXX = ±.005
.XXXX = ±.0002
ANGULAR ± 0°-30'
DO NOT SCALE DRAWING
MATERIAL ALUMINUM
FINISH –
HEAT TREATMENT –
CONTRACT NO.
B26693-12
APPROVALS
DATE
DRAWN KLR
1SEP89
CHECKED NP
2SEP89
ISSUED
–
JAN ENGINEERING
PETERBOROUGH, NEW HAMPSHIRE 03458
TITLE
SLIDE ADJUSTING
SIZE A
FSCM NO. 177-D791
DWG NO. 39812279
REV.
SCALE FULL
SHEET 1 OF 1
NEXT ASSY
USED ON
APPLICATION

THREE-VIEW DRAWINGS

A three-view drawing is similar to a two-view drawing, except there is another view added. As with any multiview drawing, the front view is the most important view and is the view with the most details. FIGURE 5-10, center, illustrates a pictorial view of an object, with an arrow indicating from which direction each surface is to be viewed: the front, the right side, and the top.

Imagine holding the object in your hand, looking directly at the front surface of the object (see FIG.5-10, lower left). Rotate the object to the right so you see the right-side view. Then, starting back at the front view, rotate the object so the top view can be seen, projected directly above the front view.

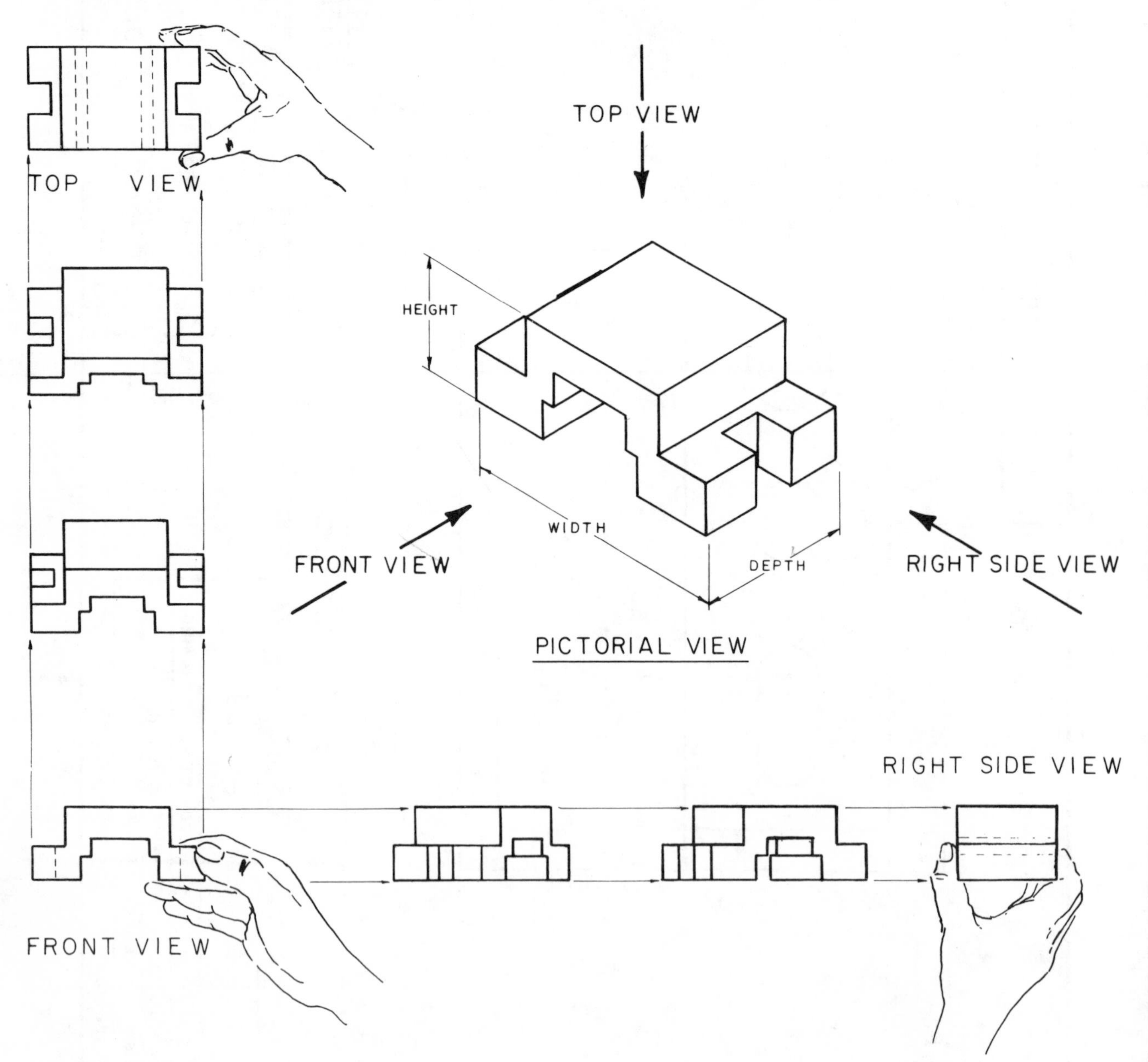

Fig. 5-10. Three-view drawings.

FIGURE 5-11 shows how a drafter would draw the front view. FIGURE 5-12 illustrates how the right side would be drawn. The rear view and left-side view are shown in FIGS. 5-13 and 5-14. (In this example the front view and the rear view are identical.) FIGURE 5-15 illustrates the top view, with its hidden surfaces noted by the four hidden lines. When viewing the bottom of the object, (FIG. 5-16), those surfaces *can* be seen; therefore, the lines are drawn as solid lines.

It is possible to have six different views of the same object: the front view, right-side view, left-side view, rear view, top view, and bottom view. The views usually drawn are the front view, right-side view, and the top view (see FIG. 5-17). (*Notice:* All views are projected from the front view.)

A pictorial view of this object is illustrated in FIG. 5-18. FIGURE 5-19 shows the front view, FIG. 5-20 shows the right- side view, and FIG. 5-21 shows the top view. FIGURE 5-22 illustrates three views with projection lines from the front view over to the right-side view, and from the front view up to the top view. These are the lines that usually do not show up on the drawing copy. In three-view drawings, projection lines also project from the right-side view up and over to the top view and vice versa.

Study FIG. 5-23. Surface "A" in the right-side view is actually the same surface "A" in the top view. To lay out projection lines from the right-side view to the top view or vice versa, draw a light line directly up from surface "A" in the right-side view. Then draw a light line from surface "A" in the top view directly to the right as shown. Where these two lines meet, point "X," draw a line at 45 degrees. Using this 45-degree projection line, all points, surfaces, or features can now be projected from the top view to the right-side view or vice versa. In order to fully understand a drawing, you should practice this many times.

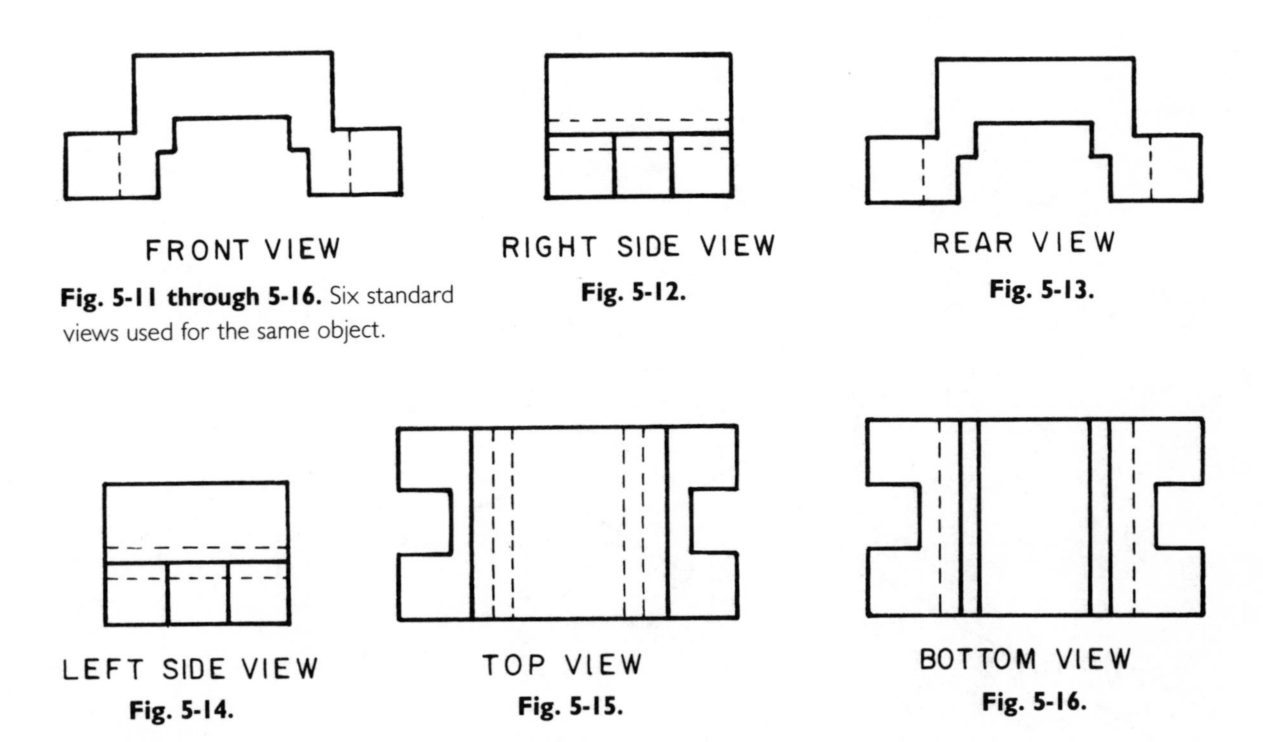

Fig. 5-11 through 5-16. Six standard views used for the same object.

Fig. 5-12.

Fig. 5-13.

Fig. 5-14.

Fig. 5-15.

Fig. 5-16.

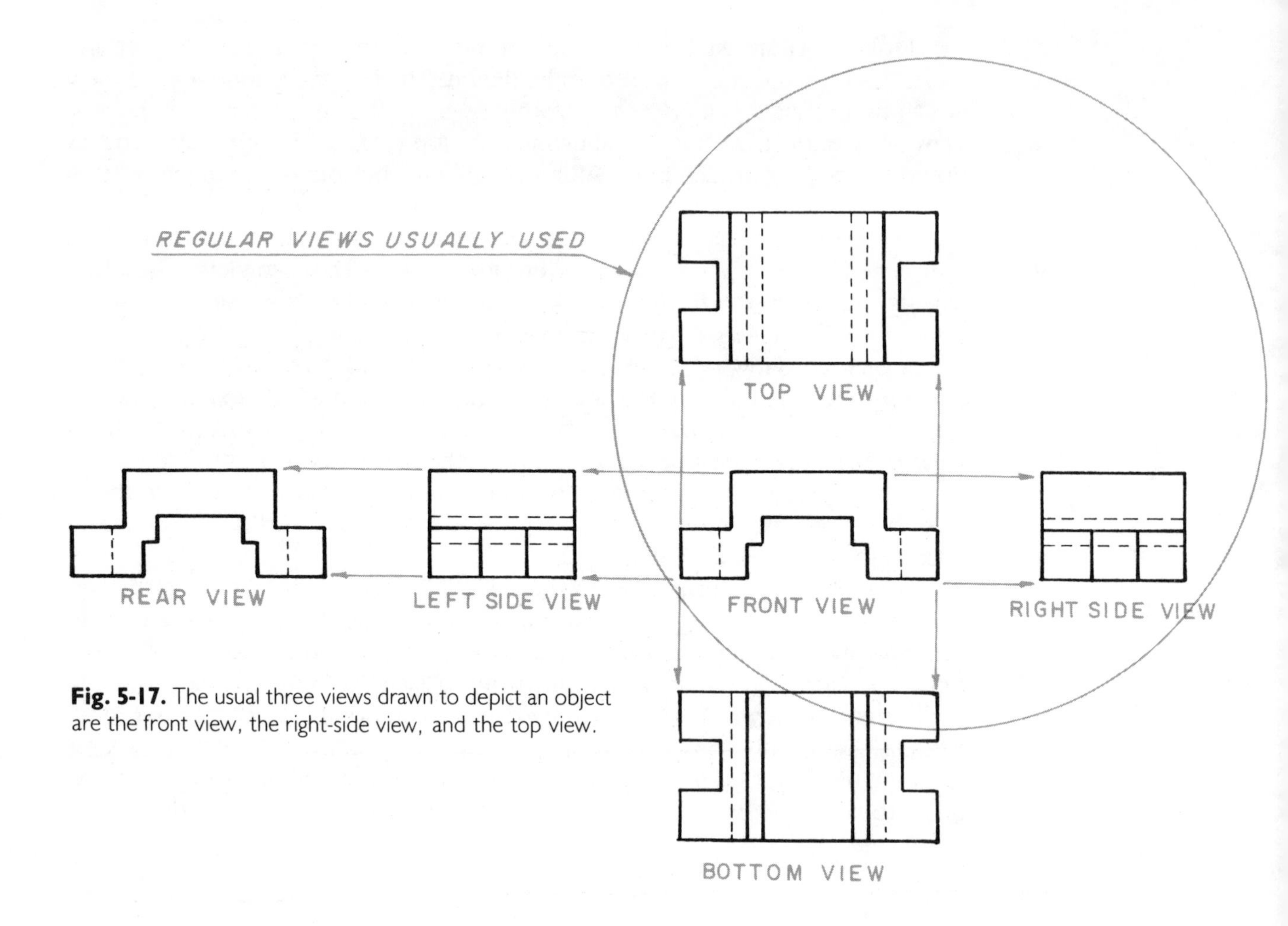

Fig. 5-17. The usual three views drawn to depict an object are the front view, the right-side view, and the top view.

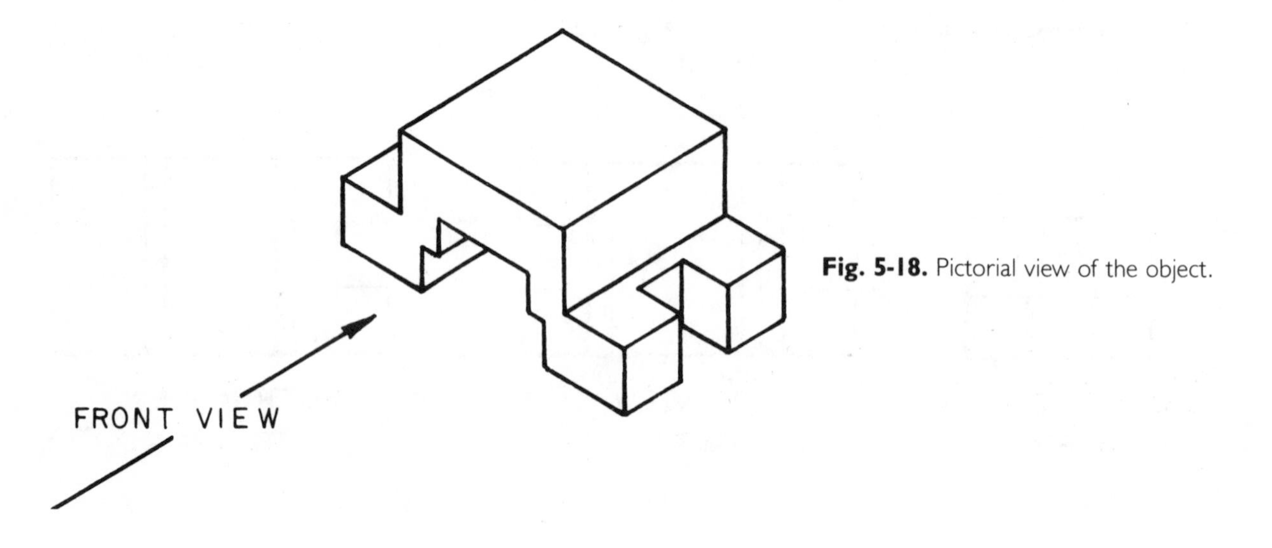

Fig. 5-18. Pictorial view of the object.

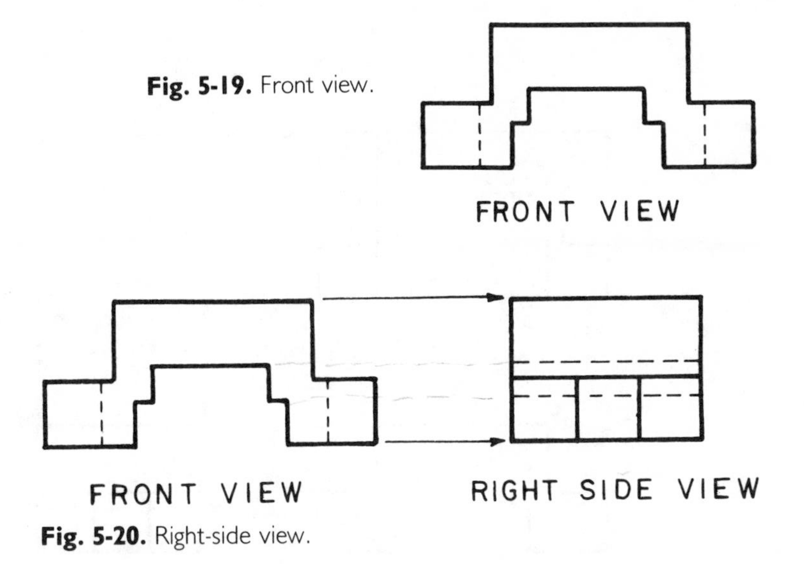

Fig. 5-19. Front view.

Fig. 5-20. Right-side view.

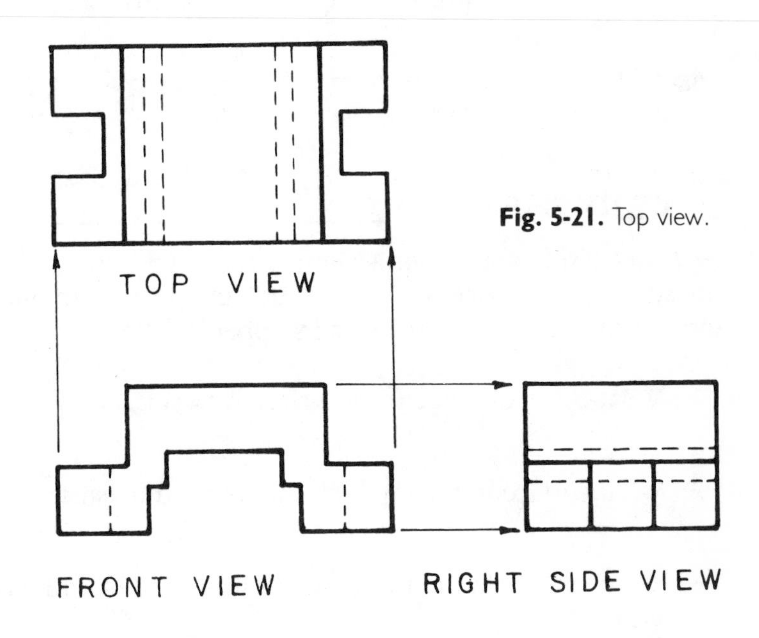

Fig. 5-21. Top view.

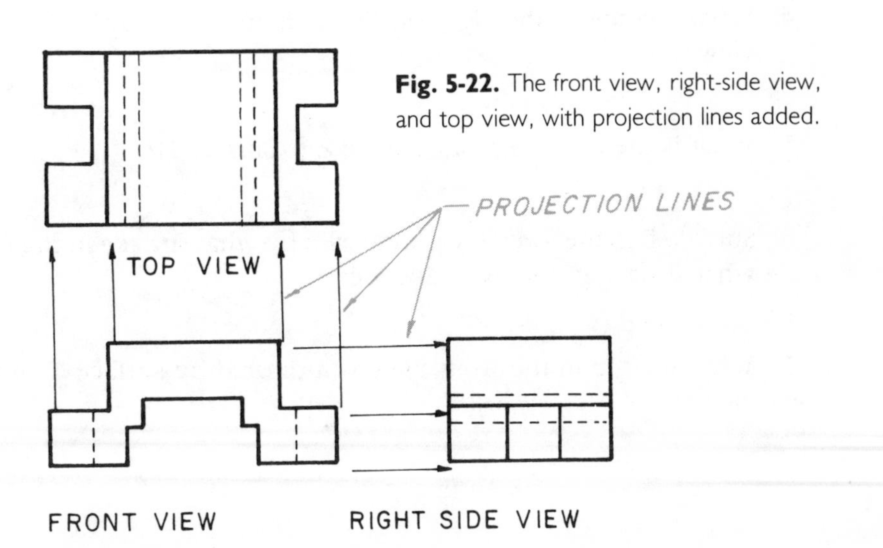

Fig. 5-22. The front view, right-side view, and top view, with projection lines added.

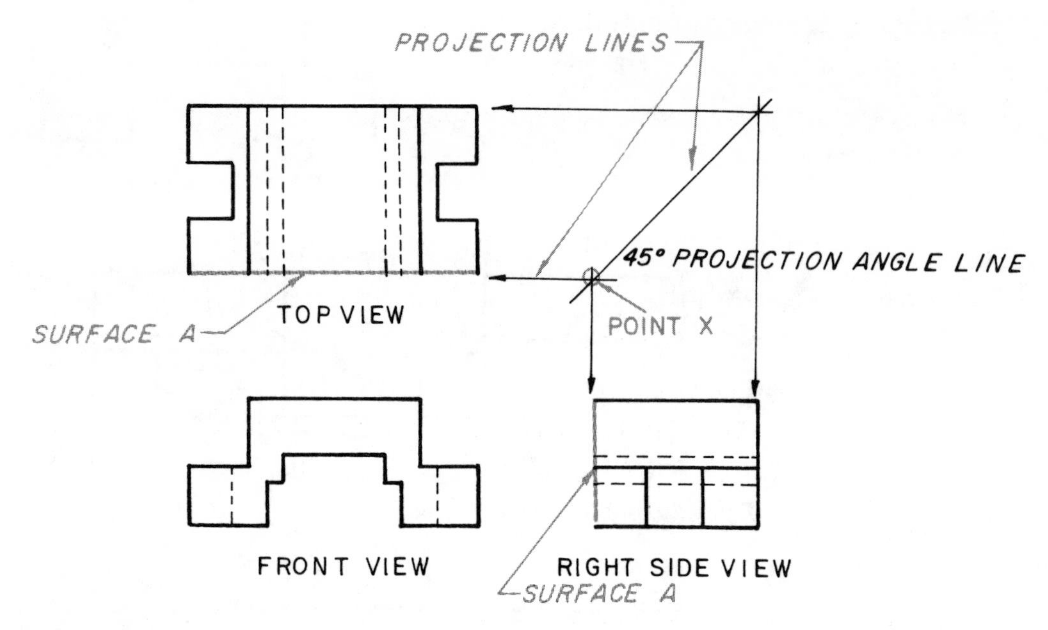

Fig. 5-23. By using the 45-degree projection line, you can project all points, surfaces, or features from the top view to the right-side view, and vice versa.

WORKSHEET 5-3

Instructions: Using Drawing Number A1237052 on p. 64, answer the following questions in the spaces provided. Add light projection lines if necessary to help interpret the drawing. (Answers in Appendix A.)

1. What is the distance from surface A to surface N?

2. What surface in the top view is indicated by surface K in the front view?

3. What surface in the front view is indicated by hidden surface C in the top view?

4. What surface in the right-side view is indicated by surface D in the top view?

5. What is the distance from surface Z to surface H?

6. Surface E in the top view is indicated by what surface in the front view? By what surface in the right-side view?

7. What surface in the front view is indicated by surface T in the right-side view?

8. What is the distance from surface X to surface S?

9. Surface P in the front view is indicated by what surface in the right-side view?

10. How far is surface V from surface A?

11. Calculate the distance from surface X to surface M.

12. Surface A in the top view is indicated by what surface in the front view?

13. What is the distance from surface E to surface S?

14. Surface G in the top view is indicated by what surface in the right-side view?

15. Calculate the distance from surface Q to surface R.

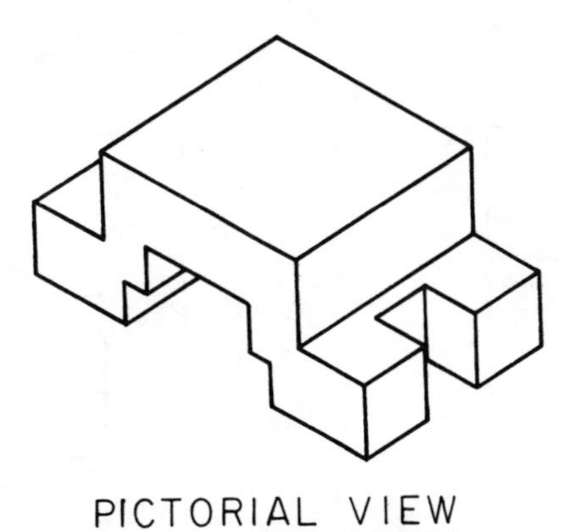

PICTORIAL VIEW

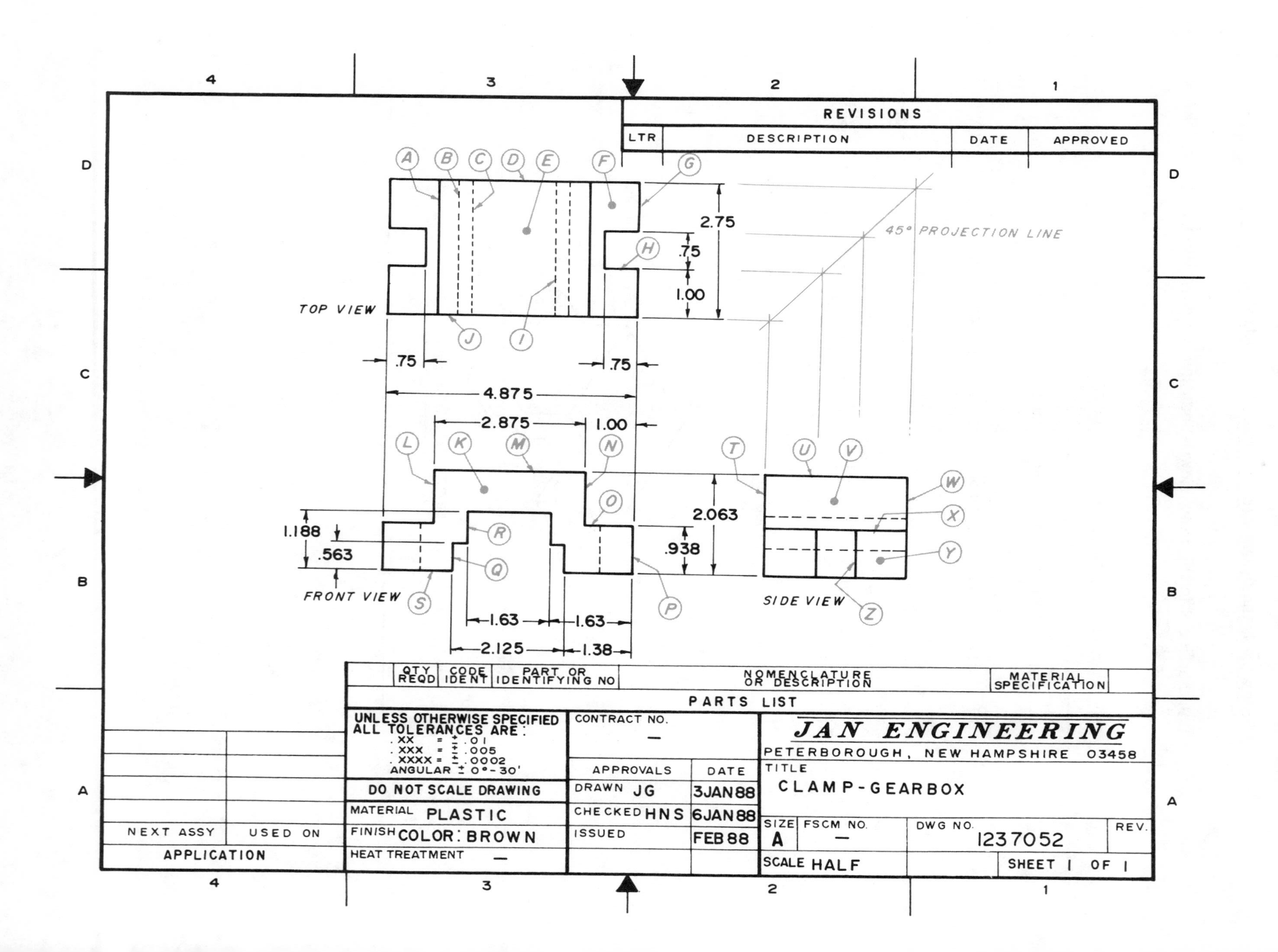
REVISIONS
LTR
DESCRIPTION
DATE
APPROVED
45° PROJECTION LINE
TOP VIEW
FRONT VIEW
SIDE VIEW
2.75
.75
1.00
.75
.75
4.875
2.875
1.00
2.063
.938
1.188
.563
1.63
1.63
2.125
1.38
A B C D E F G H I J K L M N O P Q R S T U V W X Y Z
QTY REQD
CODE IDENT
PART OR IDENTIFYING NO
NOMENCLATURE OR DESCRIPTION
MATERIAL SPECIFICATION
PARTS LIST
UNLESS OTHERWISE SPECIFIED ALL TOLERANCES ARE:
.XX = ± .01
.XXX = ± .005
.XXXX = ± .0002
ANGULAR ± 0°-30'
DO NOT SCALE DRAWING
MATERIAL PLASTIC
FINISH COLOR: BROWN
HEAT TREATMENT —
CONTRACT NO.
—
APPROVALS
DATE
DRAWN JG
3JAN88
CHECKED HNS
6JAN88
ISSUED
FEB88
JAN ENGINEERING
PETERBOROUGH, NEW HAMPSHIRE 03458
TITLE
CLAMP-GEARBOX
SIZE
A
FSCM NO.
—
DWG NO.
1237052
REV.
SCALE HALF
SHEET 1 OF 1
NEXT ASSY
USED ON
APPLICATION

WORKSHEET 5-4

Instructions: Using Drawing Number A63214-A on p. 67, answer the following questions in the spaces provided. (Answers in Appendix A.)

1. The hole indicated by the 0 is what kind of hole and what is its size?

2. Calculate the distance from surface N in the front view to surface C in the top view.

3. What is the distance between surface K and surface R in the front view?

4. What surface in the top view is indicated by surface M in the front view?

5. What size is dimension A?

6. What surface in the right-side view is indicated by surface G in the top view?

7. Calculate dimension F.

8. What surface in the top view is indicated by surface P in the front view?

9. What surface in the front view and in the right-side view is indicated by surface E in the top view?

10. What feature does V represent and what is its size?

11. What surface in the top view is indicated by surface L in the front view?

12. Calculate the distance from surface B in the top view to surface M in the front view.

13. What surface in the right-side view is indicated by surface H in the top view?

14. What does hidden line S indicate?

15. What surface in the top view is indicated by surface W in the right-side view?

16. What does the thick line under the 2.13 dimension (front view) indicate?—Explain.

17. What do the parentheses, ( ), around the (6.00) dimension mean?

18. What size is the "R" (radius) in the top view, left end?

19. What is the FULL drawing number?

20. What kind of line is line J?

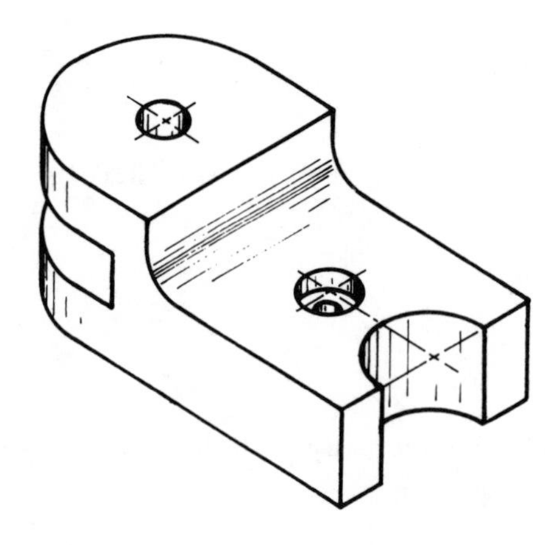

PICTORIAL VIEW

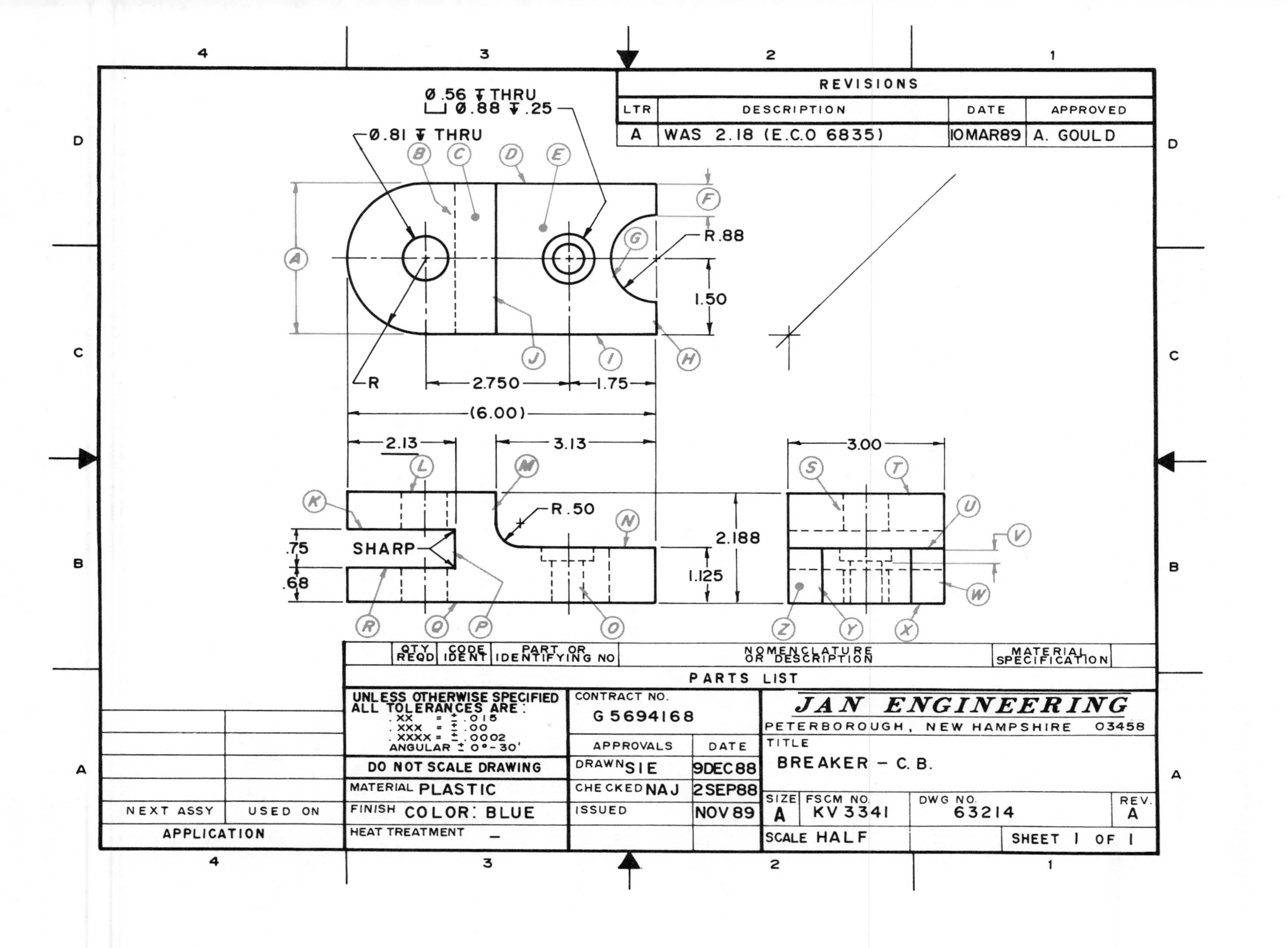
REVISIONS
LTR
DESCRIPTION
DATE
APPROVED
A
WAS 2.18 (E.C.O 6835)
10MAR89
A. GOULD
Ø.56 ↧ THRU
⌴ Ø.88 ↧ .25
Ø.81 ↧ THRU
R.88
1.50
R
2.750
1.75
(6.00)
2.13
3.13
R.50
SHARP
.75
.68
2.188
1.125
3.00
QTY REQD
CODE IDENT
PART OR IDENTIFYING NO
NOMENCLATURE OR DESCRIPTION
MATERIAL SPECIFICATION
PARTS LIST
UNLESS OTHERWISE SPECIFIED ALL TOLERANCES ARE:
.XX = ± .015
.XXX = ± .00
.XXXX = ± .0002
ANGULAR ± 0°-30'
DO NOT SCALE DRAWING
MATERIAL PLASTIC
FINISH COLOR: BLUE
HEAT TREATMENT –
CONTRACT NO.
G 5694168
APPROVALS
DATE
DRAWN SIE
9DEC88
CHECKED NAJ
2SEP88
ISSUED
NOV 89
JAN ENGINEERING
PETERBOROUGH, NEW HAMPSHIRE 03458
TITLE
BREAKER – C. B.
SIZE
A
FSCM NO.
KV 3341
DWG NO.
63214
REV.
A
SCALE HALF
SHEET 1 OF 1
NEXT ASSY
USED ON
APPLICATION

WORKSHEET 5-5

Instructions: Using Drawing Number A041237-B on p. 69, answer the following questions in the spaces provided. Add a light 45 degree project line and required projections lines to project surfaces from the top view to the right-side view and vice-versa, in order to check and verify all answers. (Answers in Appendix A.)

1. What system of dimensioning was used on this drawing?

2. What two given dimensions fall within the 3-C zone?

3. What size is dimension A?

4. What surface in the right-side view is indicated by surface E in the top view?

5. Calculate radius D.

6. What surface in the right-side view is indicated by surface G in the top view?

7. What surface in the top view is indicated by surface V in the right-side view?

8. Calculate dimension H.

9. What surface in the right-side view is indicated by surface I in the front view?

10. What surface in the top view is indicated by surface K in the front view?

11. Hole "a" is represented in the right-side view by what letter?

12. What was diameter M before it was changed? What does the thick line under the dimension mean?

13. What size is dimension S?

14. Calculate dimension P.

continued on p. 70

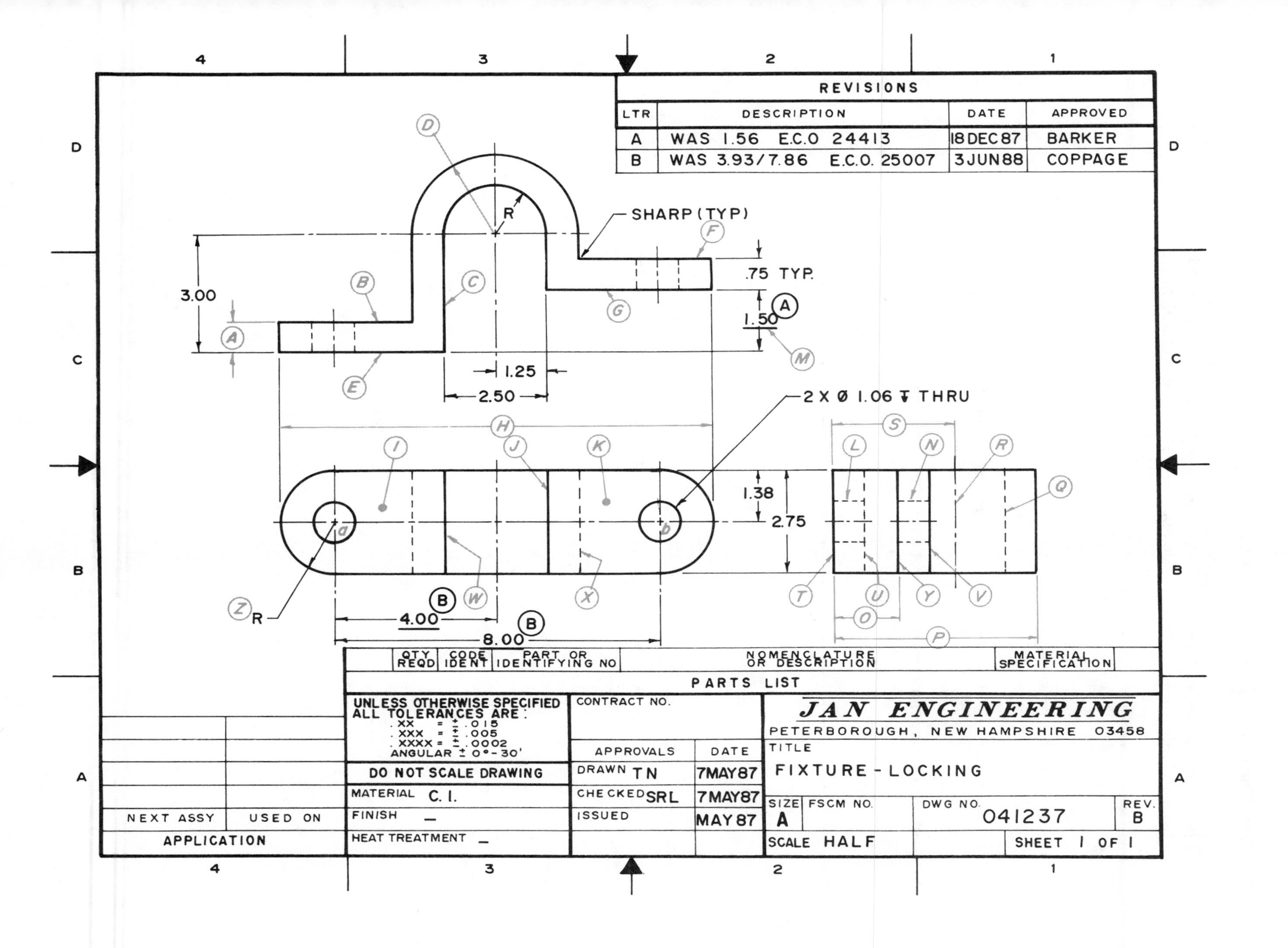
REVISIONS
LTR
DESCRIPTION
DATE
APPROVED
A
WAS 1.56 E.C.O 24413
18 DEC 87
BARKER
B
WAS 3.93/7.86 E.C.O. 25007
3 JUN 88
COPPAGE
SHARP (TYP)
.75 TYP.
1.50
3.00
1.25
2.50
2 X Ø 1.06 ↧ THRU
1.38
2.75
4.00
8.00
R
R
QTY REQD
CODE IDENT
PART OR IDENTIFYING NO
NOMENCLATURE OR DESCRIPTION
MATERIAL SPECIFICATION
PARTS LIST
UNLESS OTHERWISE SPECIFIED ALL TOLERANCES ARE:
.XX = ± .015
.XXX = ± .005
.XXXX = ± .0002
ANGULAR ± 0°-30'
DO NOT SCALE DRAWING
MATERIAL C. I.
FINISH —
HEAT TREATMENT —
CONTRACT NO.
APPROVALS
DATE
DRAWN TN
7MAY87
CHECKED SRL
7MAY87
ISSUED
MAY 87
JAN ENGINEERING
PETERBOROUGH, NEW HAMPSHIRE 03458
TITLE
FIXTURE - LOCKING
SIZE A
FSCM NO.
DWG NO. 041237
REV. B
SCALE HALF
SHEET 1 OF 1
NEXT ASSY
USED ON
APPLICATION

15. Centerline R indicates the axis of what feature?

16. What are the upper and lower limits for the 1.06 diameter holes and what is the maximum distance they can be apart and still be within tolerance?

17. What is the radius at Z?

18. Hidden line N represents what feature?

19. What scale was used by the drafter to draw this object?

20. What is the minimum distance surface I in the front view can be from surface F in the top view and be within tolerance?

Section Views

Objective: The reader will identify and understand the various kinds of section views used in industry and know the many drafting standard practices used to illustrate a part.

A section view drawing is like any multiview drawing, except one view is drawn as if it were cut in half. This is done in order to more clearly illustrate the interior features of the object. In FIG. 6-1, a front view is shown with a normal right-side view. Notice how the right side has many hidden lines and is very hard to interpret. Directly to the right of the right-side view is the *same* right-side view, except drawn as a full section. Compare this section view with the other view and you can see that the interior features are fully illustrated, without question. Any of the regular views can be made into a section view, depending on what particular features are to be illustrated.

CUTTING PLANE LINES

A line shown, but not explained, in the alphabet of lines section in Chapter 3, was the *cutting plane line.* The cutting plane line is a *thick black* line and can be drawn one of three ways (FIG. 6-2). It extends about 1/2 inch (12mm) beyond either side of the object. The arrows point in the direction the object is to be viewed, or the line of sight. The letters A-A are for identification. If more than one section view is needed, the second section view would be labeled B-B, the third C-C, and so on.

Think of the cutting plane line as the location of an actual cut through the object. If a saw was to be used to cut the object, the cutting plane line would be the guideline for the saw blade. The cutting plane lines is always placed in the view next to the section view.

SECTION LINES

Another line from the alphabet of lines is the *section line*. Section lines are *thin black* lines, usually drawn at 45 degrees and spaced about 1/16 inch (1.5 mm) to 1/8 inch (3 mm) apart. Section lines are placed only on the surfaces that the cutting plane line touches. Think of cutting an object in half, imagine painting only the surfaces that the saw blade touched. The painted surfaces are those surfaces on which the drafter would have applied section lines. In past years different section lining was used to represent different kinds of materials (see FIG. 6-3). Today, only one kind of section lining is used: an all-purpose section lining (see FIG. 6-4). Now refer to FIG. 6-5, top view. The cutting plane line did not pass through the back wall nor around the surface of the hole; therefore, section lines are not put on those surfaces. Notice: now that the front view has

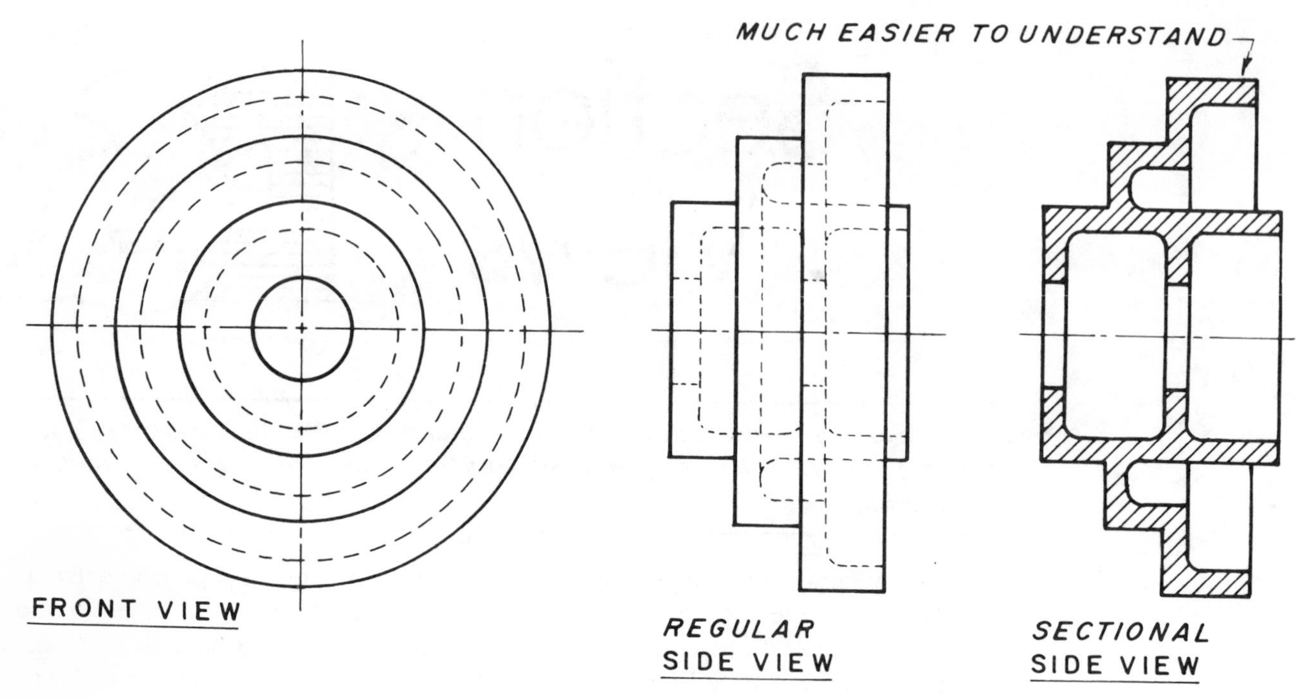

Fig. 6-1. Section views.

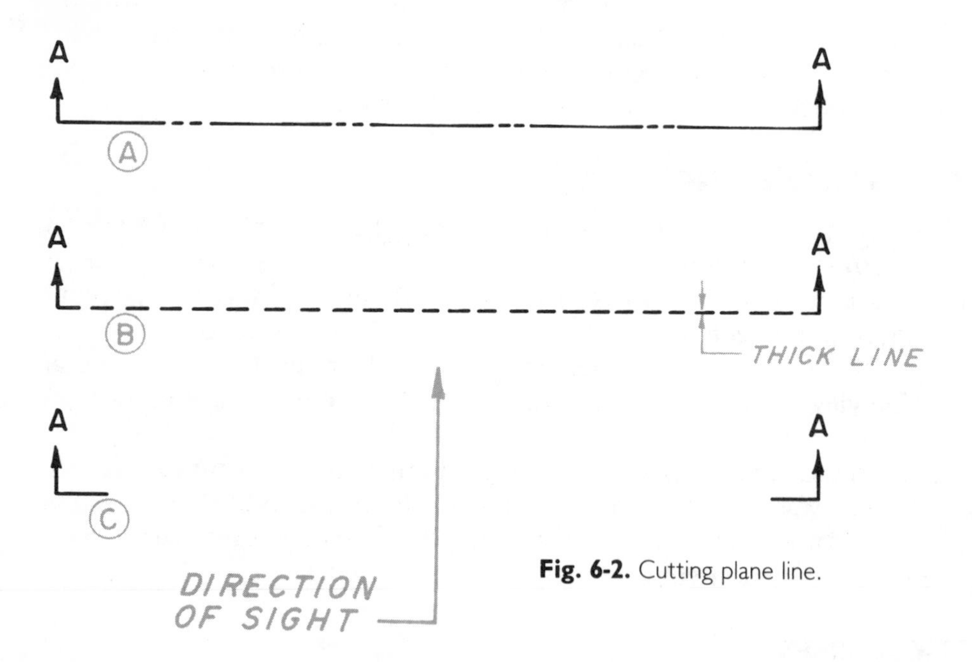

Fig. 6-2. Cutting plane line.

been cut in half, the lines that would have been hidden lines are now object lines.

KINDS OF SECTION VIEWS

There are eight major kinds of section views. Each will be fully explained and illustrated:

- Full section
- Half section
- Offset section
- Removed section
- Rotated (or revolved) section
- Broken out seciton
- Thin wall section
- Assembly section

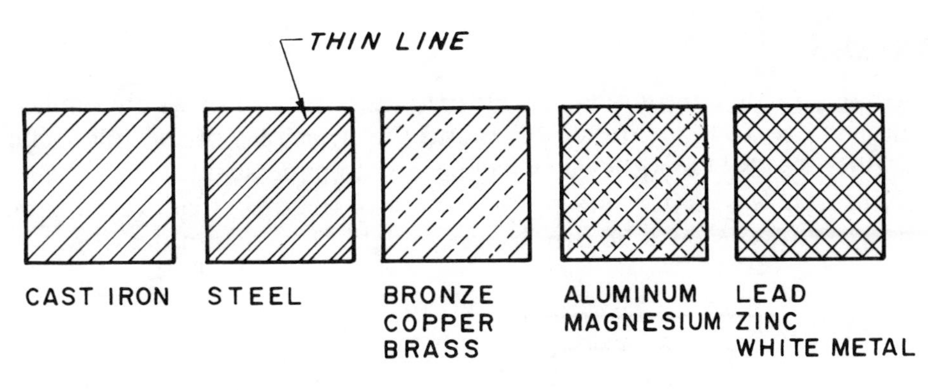

Fig. 6-3. Section lining.

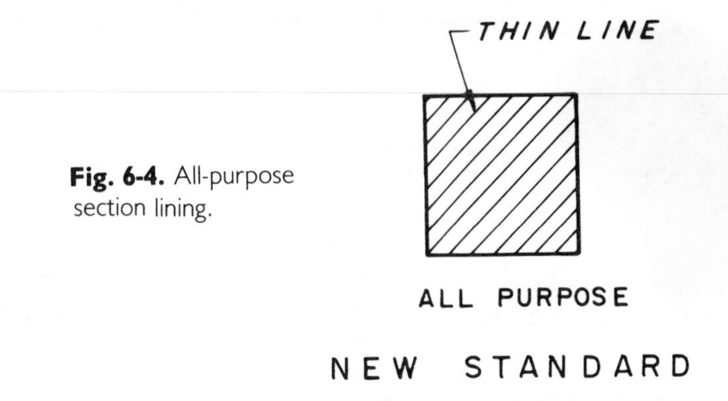

Fig. 6-4. All-purpose section lining.

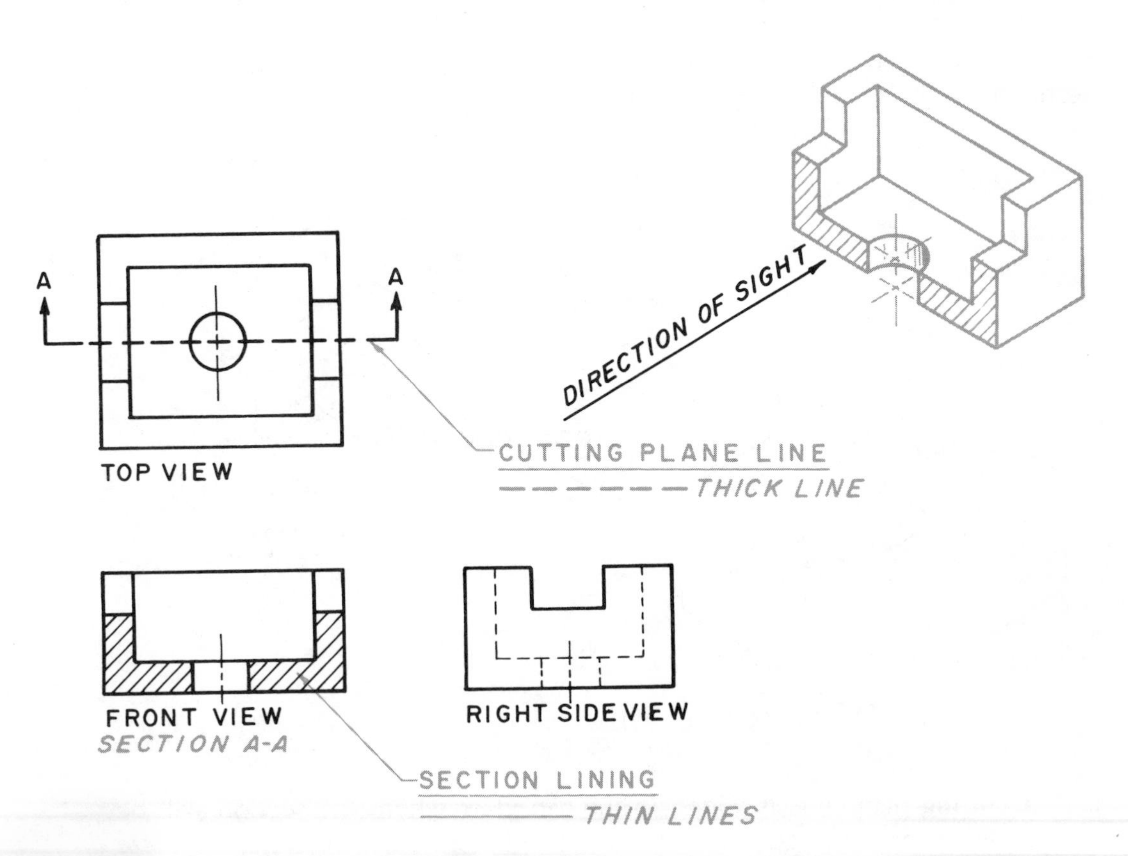

Fig. 6-5. Section lining is added only to those areas where the cutting plane line actually touches.

Full Section

Full sections are used most often. All other kinds of section views are similar to the full section, so if you fully understand the principles of the full section views, other kinds of section views will be easy to understand.

A *full section* is, as its name implies, a view of an object that has been fully cut in two. FIGURE 6-6 will be used as an example. Note that in the front view

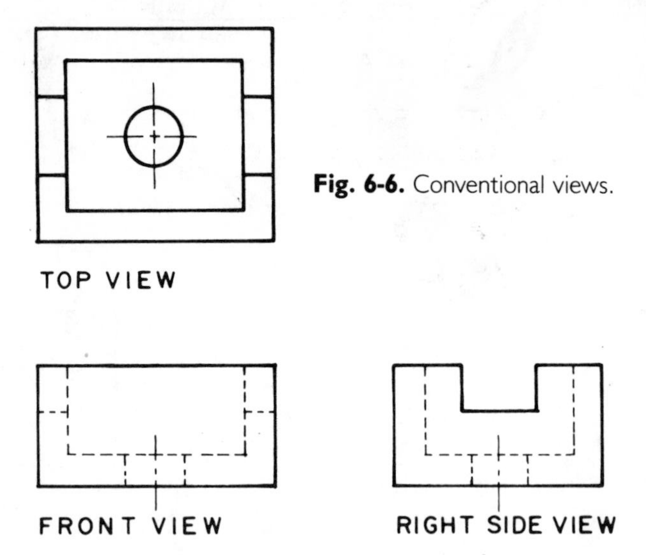

Fig. 6-6. Conventional views.

there are many hidden lines, causing some confusion concerning exactly what is needed. FIGURE 6-7 illustrates the same object as a pictorial view and how it would look when cut in half. Note the *direction of sight,* this is the direction from which the section is being viewed.

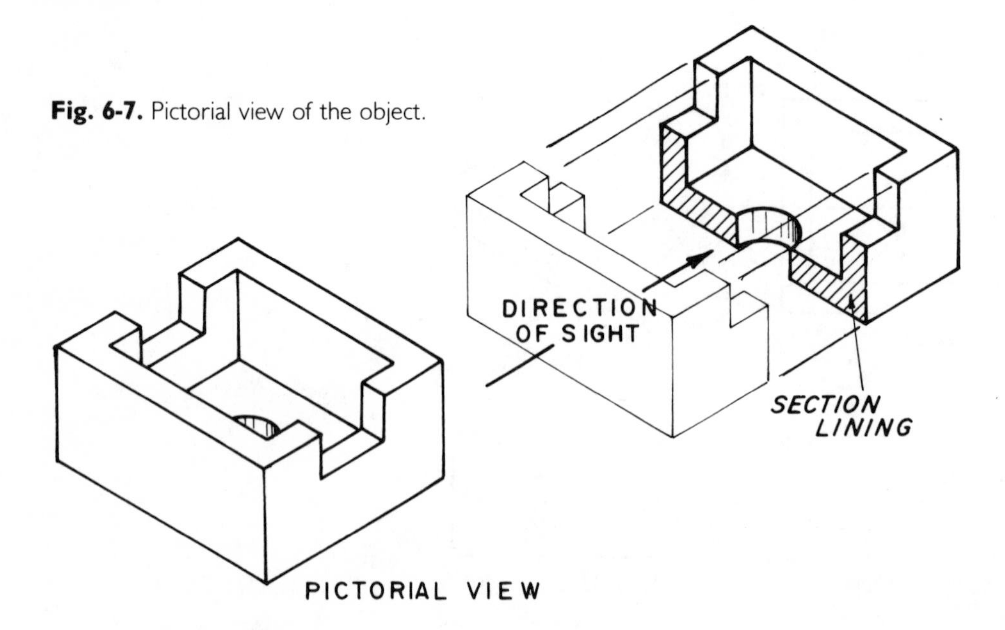

Fig. 6-7. Pictorial view of the object.

A cutting plane line is added to the top view where the section will be taken. The arrows indicate the direction of sight (see FIG. 6-8). Try to visualize the area in front of the cutting plane line (noted by the shaded lines) as being *removed*. Now look at the front view in FIG. 6-9 and try to visualize the front

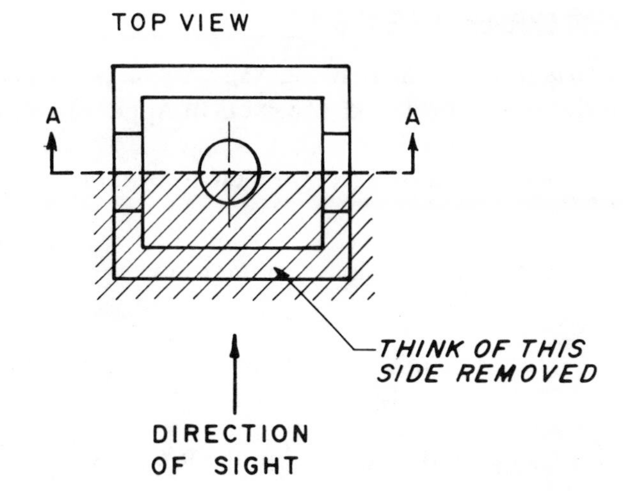

Fig. 6-8. Adding cutting plane line to the top view

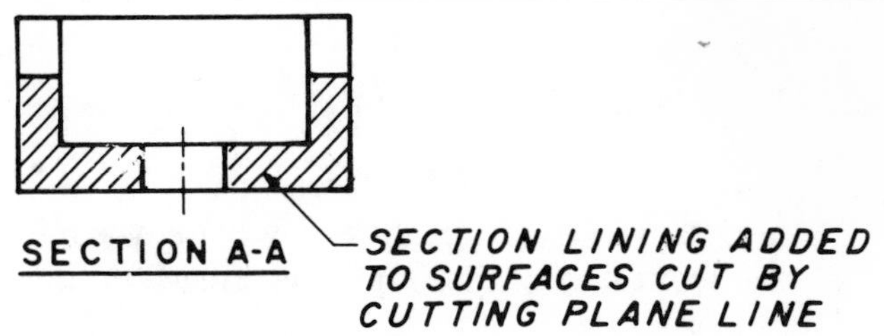

Fig. 6-9. View with front area removed.

view as it would look with the front area removed. Section lining is added to the surfaces that the cutting plane line touched.

The completed, full-section drawing is shown in FIG. 6-10. It is made up of a front view that is a section view; a top view; and a right-side view. The label under the section view identifies it with the letters of the cutting plane line.

Hidden line rule. Hidden lines are usually omitted from section views and are added only if absolutely necessary to make the view clearer.

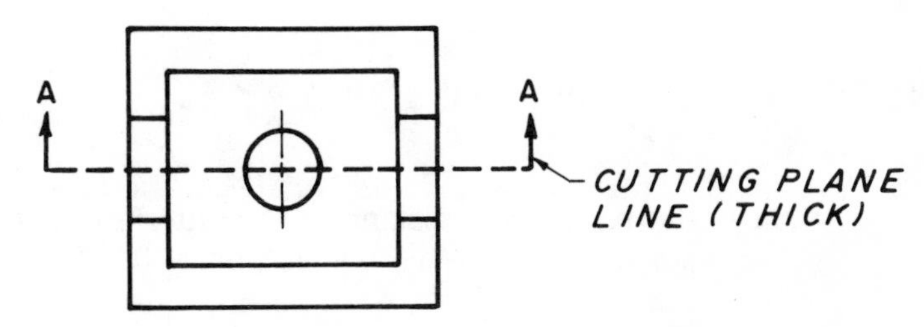

Fig. 6-10. Completed front view, full section.

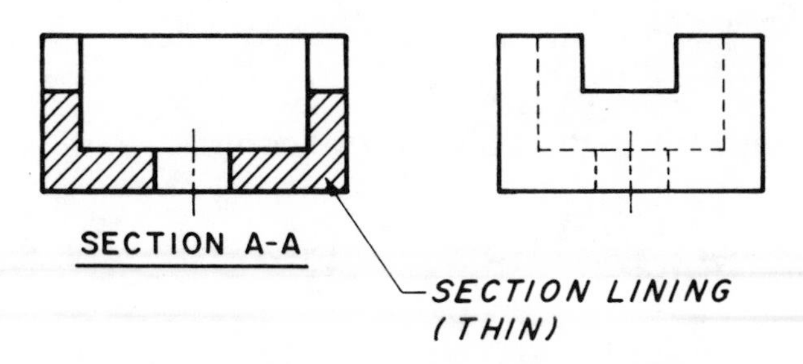

WORKSHEET 6-1

Instructions: Using drawing A337198 on p. 77, answer the following questions in the spaces provided. (Answers in Appendix A.)

1. What kind of a section view is used on this drawing?

2. What is dimension A in the top view?

3. Surface D in the top view is what surface in the right-side view?

4. Calculate dimension F in the top view.

5. What scale was used in this drawing?

6. What is dimension G in the top view?

7. Surface I in the front view is what surface in the top view? (Don't forget, the front half of the object has been removed.)

8. Calculate dimension H in the front view.

9. Surface J in the front view is what surface in the right-side view?

10. What is dimension Y in the front view and WHAT does it represent?

11. What is the *support block* made of?

12. How many .44 diameters are there?

13. What is the *maximum* distance the centerline of the .44 diameter holes can be located from surface V?

14. How far is surface J in the front view from surface Q in the right-side view?

15. Surface E in the top view is what surface in the right-side view?

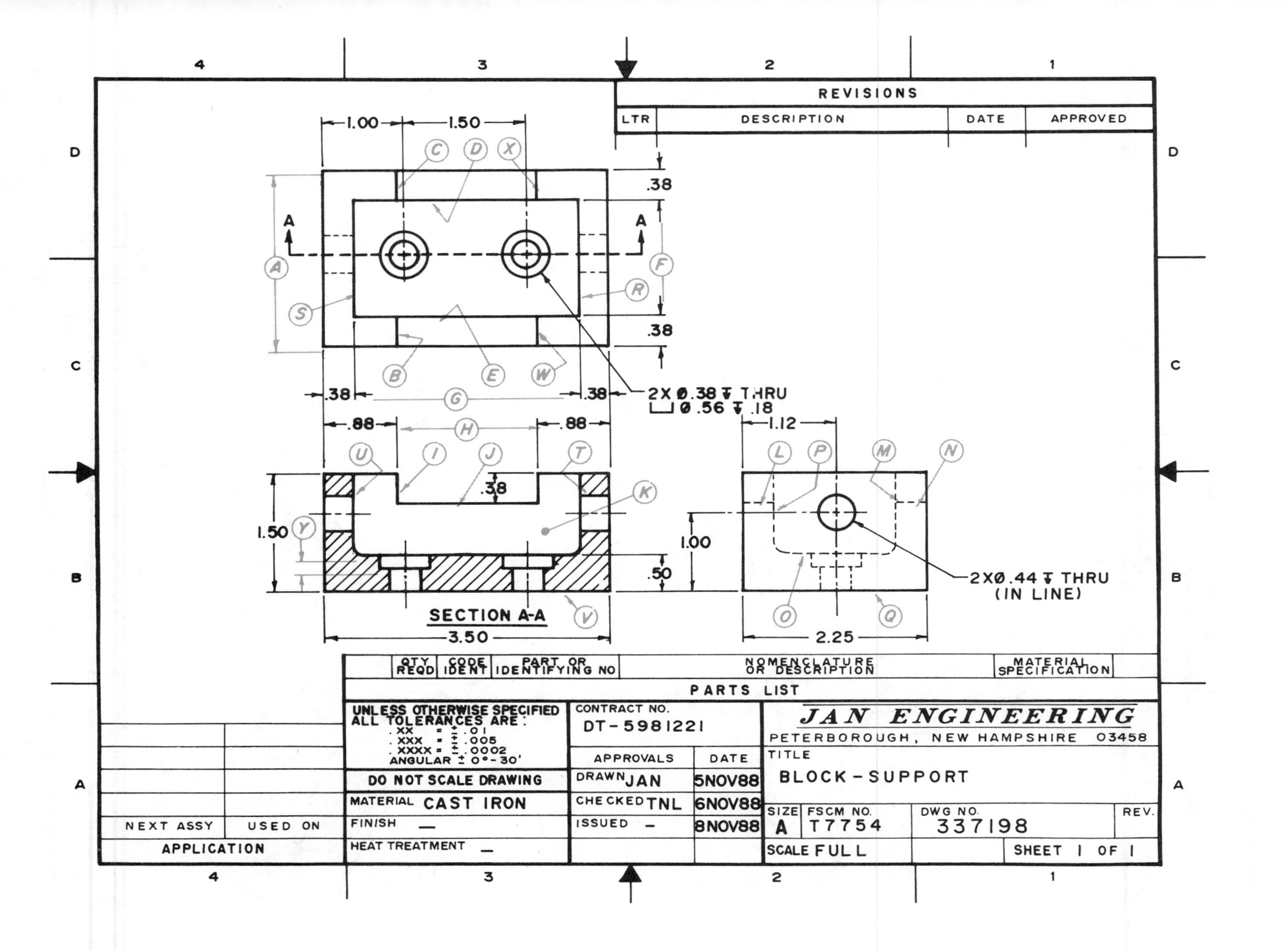

REVISIONS
LTR
DESCRIPTION
DATE
APPROVED
1.00
1.50
.38
.38
A
A
.38
.38
2X Ø.38 ↧ THRU
⌴ Ø.56 ↧ .18
.88
.88
1.12
.38
1.50
1.00
.50
2XØ.44 ↧ THRU
(IN LINE)
SECTION A-A
3.50
2.25
QTY REQD
CODE IDENT
PART OR IDENTIFYING NO
NOMENCLATURE OR DESCRIPTION
MATERIAL SPECIFICATION
PARTS LIST
UNLESS OTHERWISE SPECIFIED ALL TOLERANCES ARE:
.XX = ± .01
.XXX = ± .005
.XXXX = ± .0002
ANGULAR ± 0°-30'
CONTRACT NO.
DT-5981221
JAN ENGINEERING
PETERBOROUGH, NEW HAMPSHIRE 03458
APPROVALS
DATE
TITLE
BLOCK-SUPPORT
DO NOT SCALE DRAWING
DRAWN JAN 5NOV88
MATERIAL CAST IRON
CHECKED TNL 6NOV88
SIZE A
FSCM NO. T7754
DWG NO. 337198
REV.
NEXT ASSY
USED ON
FINISH —
ISSUED — 8NOV88
APPLICATION
HEAT TREATMENT —
SCALE FULL
SHEET 1 OF 1

Half Section

A *half section* is very similar to a full section. The only difference is that the cutting plane line of a half section passes through only *half* the object. The cutting plane line of a full section passes through and across the entire object. A half section removes only one *quarter* of the object (see FIG. 6-11). A quarter of the object is removed and taken away from the object. If you viewed the object with the quarter removed it would look like FIG. 6-12. Note the direction of sight.

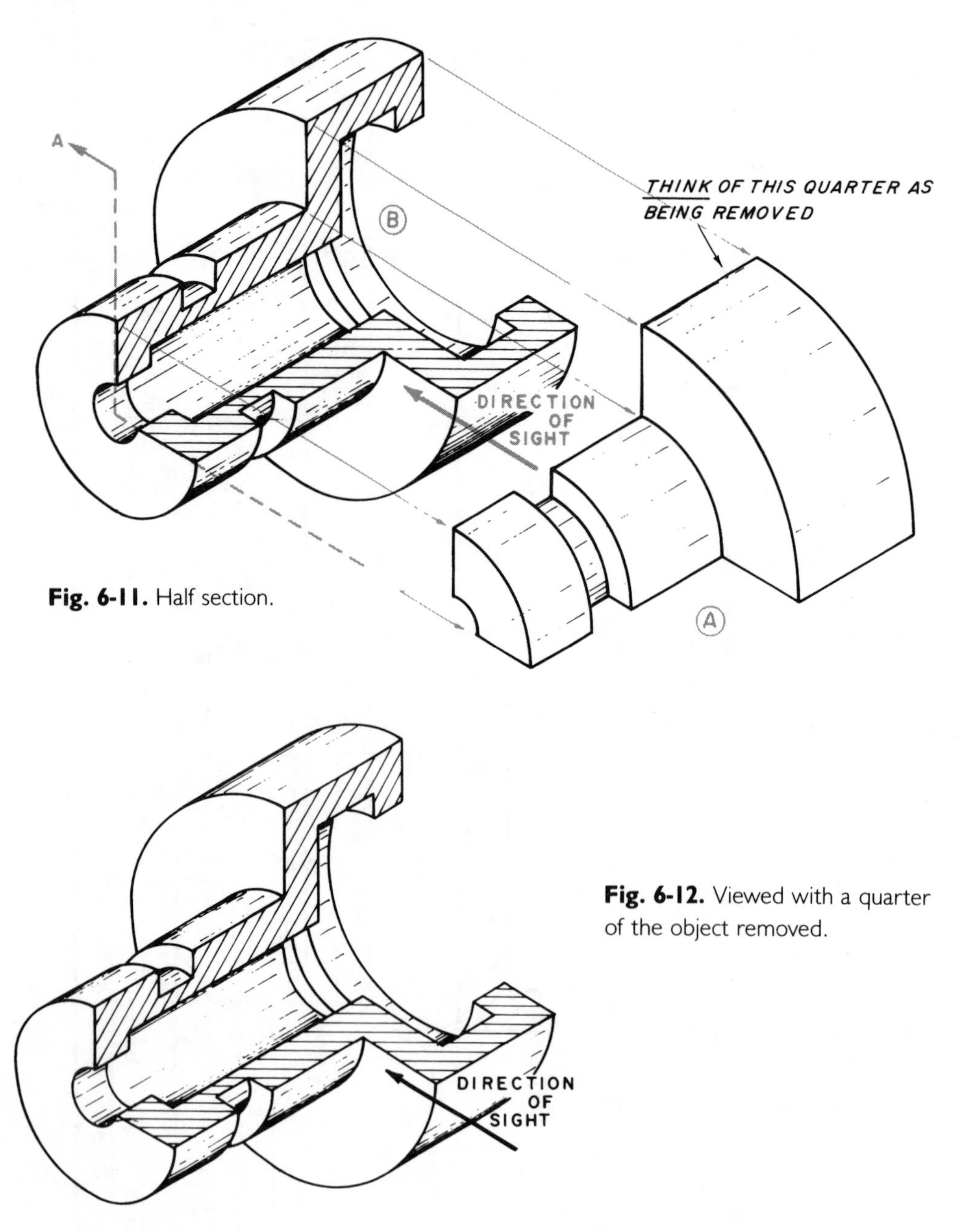

Fig. 6-11. Half section.

Fig. 6-12. Viewed with a quarter of the object removed.

Looking at a usual multiview drawing, FIG. 6-13, notice how the side view is hard to understand. Passing a cutting line through one quarter of the object making it a half section, and adding section lining to the other view makes the interior much easier to read and understand (see FIG. 6-14): a half section illustrated both the interior of the object and the exterior of the object at the same

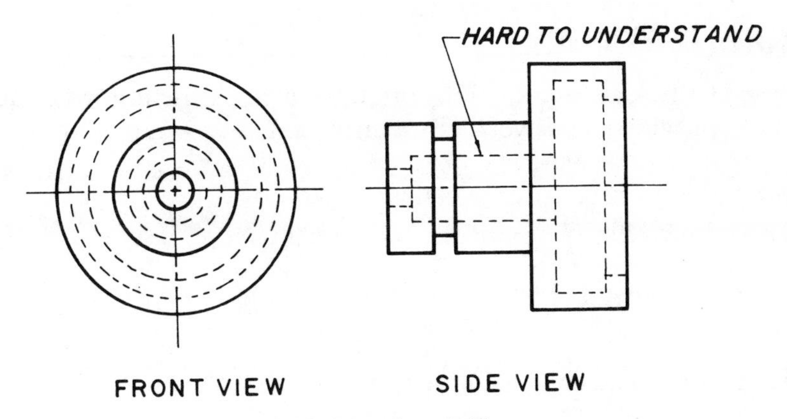

Fig. 6-13. Direction of sight.

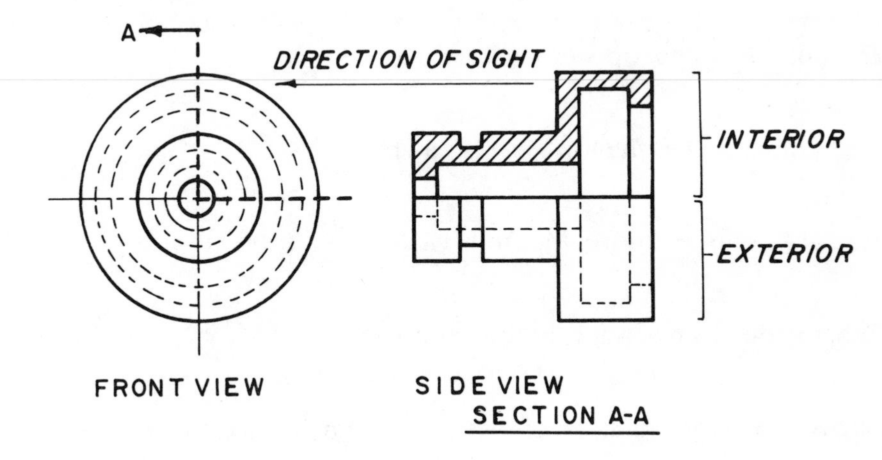

Fig. 6-14. A half section illustrates both the outside and the inside of the object in the same view.

time. Always remember the "hidden line rule": the half that is "in section" does *not* have any hidden lines added to it unless absolutely necessary. The other half, not "in section," *does* have hidden lines added to it.

WORKSHEET 6-2

Instructions: Using drawing A379664 on p. 81, answer the following questions in the spaces provided. (Answers in Appendix A.)

1. What kind of a section view is this?
2. Surface B in the top view is what surface in the front view?
3. What is the diameter of C in the top view?
4. What kind of a line is line S?
5. Diameter D in the top view is what diameter?
6. The radius at G in the top view is what?
7. Calculate dimension I in the front view.
8. What is the diameter at K in the front view?
9. Surface O in the front view is what surface in the top view?
10. Surface L in the front view is what surface in the top view?
11. Surface Q in the front view is what surface in the top view?
12. What is the *bearing guide* made of?
13. Calculate dimension P in the front view.
14. What kind of a line is line V in the top view?
15. How far is surface B in the top view from surface U in the front view?

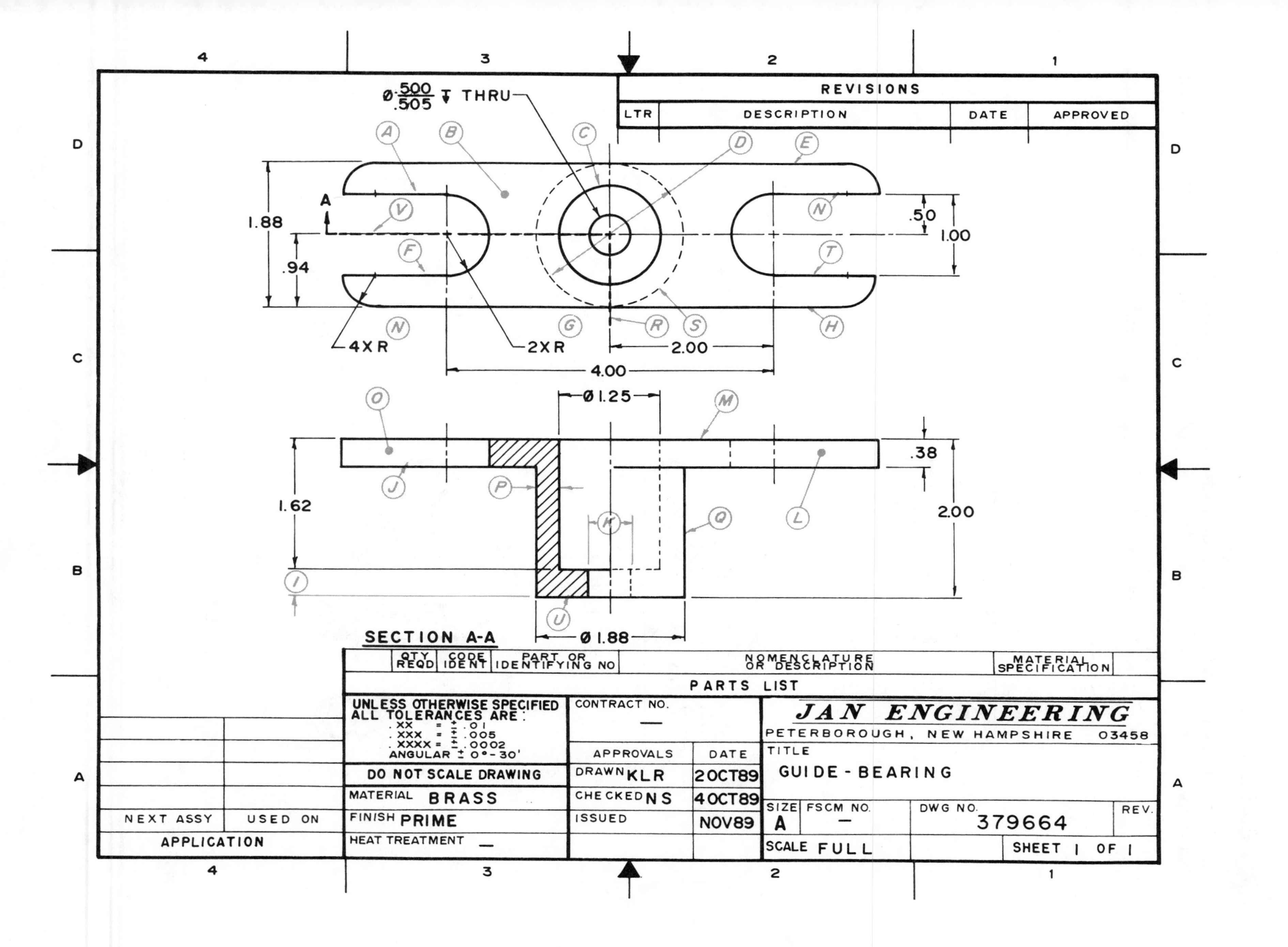
REVISIONS
LTR
DESCRIPTION
DATE
APPROVED
Ø .500/.505 ↧ THRU
A
1.88
.94
4X R
2X R
2.00
4.00
.50
1.00
Ø 1.25
.38
1.62
2.00
Ø 1.88
SECTION A-A
QTY REQD
CODE IDENT
PART OR IDENTIFYING NO
NOMENCLATURE OR DESCRIPTION
MATERIAL SPECIFICATION
PARTS LIST
UNLESS OTHERWISE SPECIFIED ALL TOLERANCES ARE:
.XX = ± .01
.XXX = ± .005
.XXXX = ± .0002
ANGULAR ± 0°-30'
DO NOT SCALE DRAWING
MATERIAL BRASS
FINISH PRIME
HEAT TREATMENT —
CONTRACT NO. —
APPROVALS
DATE
DRAWN KLR 20CT89
CHECKED NS 4OCT89
ISSUED NOV89
JAN ENGINEERING
PETERBOROUGH, NEW HAMPSHIRE 03458
TITLE
GUIDE - BEARING
SIZE A
FSCM NO. —
DWG NO. 379664
REV.
SCALE FULL
SHEET 1 OF 1
NEXT ASSY
USED ON
APPLICATION

Offset Section

An *offset section* is very similar to a full section, except the cutting plane line is not drawn in a straight line. Offset cutting plane lines make 90-degree bends in order to include important features that would have been left out with a straight line. FIGURE 6-15 illustrates a straight cutting plane line passing through the large hole (Y), but missing holes X and Z. An offset cutting plane line with two 90-degree bends through the same object will include the features "X", "Y," and "Z" (see FIG. 6-16).

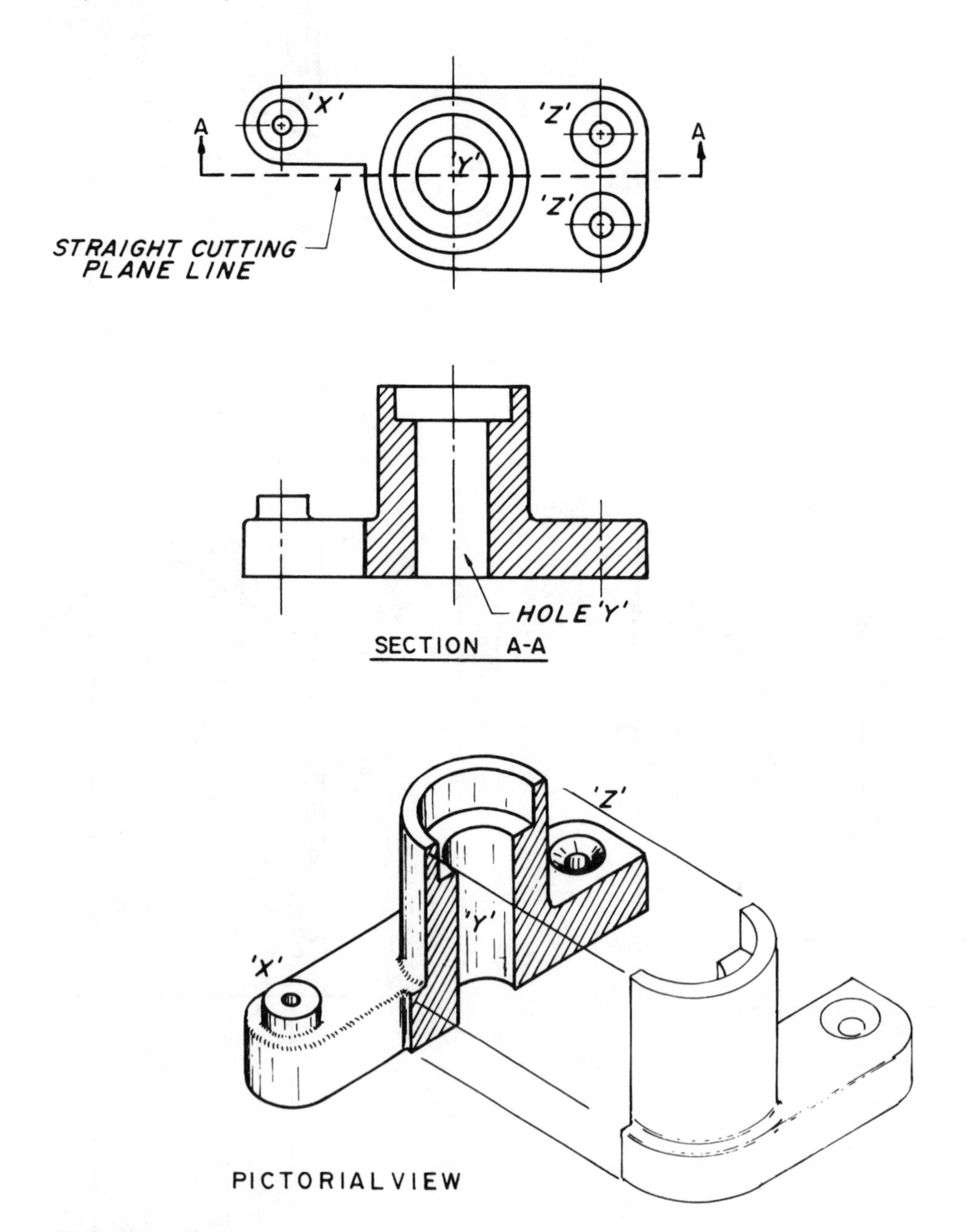

Fig. 6-15. Offset section.

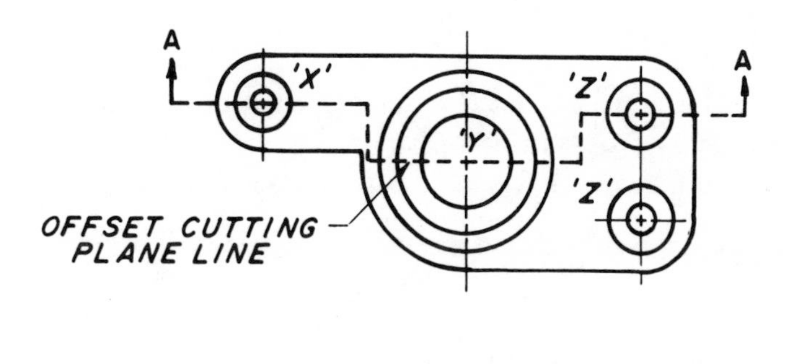

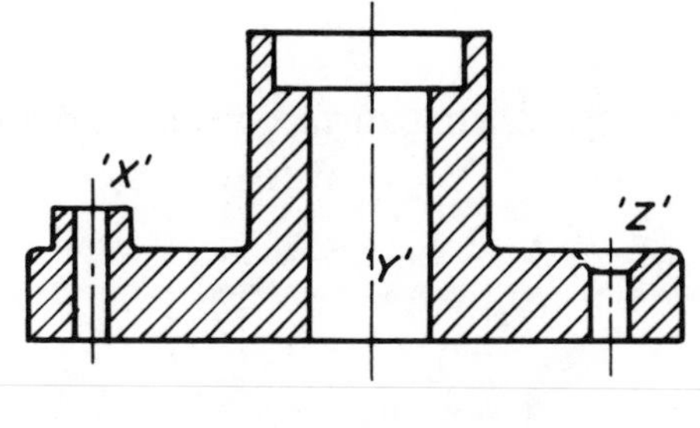

SECTION A-A

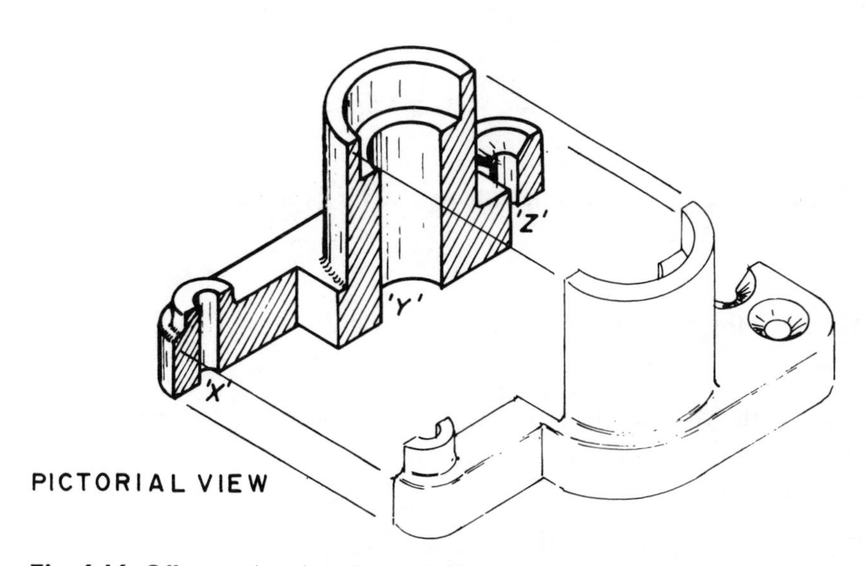

Fig. 6-16. Offset cutting plane line with 90-degree bends through the same object.

WORKSHEET 6-3

Instructions: Using drawing A1798222 on p. 85, answer the following questions in the spaces provided. (Answers in Appendix A.)

1. What kind of section view is this?

2. What is dimension A?

3. Calculate dimension B?

4. What is radius E?

5. What does surface D represent?

6. What is diameter at I?

7. Calculate the reference dimension J.

8. Surface L in the front view is what surface in the top view?

9. What is the dimension at M?

10. Surface N in the right side view represents what in the top view?

11. Surface 0 in the right side view is what surface in the top view?

12. Dimension P in the right side view is what?

13. The illustrated countersunk hole in SECTION A-A is which hole in the top view, "a" or "b"?

14. What is dimension Q in the right side view?

15. What is the distance from the top of the counterbored hole to the top of the countersunk holes?

16. The radius at K in the front view is what?

17. What is dimension H in the front view?

18. Surface C in the top view is what surface in the right-side view?

19. What is the contract number?

20. Surface T in the right-side view is what surface in the top view?

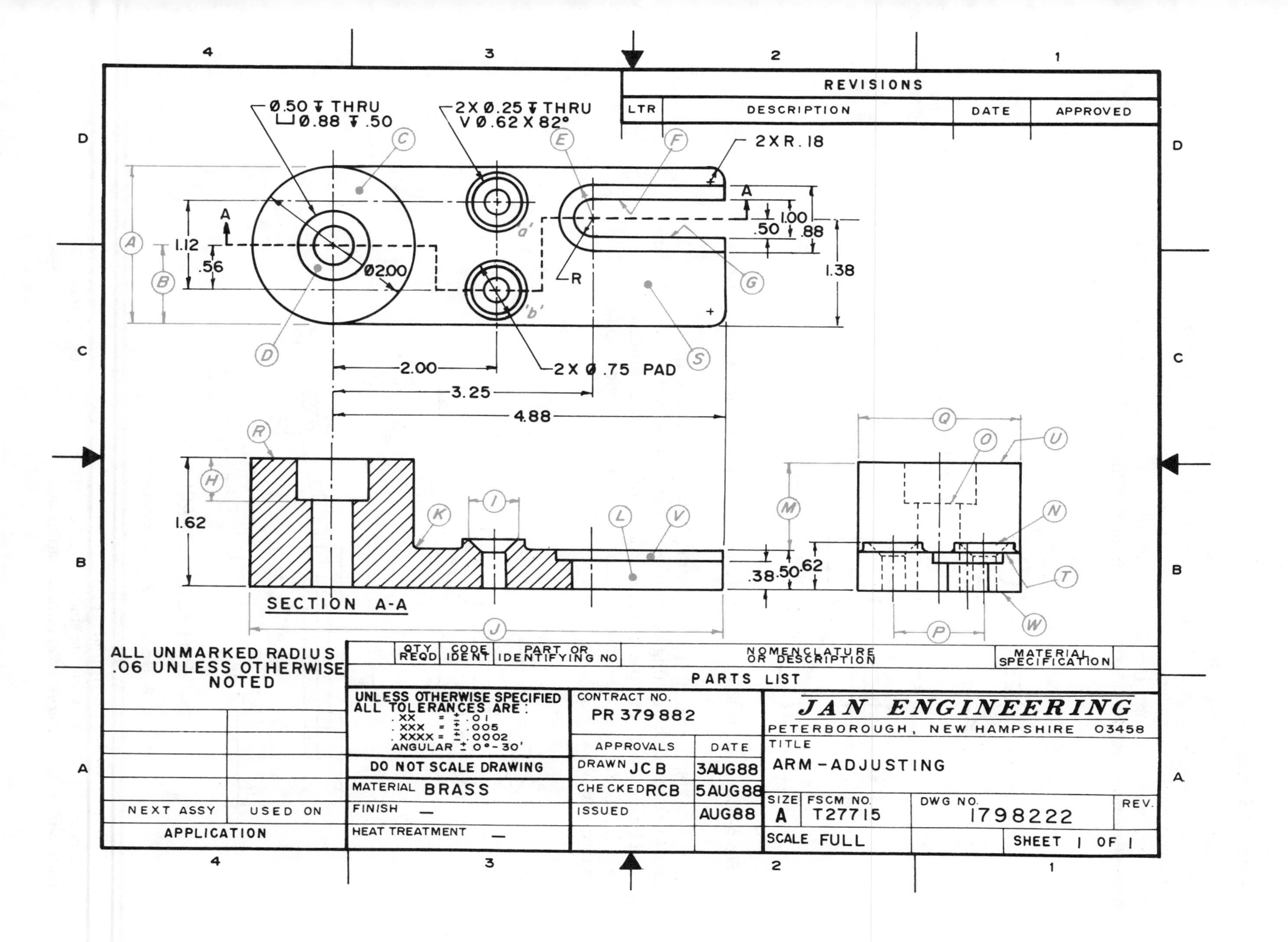
REVISIONS
LTR
DESCRIPTION
DATE
APPROVED
Ø.50 ↧ THRU
⌴ Ø.88 ↧ .50
2X Ø.25 ↧ THRU
V Ø.62 X 82°
2X R .18
Ø2.00
R
2X Ø .75 PAD
A
1.12
.56
1.00
.50
.88
1.38
2.00
3.25
4.88
1.62
.38
.50
.62
SECTION A-A
ALL UNMARKED RADIUS
.06 UNLESS OTHERWISE
NOTED
QTY REQD
CODE IDENT
PART OR IDENTIFYING NO
NOMENCLATURE OR DESCRIPTION
MATERIAL SPECIFICATION
PARTS LIST
UNLESS OTHERWISE SPECIFIED
ALL TOLERANCES ARE:
.XX = ± .01
.XXX = ± .005
.XXXX = ± .0002
ANGULAR ± 0°-30'
DO NOT SCALE DRAWING
MATERIAL BRASS
FINISH —
HEAT TREATMENT —
NEXT ASSY
USED ON
APPLICATION
CONTRACT NO.
PR 379882
APPROVALS
DATE
DRAWN JCB 3AUG88
CHECKED RCB 5AUG88
ISSUED AUG88
JAN ENGINEERING
PETERBOROUGH, NEW HAMPSHIRE 03458
TITLE
ARM-ADJUSTING
SIZE A
FSCM NO. T27715
DWG NO. 1798222
REV.
SCALE FULL
SHEET 1 OF 1

Removed Section

The section views of the full, half, and offset sections have been drawn in the usual multiview locations. The *removed section* is a section drawn away or removed from the regular views and shows only a portion of the object (see FIG. 6-17). Notice that only the portion of the object actually touched by the cutting plane line is drawn. The arrows indicate the direction of sight.

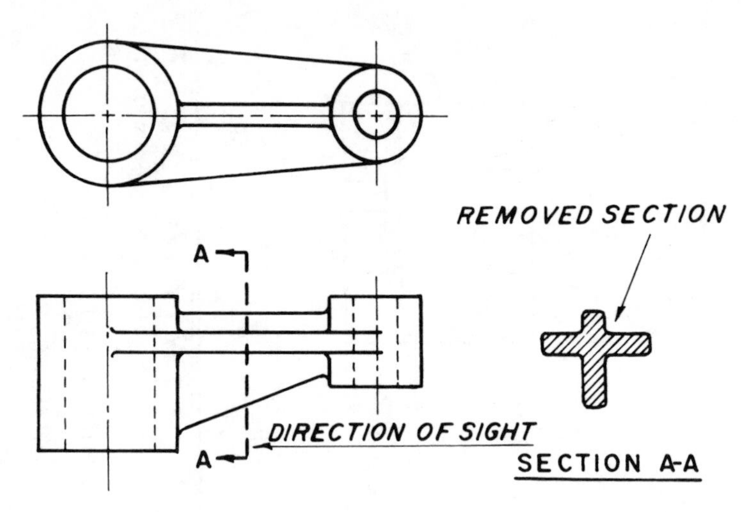

Fig. 6-17. Removed section is a section drawn away or *removed* from the regular views.

In FIG. 6-18, the height (H) and depth (D) of the section is drawn from the *exact* location of the cutting plane line. If the cutting plane line had been placed further to the left, the height and depth of the object would have been larger. In this example, dimensions X and Y remain the same throughout the length of the object.

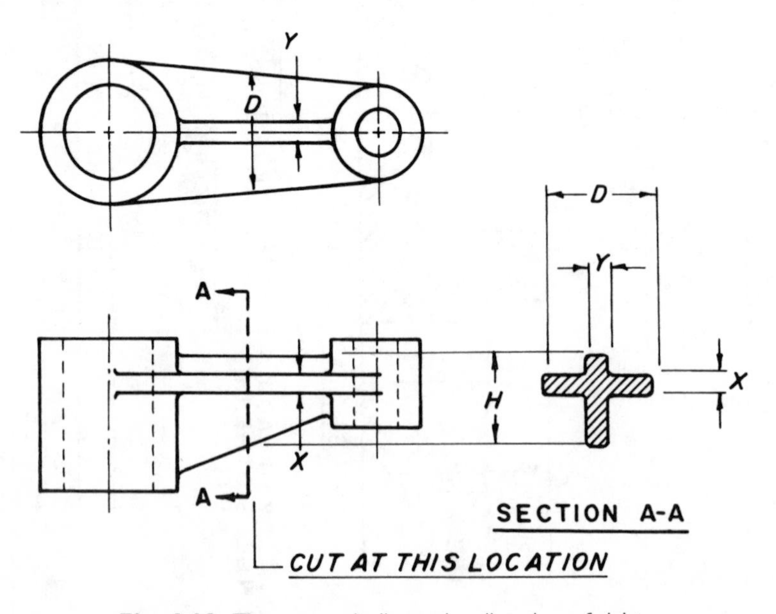

Fig. 6-18. The arrows indicate the direction of sight.

Fillets, rounds, and runouts. Fillets and rounds are terms used to describe a radius used to smooth out the sharp intersection of two surfaces. A *fillet* is an *inside* radius and a *round* is an *outside* radius.

Usually objects that are cast have fillets or rounds automatically included in their design (see FIG. 6-19). Rounds and fillets are used for three reasons:

- appearance
- strength
- to eliminate all sharp edges

Runouts are simply where one surface "runs" or blends into another.

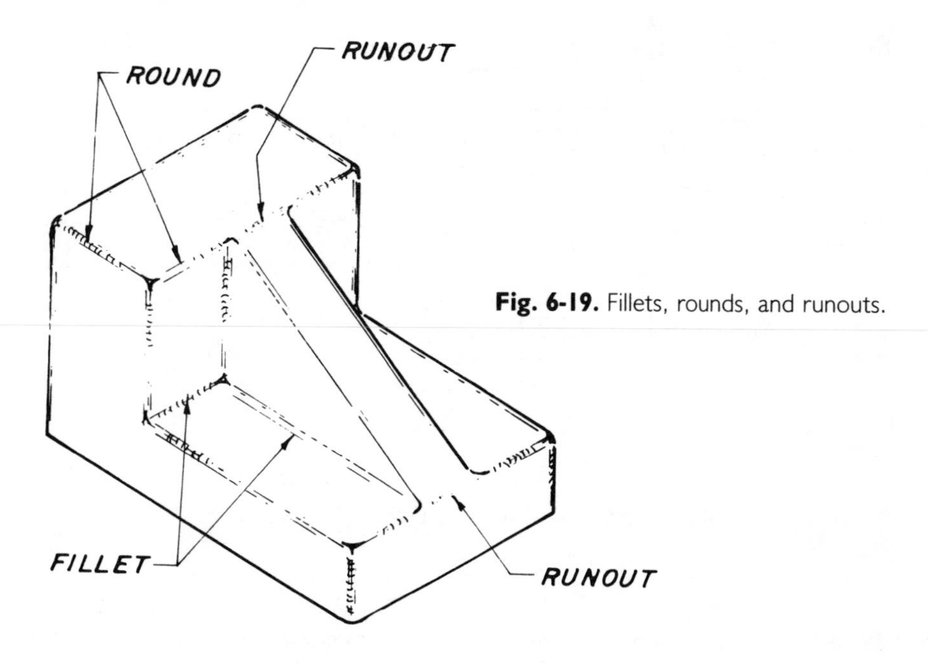

Fig. 6-19. Fillets, rounds, and runouts.

WORKSHEET 6-4

Instructions: Using drawing A0635831 on p. 89, answer the following questions in the spaces provided. (Answers in Appendix A.)

1. What kind of a section is used on this drawing?

2. Name views I, II, and III.

3. Dimension A in SECTION A-A is what?

4. Diameter B in view III is what diameter?

5. What is the *overall length* of the spacer?

6. What is dimension C in SECTION C-C?

7. The maximum size of D in SECTION D-D is what?

8. Hole E in SECTION B-B is what diameter?

9. What is the lower limit of hole Z in SECTION C-C?

10. Surface V in SECTION B-B is what surface in view II?

11. Surface Q in view II is what surface in view I?

12. Surface H in view II is what surface in SECTION C-C?

13. Surface I in view II is what surface in SECTION D-D?

14. What is dimension U in view II?

15. How far is surface V in SECTION B-B from surface G in view II?

16. What is the distance from surface R in view I to surface M in SECTION D-D?

17. What is the size of the two holes in SECTION A-A?

18. How deep are the four notches in SECTION D-D?

19. What is the maximum distance surface H in view II is from surface 0 in SECTION C-C?

20. The overall diameter of the *spacer* is what?

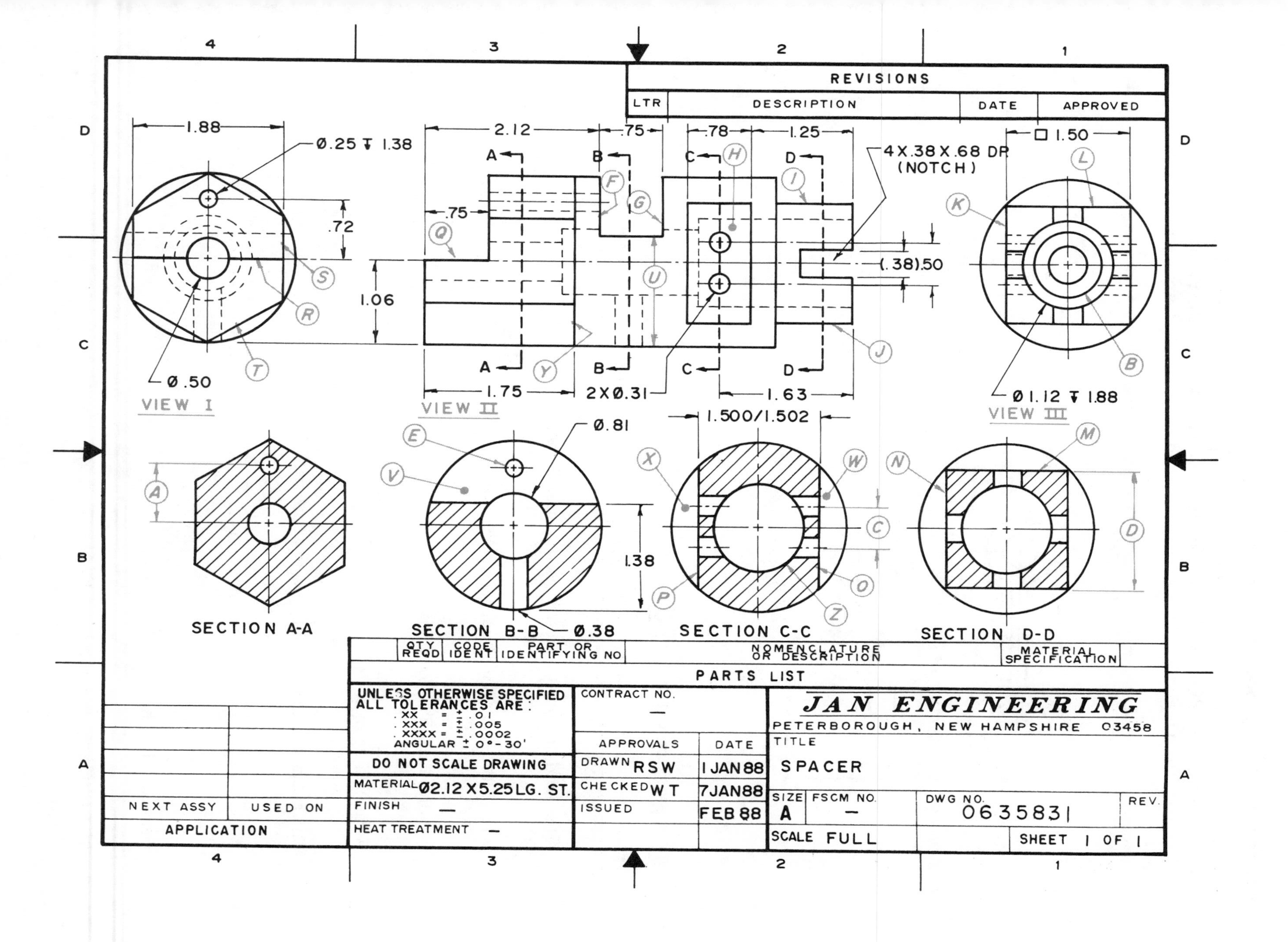
REVISIONS
LTR
DESCRIPTION
DATE
APPROVED
1.88
Ø.25 ↧ 1.38
.72
1.06
Ø.50
VIEW I
2.12
.75
.75
.78
1.25
1.75
2XØ.31
1.63
VIEW II
4X.38X.68 DP (NOTCH)
(.38).50
□ 1.50
Ø1.12 ↧ 1.88
VIEW III
1.500/1.502
Ø.81
1.38
Ø.38
SECTION A-A
SECTION B-B
SECTION C-C
SECTION D-D
QTY REQD
CODE IDENT
PART OR IDENTIFYING NO
NOMENCLATURE OR DESCRIPTION
MATERIAL SPECIFICATION
PARTS LIST
UNLESS OTHERWISE SPECIFIED ALL TOLERANCES ARE:
.XX = ±.01
.XXX = ±.005
.XXXX = ±.0002
ANGULAR ± 0°-30'
DO NOT SCALE DRAWING
MATERIAL Ø2.12 X 5.25 LG. ST.
FINISH —
HEAT TREATMENT —
CONTRACT NO. —
APPROVALS
DATE
DRAWN RSW 1 JAN 88
CHECKED WT 7 JAN 88
ISSUED FEB 88
JAN ENGINEERING
PETERBOROUGH, NEW HAMPSHIRE 03458
TITLE
SPACER
SIZE A
FSCM NO. —
DWG NO. 0635831
REV.
SCALE FULL
SHEET 1 OF 1
NEXT ASSY
USED ON
APPLICATION

Rotated (or revolved) Section

Rotated sections are like the removed section, except the section, instead of being removed, is simply *rotated* in its place (see FIG. 6-20). If the cutting plane line had been taken to the left slightly, the height and depth would have been larger.

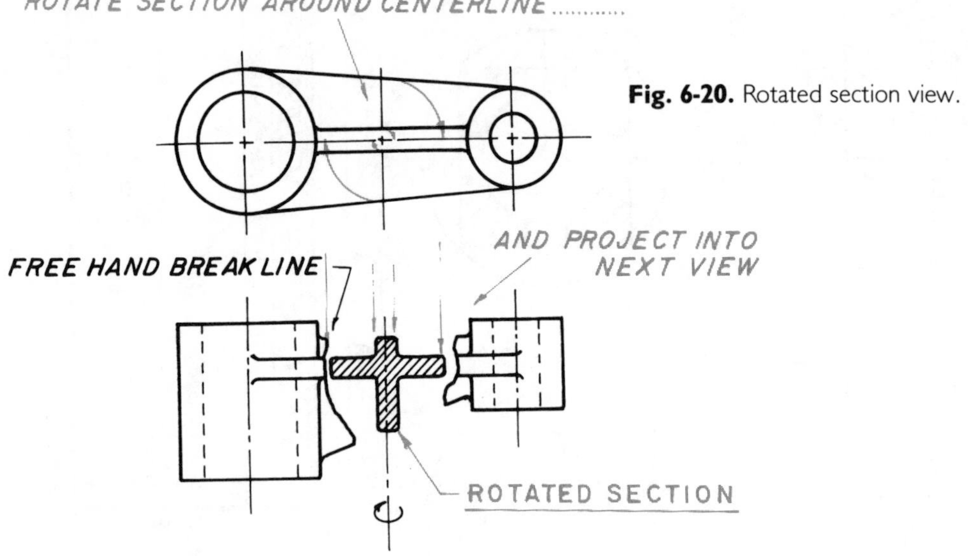

Fig. 6-20. Rotated section view.

WORKSHEET 6-5

Instruction: Using drawing A1798003 on p. 92, answer the following questions in the spaces provided. (Answers in Appendix A.)

1. What kind of a section view is used on this drawing?

2. What is the lower limit of diameter A?

3. Is radius E a "fillet" or a "round"?

4. What is dimension B?

5. The radius at U in the front view is what?

6. What is the minimum size dimension G can be and still be within tolerance?

7. Diameter H in the front view is what?

8. Surface I in the front view is what surface in the top view?

9. Dimension J in the front view is what?

10. Surface K in the front view is what surface in the top view?

11. What kind of a line is line M in the front view, other than a object line?

12. What is the maximum diameter of L?

13. What is dimension N?

14. What does the thick line T, under the 9.500 dimension indicate?

15. What scale was used on this drawing?

16. How far is surface D in the top view from surface S in the front view?

17. Calculate dimension 0.

18. What is the maximum distance, center to center, the two holes can be?

19. What was the 9.500 dimension when the drawing was first issued?

20. Surface C in the top view is what surface in the front view.

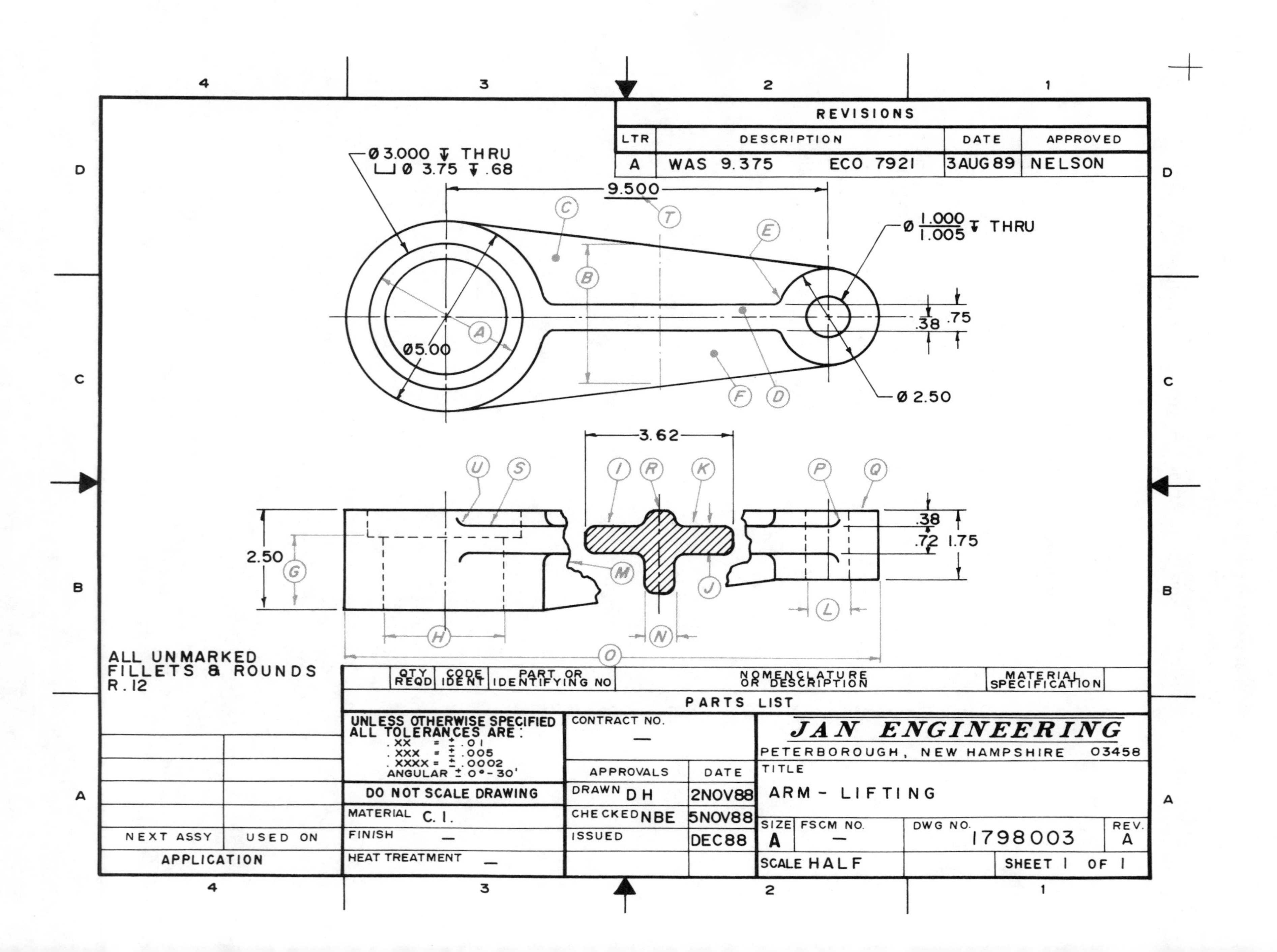
REVISIONS
LTR | DESCRIPTION | DATE | APPROVED
A | WAS 9.375 ECO 7921 | 3AUG89 | NELSON
Ø3.000 ↧ THRU
⌴ Ø 3.75 ↧ .68
9.500
Ø 1.000/1.005 ↧ THRU
.75
.38
Ø5.00
Ø 2.50
3.62
.38
.72
1.75
2.50
ALL UNMARKED FILLETS & ROUNDS R .12
QTY REQD | CODE IDENT | PART OR IDENTIFYING NO | NOMENCLATURE OR DESCRIPTION | MATERIAL SPECIFICATION
PARTS LIST
UNLESS OTHERWISE SPECIFIED ALL TOLERANCES ARE:
.XX = ± .01
.XXX = ± .005
.XXXX = ± .0002
ANGULAR ± 0°-30'
DO NOT SCALE DRAWING
MATERIAL C.I.
FINISH —
HEAT TREATMENT —
NEXT ASSY | USED ON
APPLICATION
CONTRACT NO. —
APPROVALS | DATE
DRAWN DH 2NOV88
CHECKED NBE 5NOV88
ISSUED DEC88
JAN ENGINEERING
PETERBOROUGH, NEW HAMPSHIRE 03458
TITLE ARM - LIFTING
SIZE A | FSCM NO. — | DWG NO. 1798003 | REV. A
SCALE HALF | SHEET 1 OF 1

Broken-Out Section

Many times only a small area needs to be drawn in section to make the drawing clear. For this occasion, the drafter will use a *broken-out section* (refer to FIG. 6-21). A cutting plane line is *not* used in a broken-out section. A thick, wavy object line around the area or feature to be detailed is drawn to illustrate that a piece has been broken out. The features or details will now be seen as solid object lines.

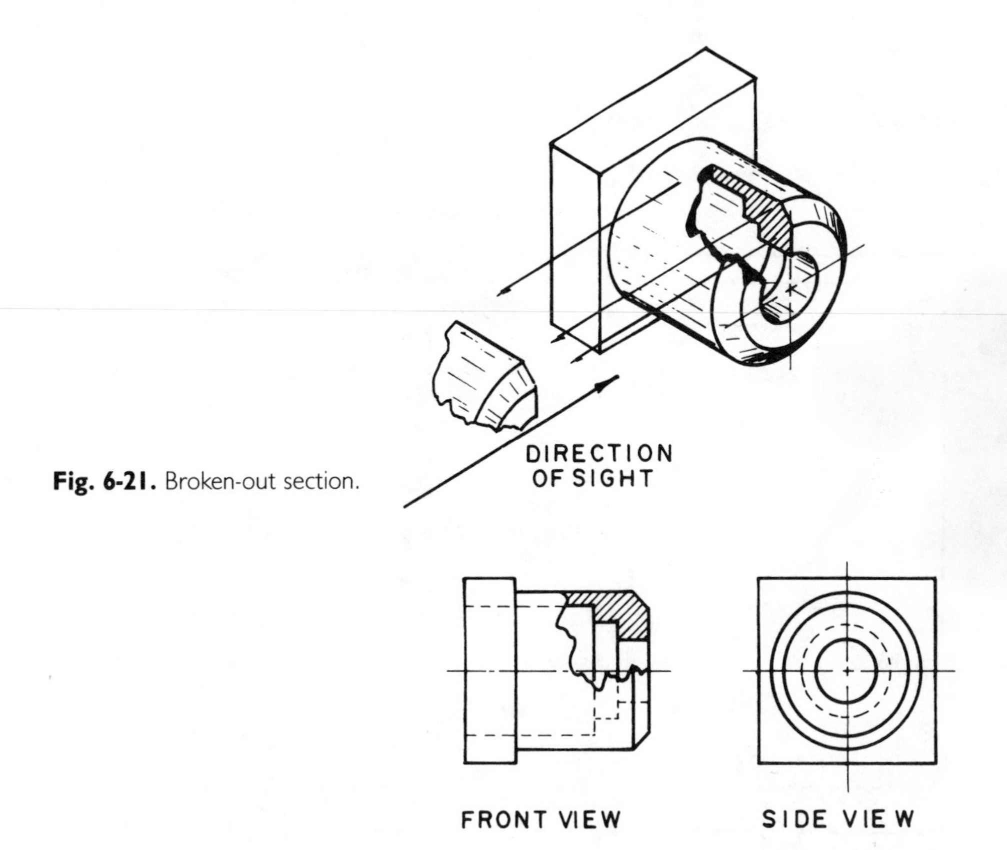

Fig. 6-21. Broken-out section.

Thin Wall Section

Section lining cannot be added to thin objects such as sheets of metal, gaskets, or shims; therefore, they are simply drawn and filled in *solid black* (see FIG. 6-22). If two or more thin pieces that are filled in solid black touch each other, a space is left between the parts (see FIG. 6-23).

Shafts and fasteners in section. If a cutting plane line passes through the *long* dimension of a shaft or any kind of fastener, section lining is *not* added (see FIG. 6-24). Section lining on these parts that have no interior detail would serve no purpose. If, however, the cutting plane line passes through or *across* the shaft or fastener, section lining *is* added (see FIG. 6-25).

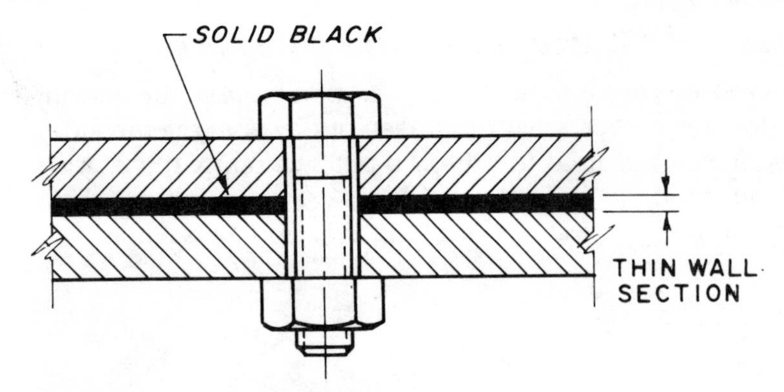

Fig. 6-22. Thin wall section.

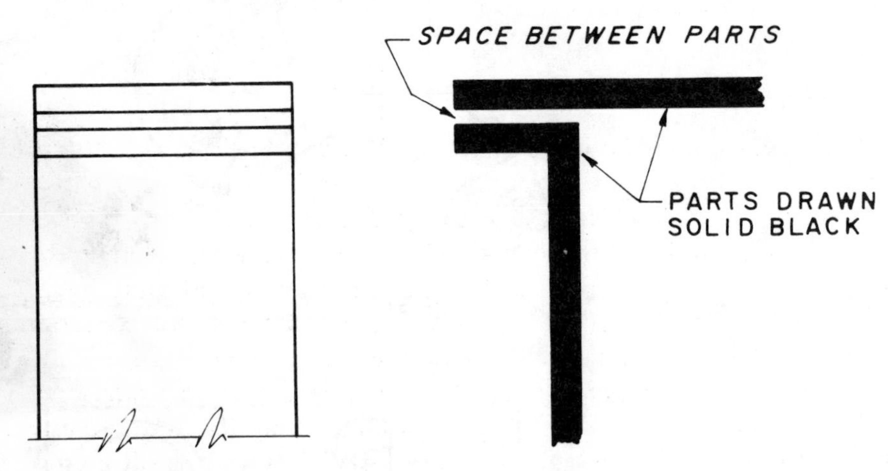

Fig. 6-23. If two or more thin pieces touch each other, a space is left between them.

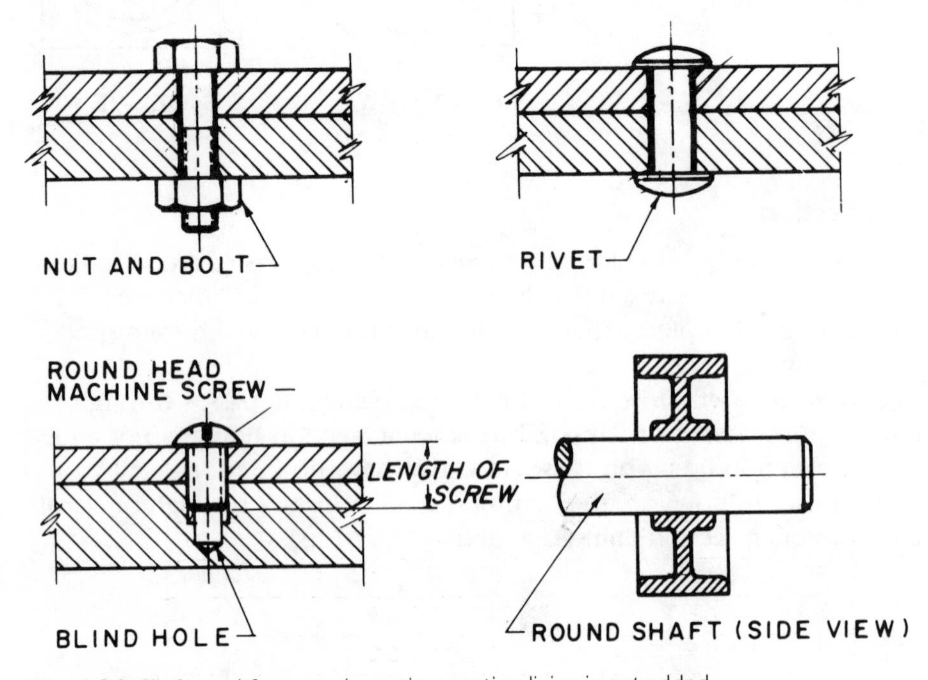

Fig. 6-24. Shafts and fasteners in section; section lining is *not* added.

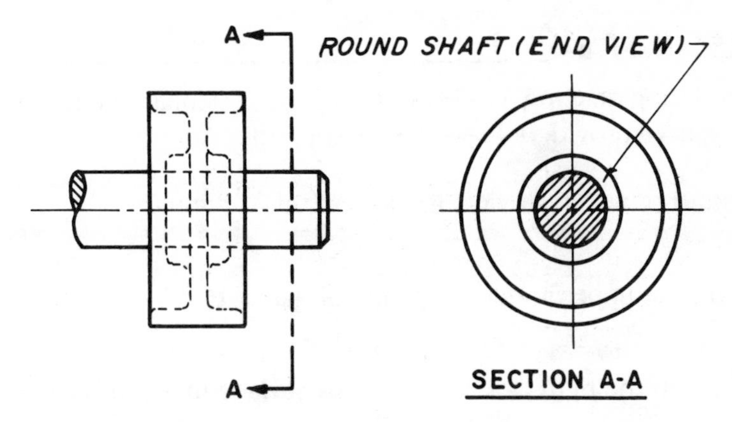

Fig. 6-25. If the cutting plane line passes through or across the shaft or fastener, section lining *is* added.

Assembly Section

An assembly drawing shows how the many parts of an object fit together to make up the object. This kind of a section view also shows where each part is located. Each part of the assembly is also labeled with its *name, part number,* and *quantity required.* If the assembly is small, this information is given on the drawing by a note next to each part. If the assembly is large and there is not enough room for the many notes, each part is identified by a number or letter within a circle with a leader line touching each part. The circle with the number or letter in it is called a *balloon* (see FIG. 6-26). A table, known as a *parts list,* must be added to the drawing to give the name, part number, and quantity required of each part shown in a balloon. If the parts list is long, it is put on a separate sheet, rather than on the drawing (see FIG. 6-27).

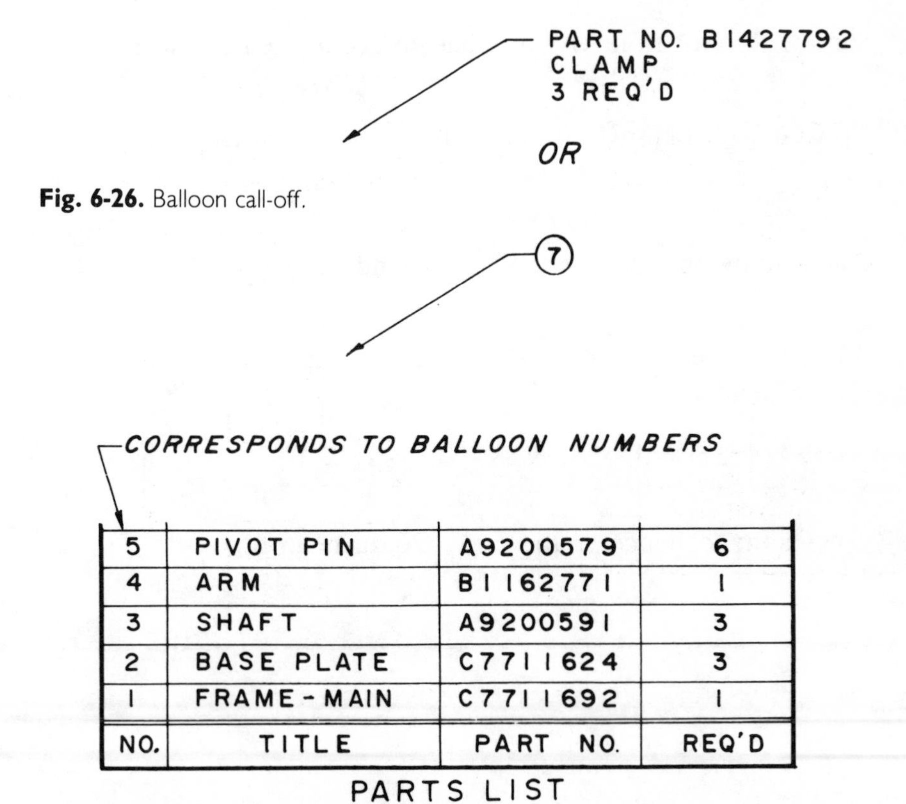

Fig. 6-26. Balloon call-off.

NO.	TITLE	PART NO.	REQ'D
5	PIVOT PIN	A9200579	6
4	ARM	B1162771	1
3	SHAFT	A9200591	3
2	BASE PLATE	C7711624	3
1	FRAME - MAIN	C7711692	1

PARTS LIST

Fig. 6-27. If the parts list is long it is sometimes put on a separate sheet.

WORKSHEET 6-6

Instructions: Using drawing A19732114 on p. 97, calculate the following questions in the spaces provided. (Answers in Appendix A.)

1. What kind of a section view is used on this drawing?

2. What are the upper and lower limits of diameter A?

3. Surface K in the right-side view is what surface in the front view?

4. Line B is what kind of a line, other than an object line?

5. How far is it from surface Q in the front view to surface M?

6. What was the part made of when it was first made?

7. To what scale is this drawing drawn to?

8. What is the diameter of J?

9. Surface P in the front view is what surface in the right-side view?

10. Surface N in the front view is what surface in the right-side view?

11. How far is it from surface L in the front view to surface K in the right-side view?

12. Calculate the maximum size R can be and still be within limits.

13. What does the thick line under the 1.500 dimension mean?

14. When was this part last changed?

15. What is the tolerance on the .625/.626 diameter hole?

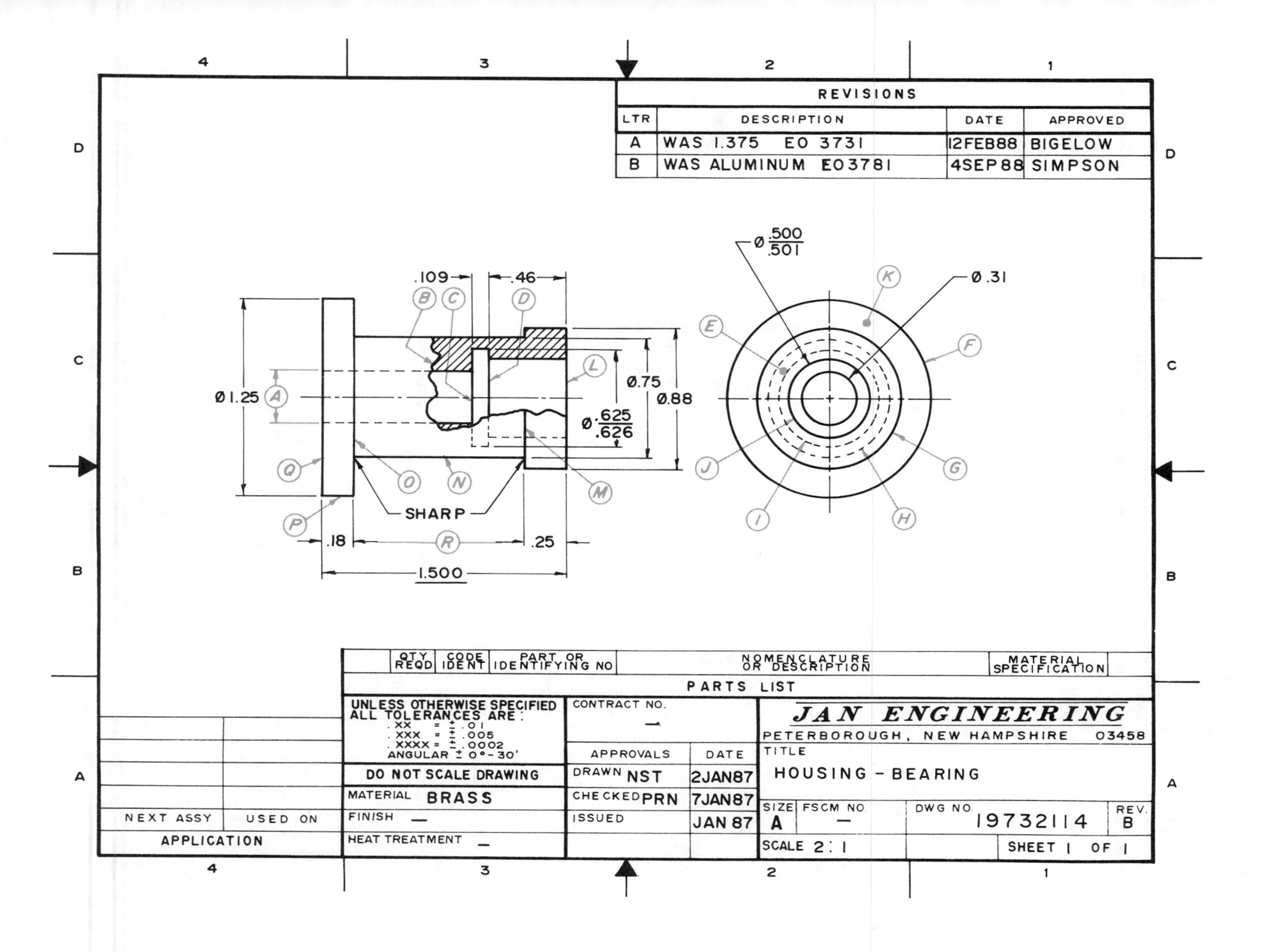
REVISIONS
LTR
DESCRIPTION
DATE
APPROVED
A
WAS 1.375 EO 3731
12FEB88
BIGELOW
B
WAS ALUMINUM EO3781
4SEP88
SIMPSON
Ø.500/.501
Ø.31
.109
.46
Ø1.25
Ø.75
Ø.88
Ø.625/.626
SHARP
.18
.25
1.500
QTY REQD
CODE IDENT
PART OR IDENTIFYING NO
NOMENCLATURE OR DESCRIPTION
MATERIAL SPECIFICATION
PARTS LIST
UNLESS OTHERWISE SPECIFIED ALL TOLERANCES ARE:
.XX = ± .01
.XXX = ± .005
.XXXX = ± .0002
ANGULAR ± 0°-30'
DO NOT SCALE DRAWING
MATERIAL BRASS
FINISH —
HEAT TREATMENT —
CONTRACT NO. —
APPROVALS
DATE
DRAWN NST 2JAN87
CHECKED PRN 7JAN87
ISSUED JAN 87
JAN ENGINEERING
PETERBOROUGH, NEW HAMPSHIRE 03458
TITLE
HOUSING - BEARING
SIZE A
FSCM NO. —
DWG NO. 19732114
REV. B
SCALE 2:1
SHEET 1 OF 1
NEXT ASSY
USED ON
APPLICATION

FIGURE 6-28 is an example of an assembly drawing. Notice no section lining is added to the *long* direction of the two shafts and none added to the fasteners. There is also a thin wall section used in this assembly drawing—a thin gasket. Hidden lines are *not* used on assembly drawings unless absolutely necessary. Only overall reference dimensions and those required to locate the position of the parts are used on an assembly drawing. Remember, each and every part has its own full-dimensioned drawing; therefore, dimensions are not needed. In order to help identify each individual part of the assembly drawing, section lining of each part slants in different directions.

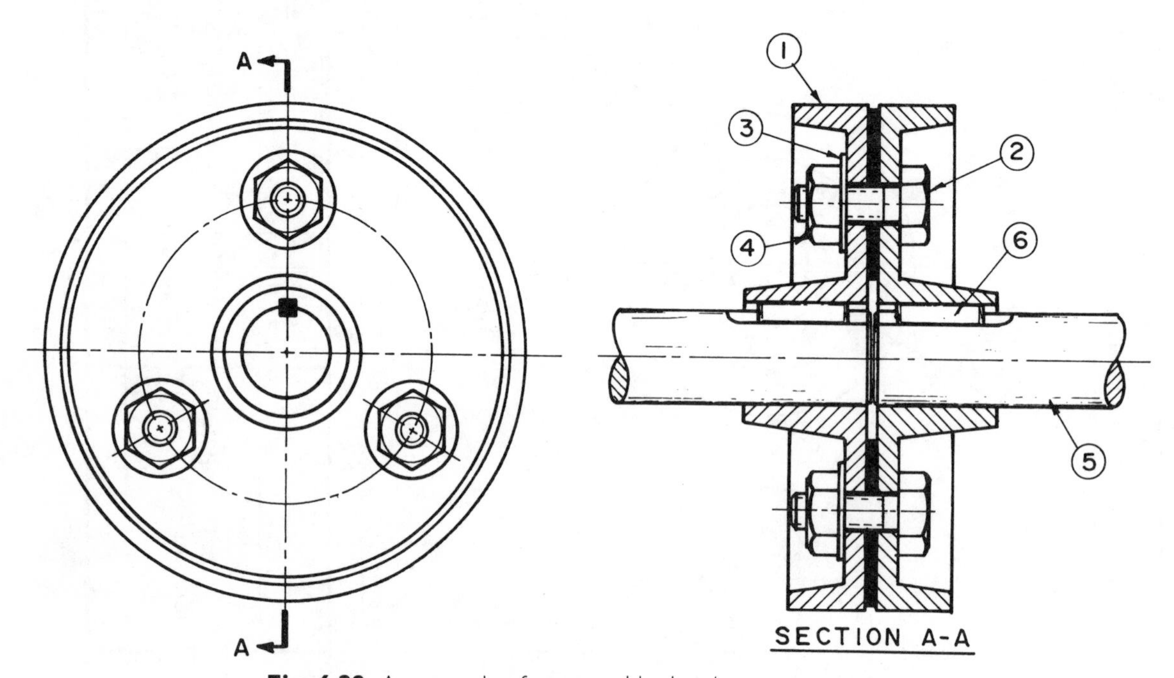

Fig. 6-28. An example of an assembly drawing.

CONVENTIONAL PRACTICES

Sometimes a view is actually *drawn incorrectly* in order to avoid confusion. There are standard ways of drawing certain features that, while *not accurate,* allow everyone reading the drawing to understand exactly what is shown. These standard ways of drawing certain features are known as *conventional practices.* Conventional practices apply to ribs, webs, holes, spokes, and keyways.

Ribs and webs. FIGURE 6-29 shows an object with four supporting ribs placed 90 degrees apart (see the pictorial view). The two given regular views illustrate the part. Directly below are the same two views, except the front view has been made a full section view. If the cutting plane line passes through the center, as shown, the front section view is correct and will look *exactly* as illustrated. This gives a false impression and implies the part is to look like the lower pictorial view—that is, a solid cone-shaped object. To avoid this problem, the conventional method would be to draw the cutting plane line as shown but *not* add section lining to the ribs or webs (see FIG. 6-30).

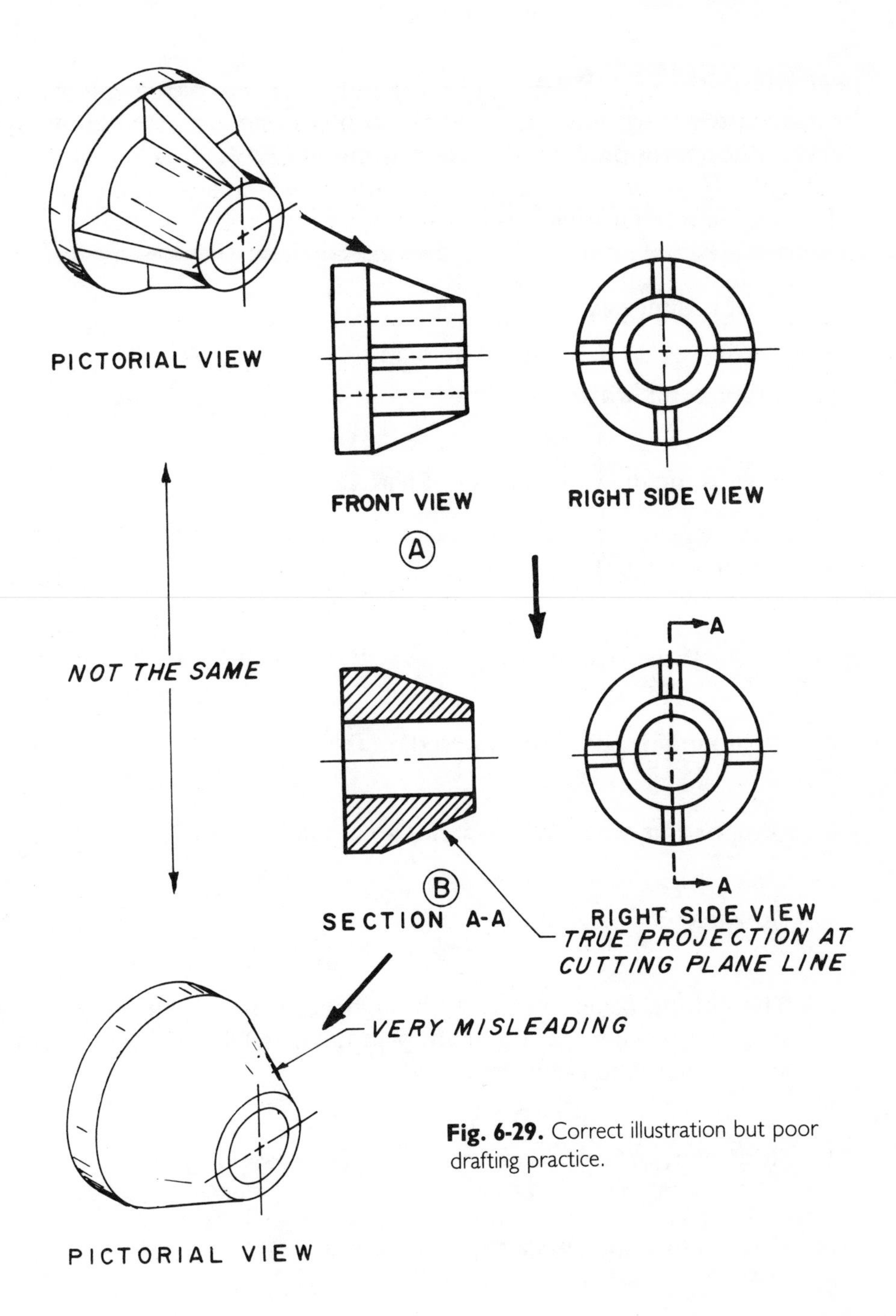

Fig. 6-29. Correct illustration but poor drafting practice.

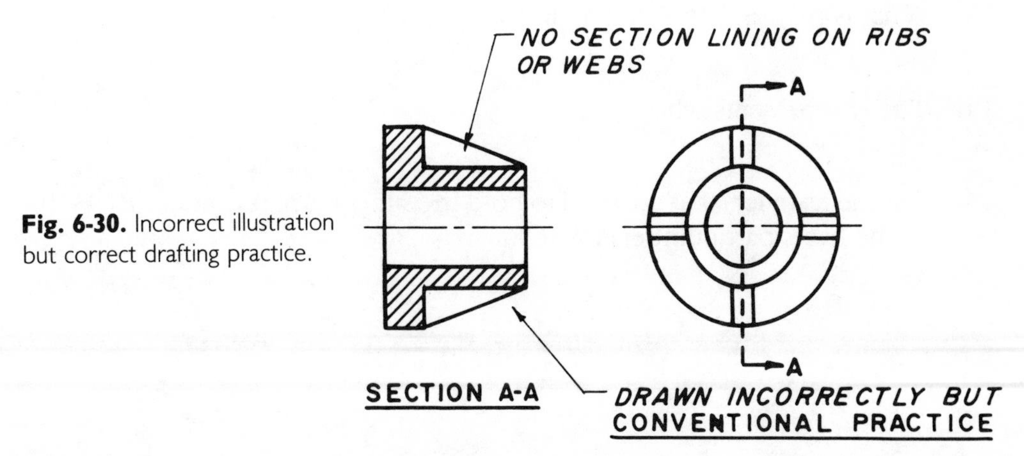

Fig. 6-30. Incorrect illustration but correct drafting practice.

WORKSHEET 6-7

Instructions: Using drawing A193505 on p. 101, answer the following questions in the spaces provided. (Answers in Appendix A.)

1. What kind of a section is this?

2. What is dimension A?

3. B represents what?

4. What is the drawing number of part C?

5. Part D is what plan number?

6. E represents what drawing number and how many are required?

7. What kind of a section was used on the washer at F?

8. G represents what? Why is it *not* sectioned?

9. How may 1/4″ washers are used?

10. The cutting plane lines pass through the center of the front view and through the center of the shaft, plan number B193506, why doesn't this shaft have section lining applied to it?

11. How many *different* parts are used on this assembly?

12. How many *total* parts are used on this assembly?

13. What is the F.S.C.M. number and what does it mean?

14. The H represents what?

15. List the fasteners that are used to hold the support, part number B193504, to the base, part number A193606.

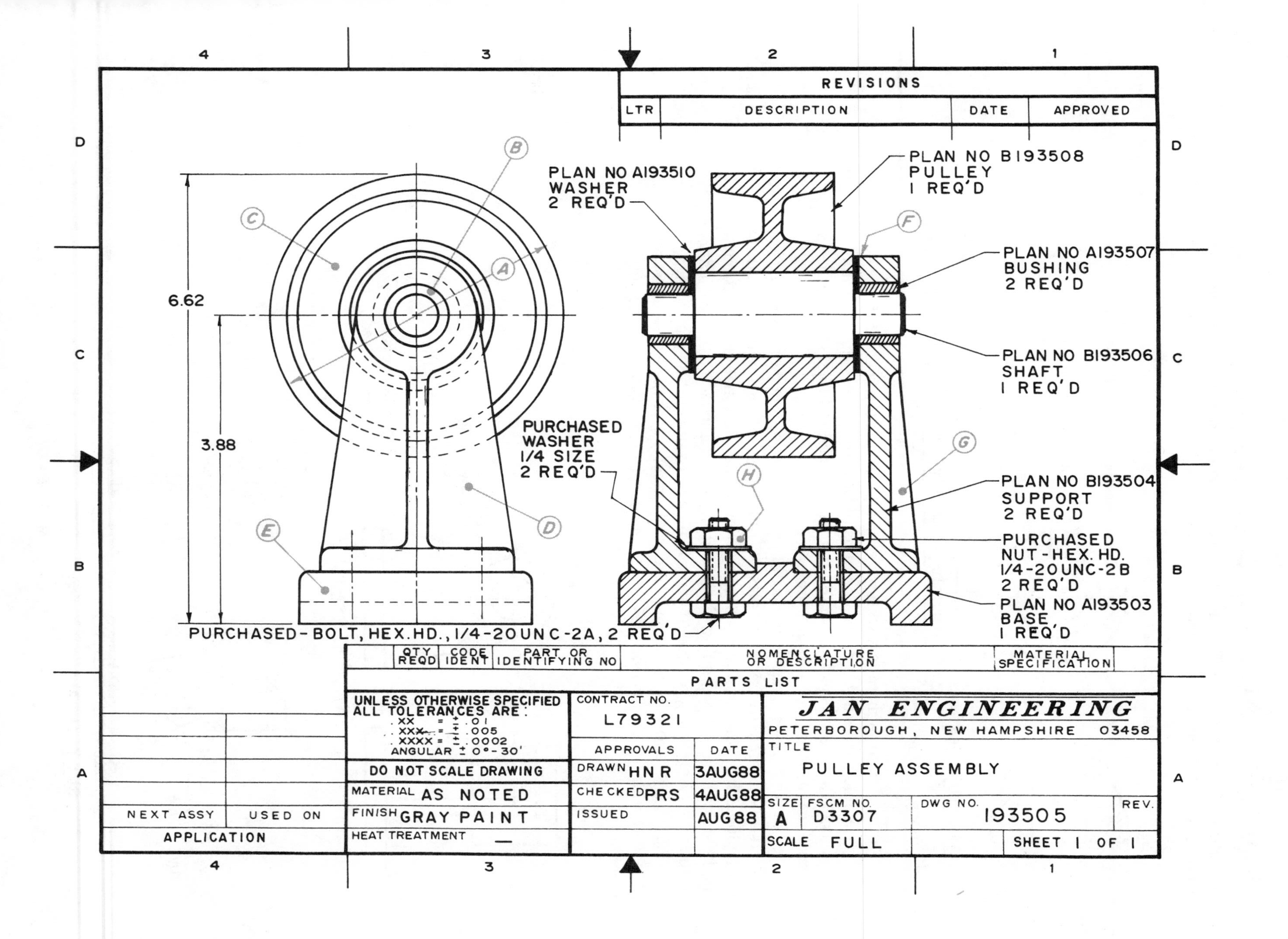
REVISIONS
LTR
DESCRIPTION
DATE
APPROVED
PLAN NO A193510
WASHER
2 REQ'D
PLAN NO B193508
PULLEY
1 REQ'D
PLAN NO A193507
BUSHING
2 REQ'D
PLAN NO B193506
SHAFT
1 REQ'D
PURCHASED
WASHER
1/4 SIZE
2 REQ'D
PLAN NO B193504
SUPPORT
2 REQ'D
PURCHASED
NUT-HEX. HD.
1/4-20UNC-2B
2 REQ'D
PLAN NO A193503
BASE
1 REQ'D
PURCHASED-BOLT, HEX.HD., 1/4-20UNC-2A, 2 REQ'D
A
B
C
D
E
F
G
H
6.62
3.88
QTY REQD
CODE IDENT
PART OR IDENTIFYING NO
NOMENCLATURE OR DESCRIPTION
MATERIAL SPECIFICATION
PARTS LIST
UNLESS OTHERWISE SPECIFIED
ALL TOLERANCES ARE:
.XX = ± .01
.XXX = ± .005
.XXXX = ± .0002
ANGULAR ± 0°-30'
CONTRACT NO.
L79321
JAN ENGINEERING
PETERBOROUGH, NEW HAMPSHIRE 03458
APPROVALS
DATE
DRAWN HNR
3AUG88
CHECKED PRS
4AUG88
ISSUED
AUG88
TITLE
PULLEY ASSEMBLY
DO NOT SCALE DRAWING
MATERIAL AS NOTED
FINISH GRAY PAINT
HEAT TREATMENT —
SIZE
A
FSCM NO.
D3307
DWG NO.
193505
REV.
SCALE FULL
SHEET 1 OF 1
NEXT ASSY
USED ON
APPLICATION

A practice used by some companies is a combination of both practices. Section lining *is* added to the ribs and webs but double-spaced through the web or rib, as illustrated in FIG. 6-31.

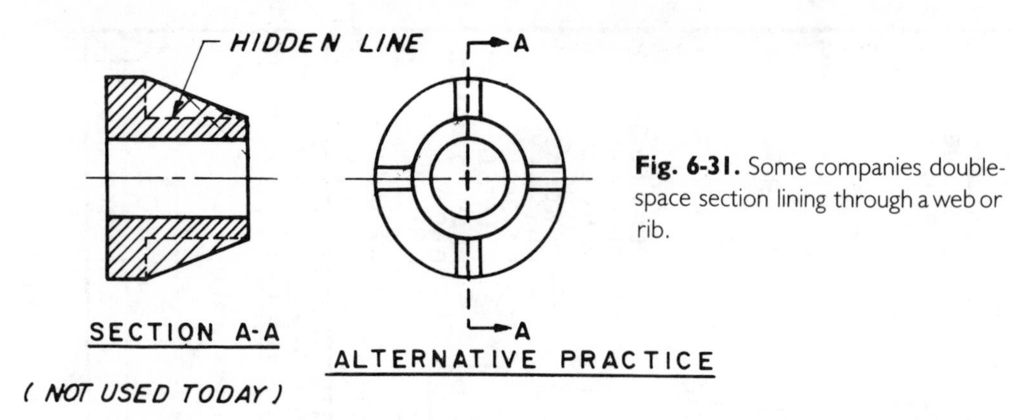

Fig. 6-31. Some companies double-space section lining through a web or rib.

FIGURE 6-32 show an object with webs to support the top section. If the cutting plane line passes through the center as illustrated, it gives the appearance the object is a massive solid object. By *not* adding section lining to the webs, a closer illustration is given of what the object is to look like.

Holes and spokes. Many circular objects have a number of holes spaced around a bolt circle (B.C.). The holes do not always line up with the cutting plane line. The conventional practice is to draw the sectional view as though the holes were actually in line with the cutting plane line (see FIG. 6-33). Think of the lower hole being rotated to make it align with the cutting plane line.

Figure 6-34 shows both the incorrect and correct method used to illustrate an object with both holes *and* ribs being rotated to align them with the cutting plane line.

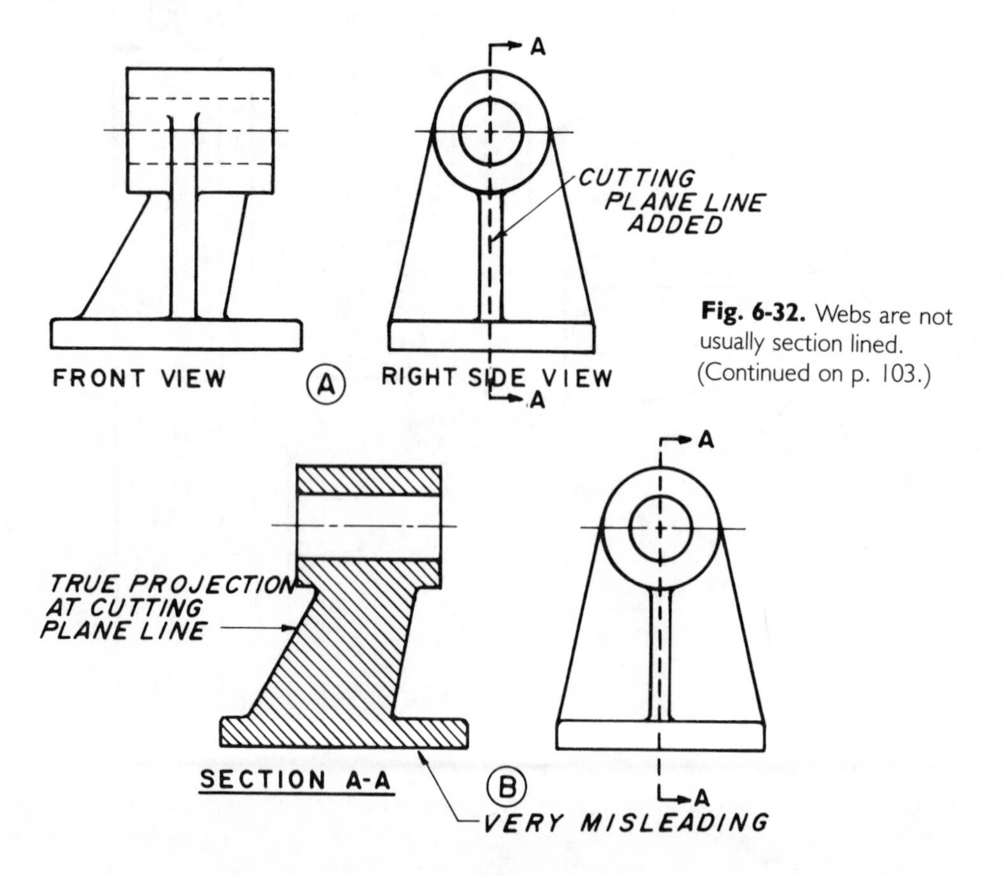

Fig. 6-32. Webs are not usually section lined. (Continued on p. 103.)

NO SECTION LINING ON RIBS OR WEBS

CONVENTIONAL PRACTICE

SECTION A-A

(C)

Fig. 6-32. Continued.

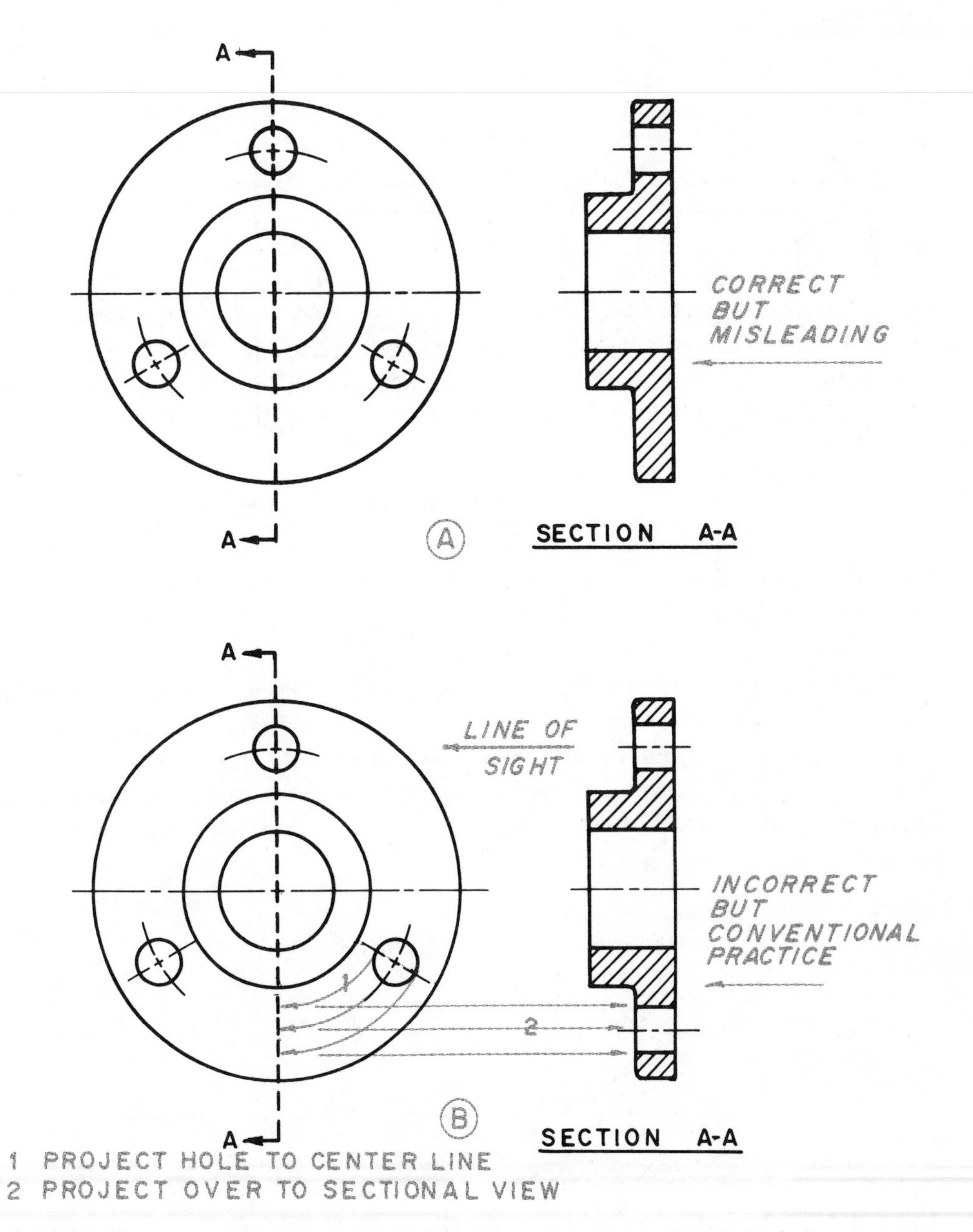

Fig. 6-33. The conventional practice is to draw the section view as though the holes were actually in line.

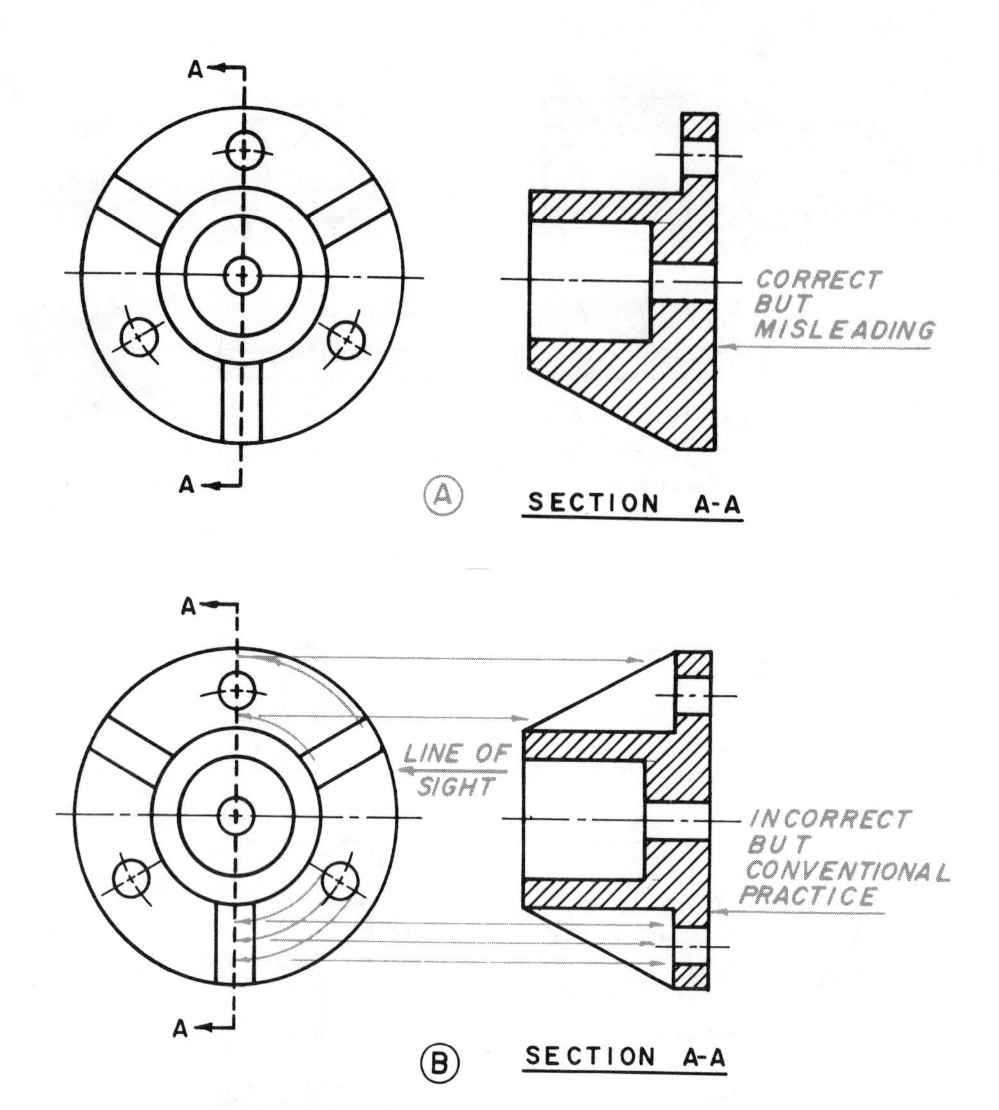

Fig. 6-34. Incorrect and correct methods used to illustrate an object with both holes and ribs being rotated to align with the cutting plane line.

Spokes of a wheel or pulley are also confusing if drawn correctly in section (see FIG. 6-35). The conventional practice is to rotate the spokes similar to the way holes are rotated, and draw them as illustrated. Note spoke "A" is rotated up and in line with the cutting plane line. Section lining is *not* added to spokes. Keyways are also rotated around in order to be in line with the cutting plane line.

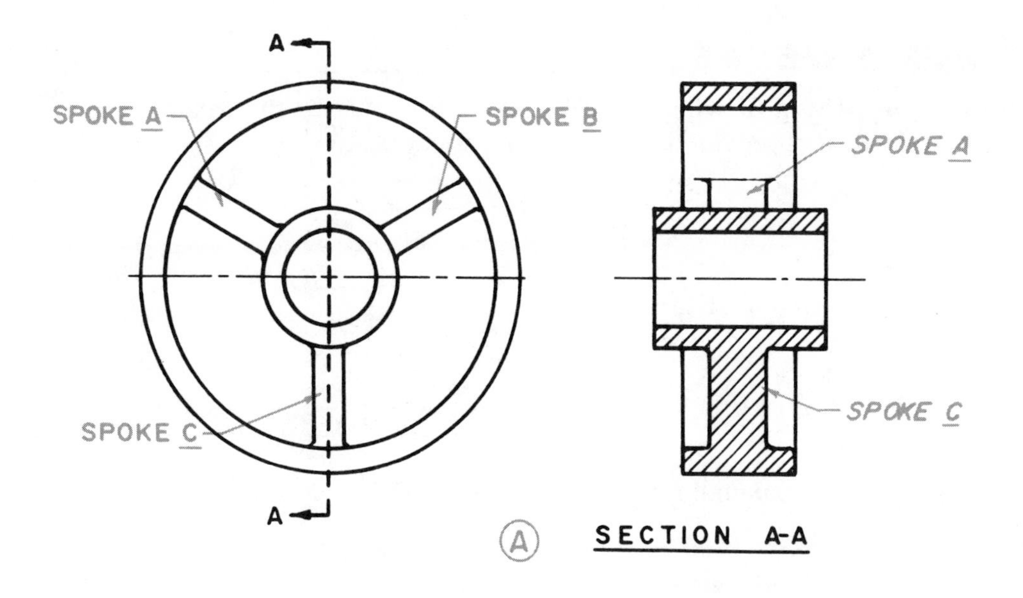

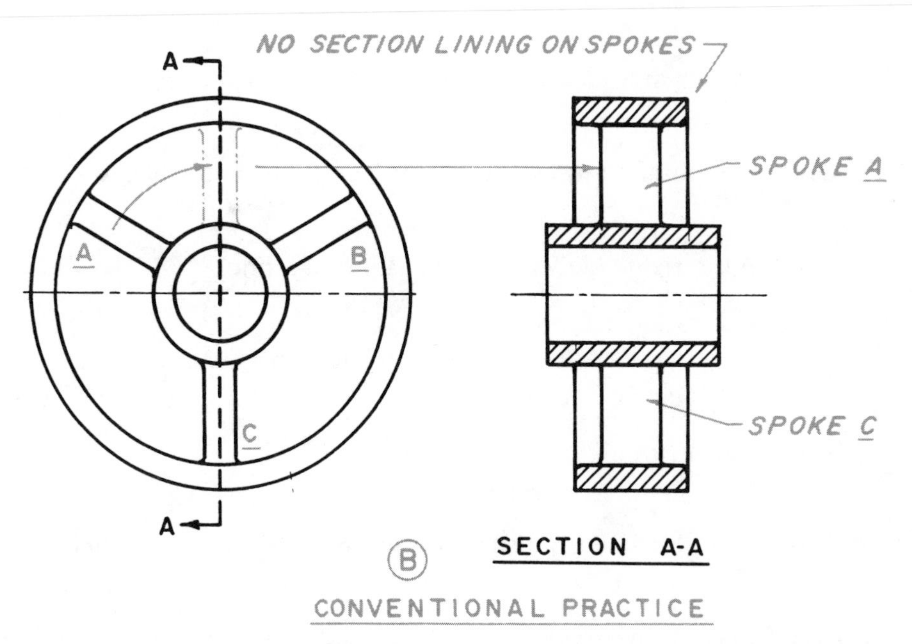

Fig. 6-35. Spokes are rotated similar to holes and drawn as illustrated.

WORKSHEET 6-8

Instructions: Using drawing A179954 on p. 107, answer the following questions in the spaces provided. (Answers in Appendix A.)

1. What kind of a section was used on this drawing?

2. Line A is what kind of a line?

3. Dimension B is what?

4. What is the maximum depth of dimension C?

5. What is the diameter at D?

6. The diameter indicated by the letter E is what diameter?

7. What is the minimum size dimension I could be and still be within limits?

8. Line J in the right-side view is what kind of a line?

9. Diameter K in the right-side view is what diameter?

10. What is diameter L?

11. What is the distance from surface A in the side view to surface O?

12. How far is it from surface M in the right-side view to surface Q in the front view?

13. How far apart are surfaces R to S, in the front view?

14. What are the high and low *limits* of the four .438 diameter holes?

15. What is the *holder* made of?

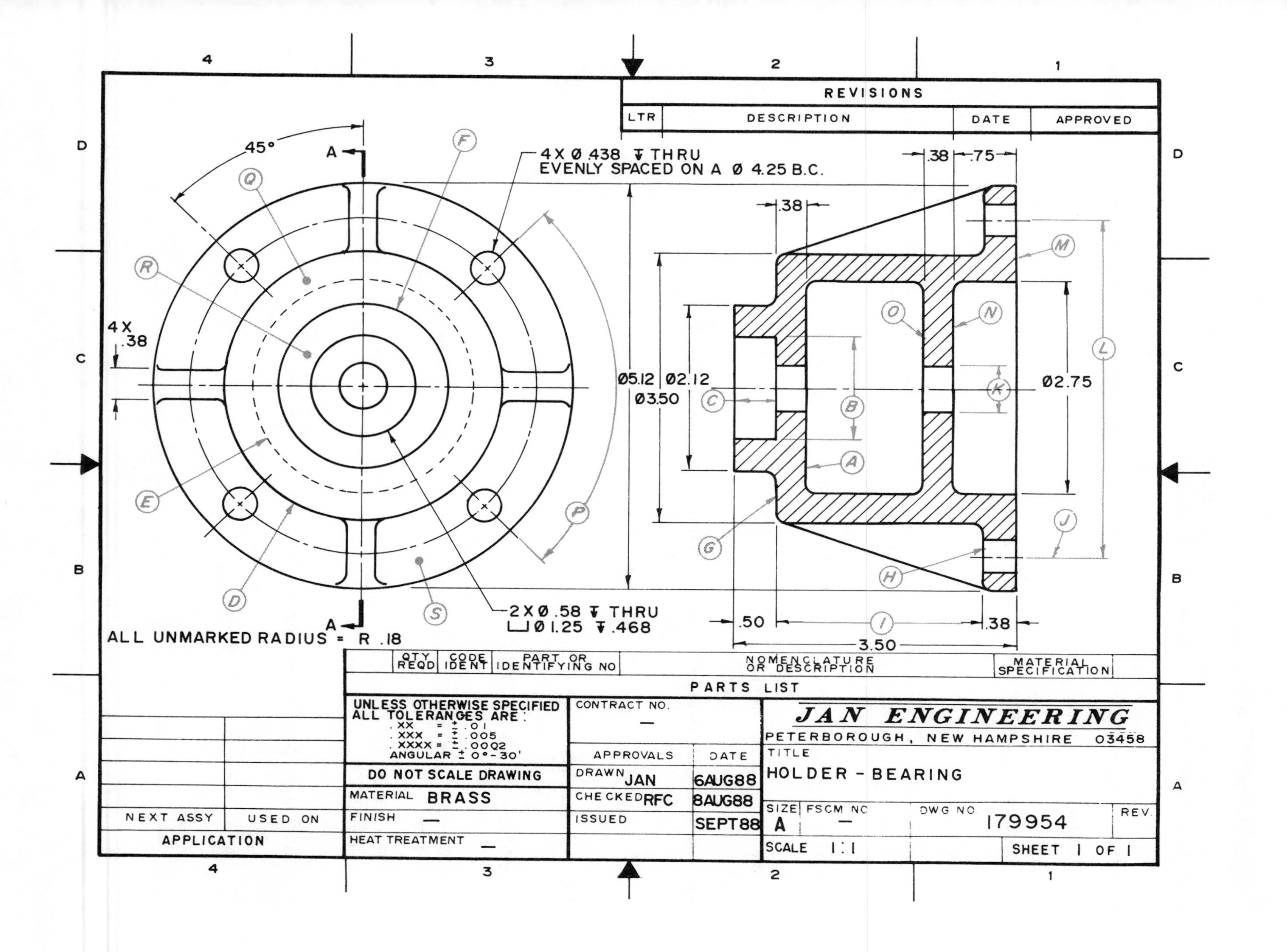
REVISIONS
LTR
DESCRIPTION
DATE
APPROVED
4X Ø .438 ↧ THRU
EVENLY SPACED ON A Ø 4.25 B.C.
45°
4X .38
.38
.75
.38
Ø5.12
Ø2.12
Ø3.50
Ø2.75
2X Ø .58 ↧ THRU
⌴ Ø 1.25 ↧ .468
ALL UNMARKED RADIUS = R .18
.50
.38
3.50
A
A
QTY REQD
CODE IDENT
PART OR IDENTIFYING NO
NOMENCLATURE OR DESCRIPTION
MATERIAL SPECIFICATION
PARTS LIST
UNLESS OTHERWISE SPECIFIED ALL TOLERANCES ARE:
.XX = ±.01
.XXX = ±.005
.XXXX = ±.0002
ANGULAR ± 0°-30'
DO NOT SCALE DRAWING
MATERIAL BRASS
FINISH —
HEAT TREATMENT —
CONTRACT NO. —
APPROVALS DATE
DRAWN JAN 6AUG88
CHECKED RFC 8AUG88
ISSUED SEPT88
JAN ENGINEERING
PETERBOROUGH, NEW HAMPSHIRE 03458
TITLE
HOLDER - BEARING
SIZE A
FSCM NO —
DWG NO 179954
REV.
SCALE 1:1
SHEET 1 OF 1
NEXT ASSY
USED ON
APPLICATION

WORKSHEET 6-9

Instructions: Using drawings A193586 on p. 109, answer the following questions in the spaces provided. (Answers in Appendix A.)

1. What kind of a section is this?
2. Surface K in the right-side view is what surface in the front view?
3. Surface N in the right-side view is what surface in the front view?
4. What is diameter T?
5. Diameter R is what diameter?
6. Hole I in the right-side view is what hole in the front view?
7. Calculate dimension P.
8. Spoke L in the right-side view is which spoke in the front view?
9. Spoke M in the right-side view is which spoke in the front view?
10. What is diameter Q?
11. Dimension S is what size?
12. What is the radius at 0?
13. How far is it from surface C in the front view to surface W in the right-side view?
14. How many degrees *clockwise* is it from hole A to hole F?
15. What is the O.D. of the wheel?

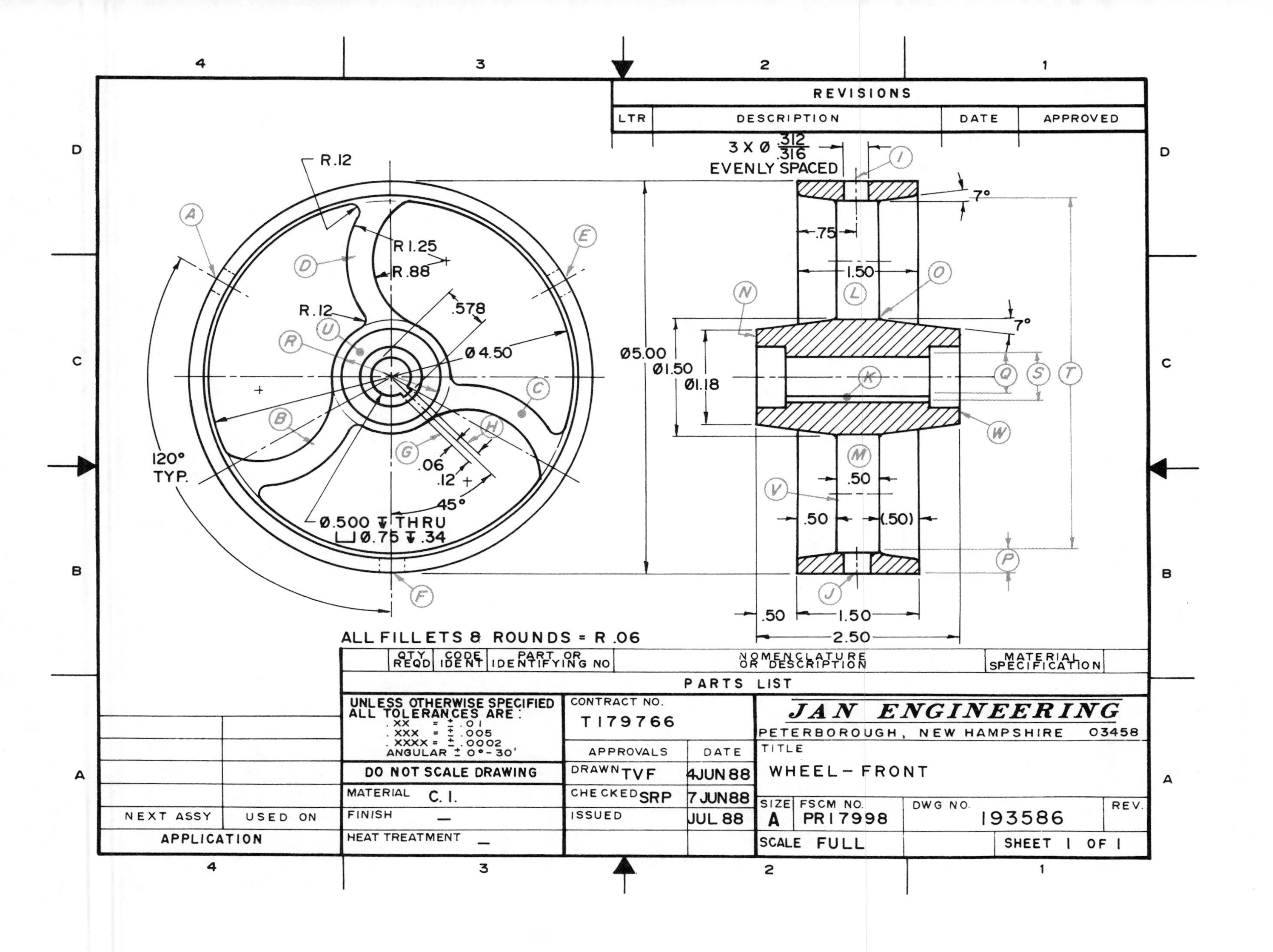
REVISIONS
LTR
DESCRIPTION
DATE
APPROVED
3 X Ø .312 .316
EVENLY SPACED
R.12
R 1.25
R.88
R.12
.578
Ø 4.50
Ø5.00
Ø1.50
Ø1.18
.75
1.50
7°
7°
.50
.50
(.50)
.06
.12
45°
120° TYP.
Ø.500 ↧ THRU
⌴ Ø.75 ↧ .34
.50
1.50
2.50
ALL FILLETS & ROUNDS = R .06
QTY REQD
CODE IDENT
PART OR IDENTIFYING NO
NOMENCLATURE OR DESCRIPTION
MATERIAL SPECIFICATION
PARTS LIST
UNLESS OTHERWISE SPECIFIED ALL TOLERANCES ARE:
.XX = ± .01
.XXX = ± .005
.XXXX = ± .0002
ANGULAR ± 0°-30'
DO NOT SCALE DRAWING
MATERIAL C. I.
FINISH —
HEAT TREATMENT —
CONTRACT NO.
T179766
APPROVALS
DATE
DRAWN TVF 4JUN88
CHECKED SRP 7JUN88
ISSUED JUL 88
JAN ENGINEERING
PETERBOROUGH, NEW HAMPSHIRE 03458
TITLE
WHEEL - FRONT
SIZE A
FSCM NO. PR17998
DWG NO. 193586
REV.
SCALE FULL
SHEET 1 OF 1
NEXT ASSY
USED ON
APPLICATION

Auxiliary Views

Objective: The reader will identify and understand the various kinds of auxiliary views used in industry and know the many drafting standard practices used to illustrate a part.

Up to this point, all the drawings and problems presented had objects with surfaces that were either parallel or at right angles to each other. They were represented by one, two, or more views in the standard multiview format (see FIG. 7-1). The surfaces of these objects can be projected in *true size* and *true shape* to the other views.

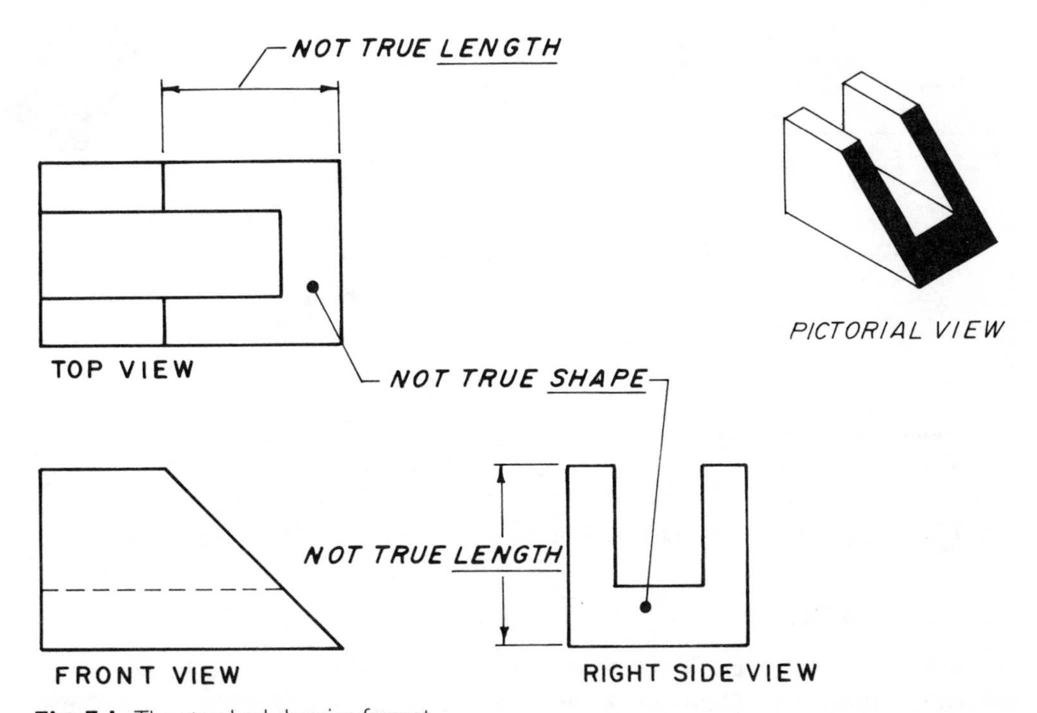

Fig. 7-1. The standard drawing format.

Some objects have one or more surfaces that slant in a direction other than the usual vertical or horizontal direction. The true size and true shape of these slanted or inclined surfaces are not shown in the usual views, therefore an *auxiliary view* must be used (see FIG. 7-2). The front view shows the slanted or inclined surface as only an edge, the top view and right-side view shows the surface, but *not* it's true size or shape. If the true size and true shape is to be illustrated, an auxiliary view must be added to the regular views. The auxiliary view is projected 90 degrees from the edge view (see FIG. 7-3).

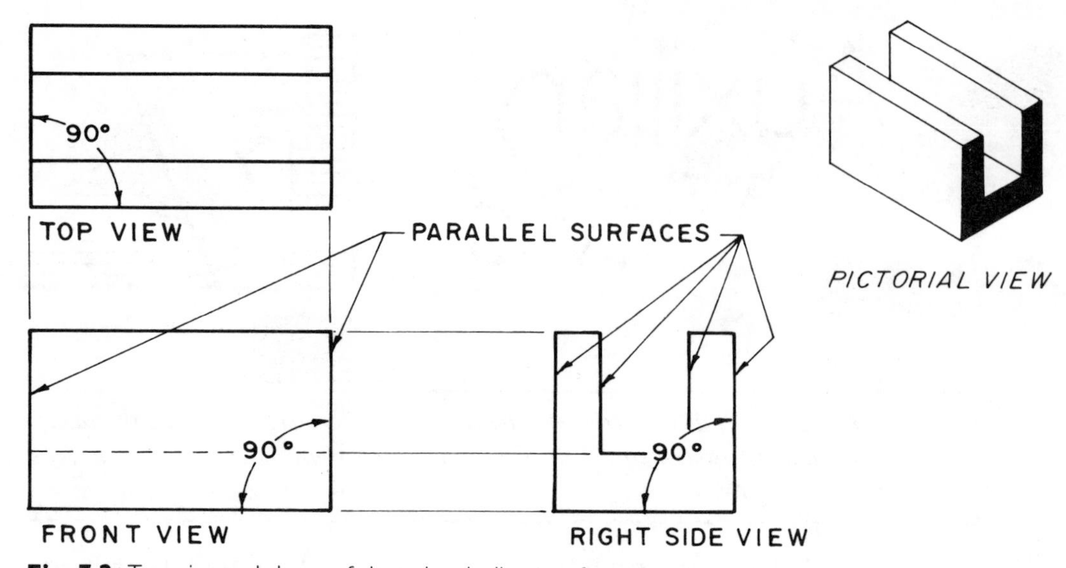

Fig. 7-2. True size and shape of slanted or inclined surfaces are viewed in the auxiliary view.

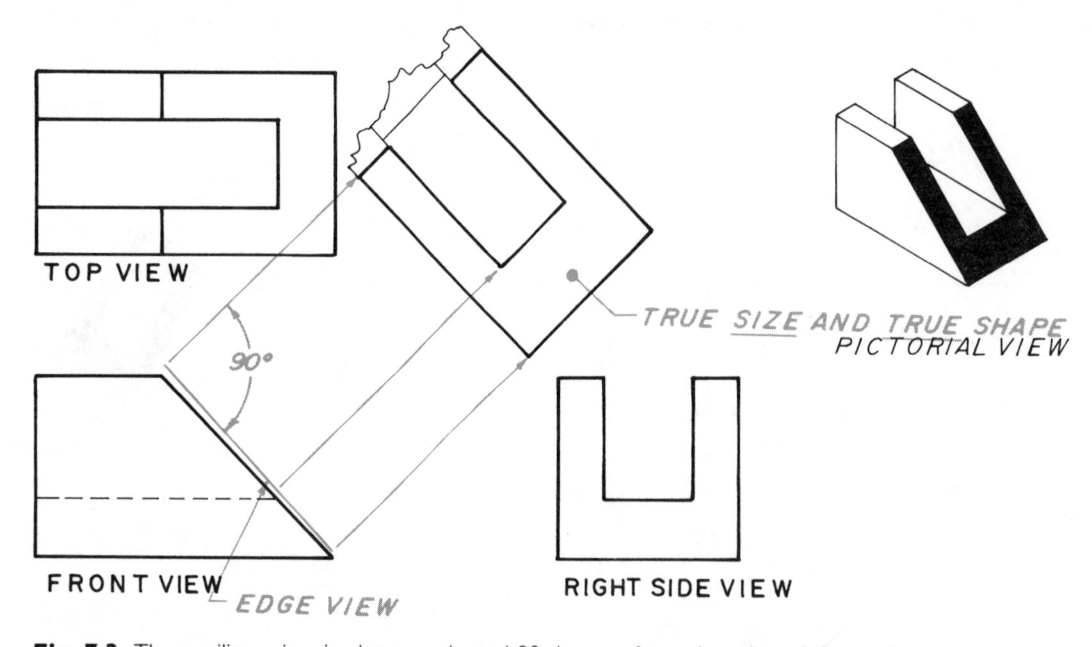

Fig. 7-3. The auxiliary view is *always* projected 90 degrees from the edge of the surface.

CLASSIFICATION

Auxiliary views are classified according to the slanted or inclined surface from which it is projected. There are three kinds of auxiliary views: a front view auxiliary (FIG. 7-4), a top view auxiliary (FIG. 7-5), and a side view auxiliary (FIG. 7-6). Regardless of which kind of auxiliary view is used, the auxiliary view is projected at 90 degrees from the *edge* view (see FIG. 7-7).

PARTIAL AUXILIARY VIEW

Background details are usually omitted on the auxiliary view, in order to simplify the drawing and avoid confusion. A *break line* is used to note the break in an incompleted view (see FIG. 7-8).

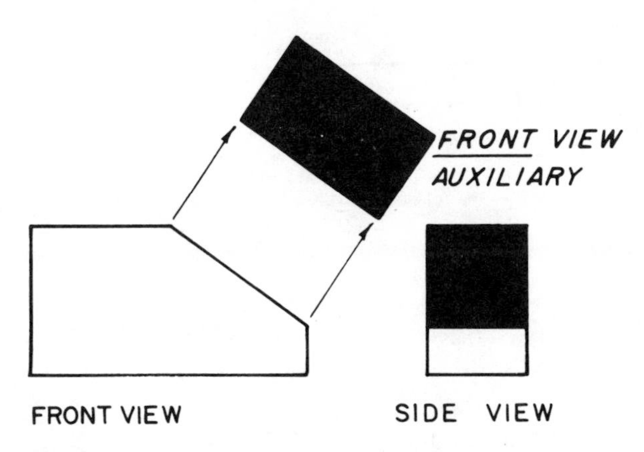

Fig. 7-4. Front view auxiliary.

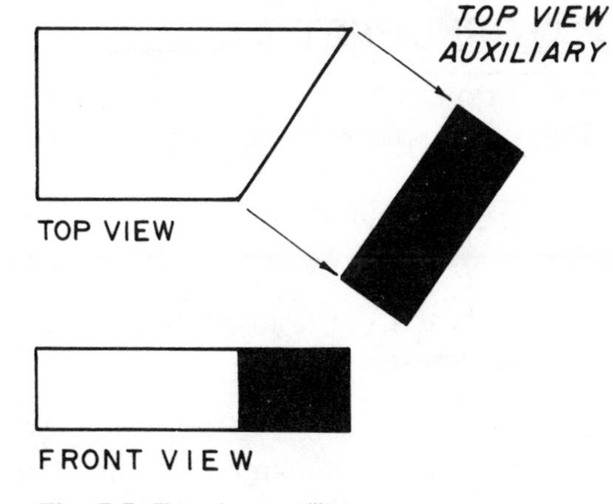

Fig. 7-5. Top view auxiliary.

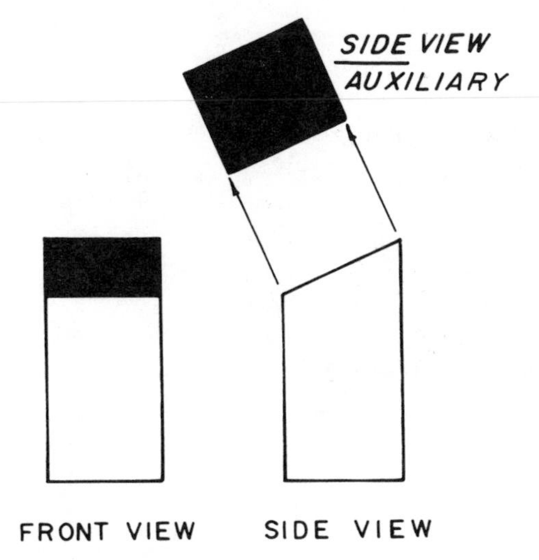

Fig. 7-6. Side auxiliary view.

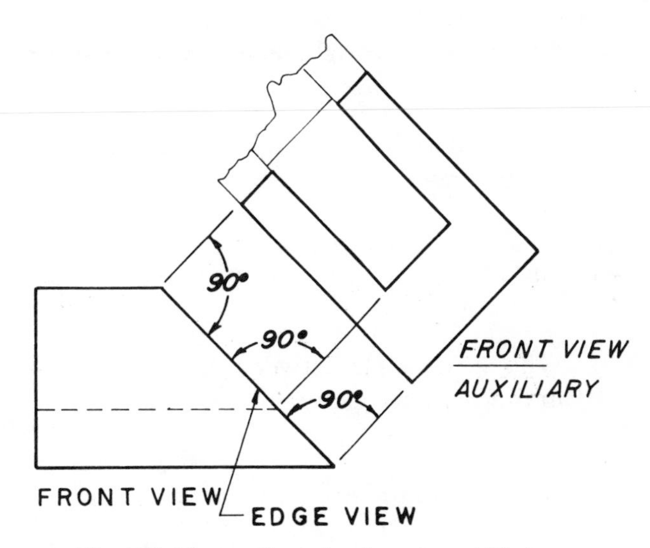

Fig. 7-7. The auxiliary view is projected 90 degrees from an edge view.

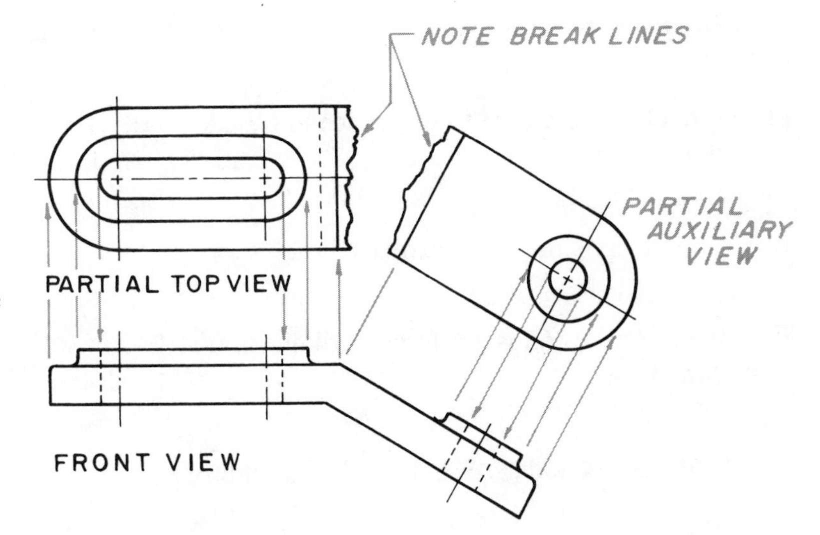

Fig. 7-8. A break line is used to indicate the break in an incomplete view.

WORKSHEET 7-1

Instructions: Using plan number A71591 on p. 115, answer the following questions in the spaces provided. (Answers in Appendix A.)

1. List the four given views.

2. Surface A in the auxiliary view is what surface in the right-side view—in the left-side view?

3. What is the minimum distance between surface P in the auxiliary view and surface V in the right-side view?

4. Surface X in the right-side view is what surface in the front view?

5. Surface 0 in the front view is what surface in the left-side view, in the auxiliary view?

6. Surface C in the auxiliary view is what surface in the front view?

7. What is the angle between surface U in the right-side view and surface D in the auxiliary view?

8. Surface R in the front view is what surface in the right-side view?

9. Surface Q in the front view is what surface in the right-side view?

10. Surface E in the auxiliary view is which surface in the left-side view?

11. What is the distance from surface I in the left-side view to surface N in the right-side view?

12. How deep is the 1.00 diameter hole?

13. What dimension is a reference dimension and why is it only a reference dimension?

14. There is *one* radius. What size is this radius?

15. Give the overall size of this object. (Width, Depth and Height)

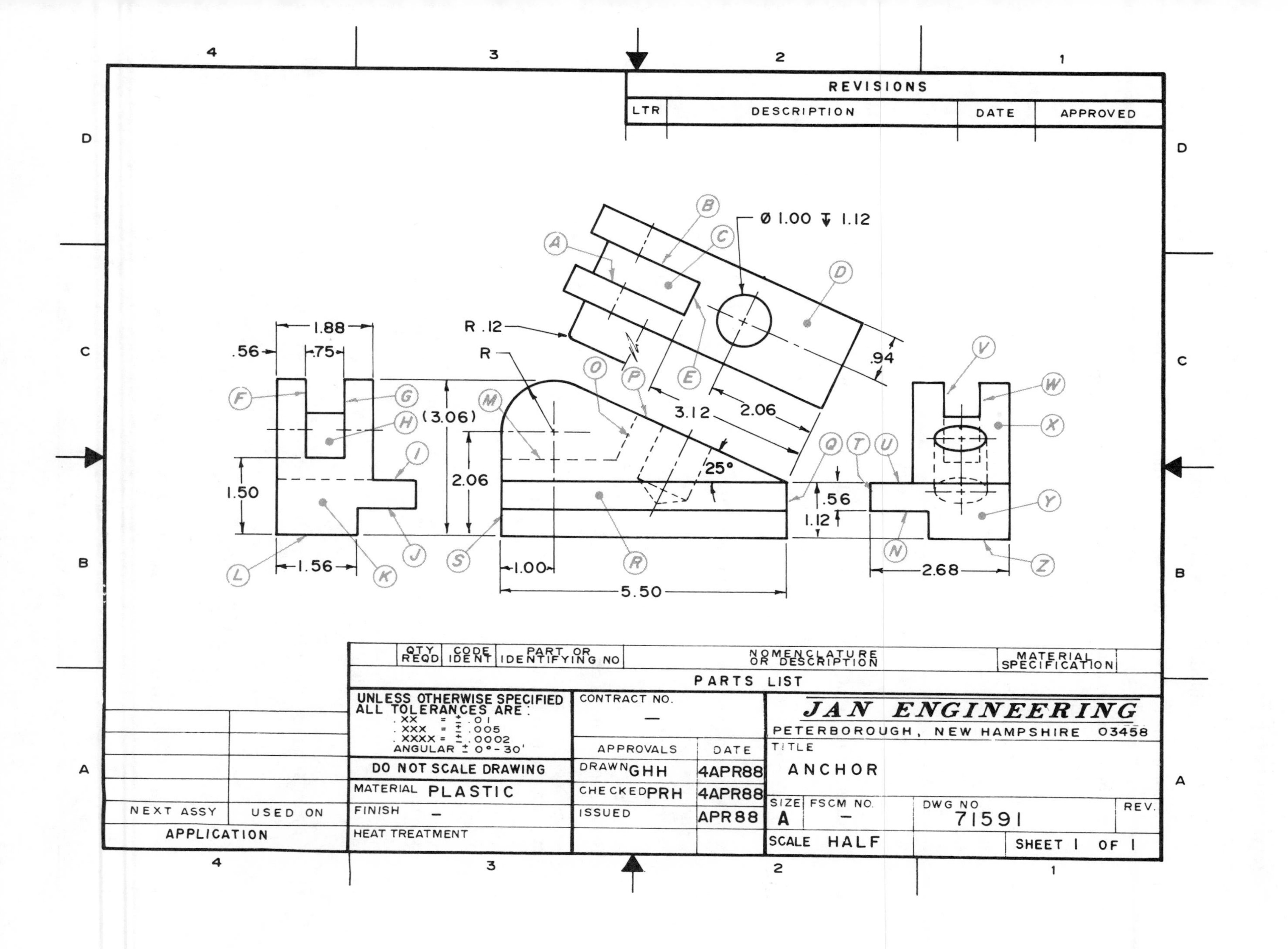
REVISIONS
LTR
DESCRIPTION
DATE
APPROVED
Ø 1.00 ↧ 1.12
.94
3.12
2.06
25°
R .12
R
(3.06)
2.06
1.88
.75
.56
1.50
1.56
1.00
5.50
.56
1.12
2.68
QTY REQD
CODE IDENT
PART OR IDENTIFYING NO
NOMENCLATURE OR DESCRIPTION
MATERIAL SPECIFICATION
PARTS LIST
UNLESS OTHERWISE SPECIFIED ALL TOLERANCES ARE:
.XX = ± .01
.XXX = ± .005
.XXXX = ± .0002
ANGULAR ± 0° - 30'
DO NOT SCALE DRAWING
MATERIAL PLASTIC
FINISH -
HEAT TREATMENT
CONTRACT NO. -
APPROVALS
DATE
DRAWN GHH 4APR88
CHECKED PRH 4APR88
ISSUED APR88
JAN ENGINEERING
PETERBOROUGH, NEW HAMPSHIRE 03458
TITLE
ANCHOR
SIZE A
FSCM NO. -
DWG NO 71591
REV.
SCALE HALF
SHEET 1 OF 1
NEXT ASSY
USED ON
APPLICATION

WORKSHEET 7-2

Instructions: Using drawing 18875541 on p. 117, answer the following questions in the spaces provided. (Answers in Appendix A.)

1. Surface A in the top view is what surface in the auxiliary view?
2. What kind of an auxiliary view is shown on this drawing?
3. What is the size of radius B?
4. Surface P in the auxiliary view is what surface in the top view—in the right-side view?
5. What size is the dimension A-A in the top view?
6. What is the radius at G? What does this radius represent?
7. Calculate dimension H.
8. What is the maximum size of dimension I?
9. What size is dimension J?
10. What size is dimension S?
11. What is the true length of dimension T?
12. What is the true length of dimension U?
13. Surface V in the right-side view is what surface in the auxiliary view?
14. What size is dimension Y?
15. Surface X in the right-side view is at what angle from surface A in the top view?
16. Surface M in the auxiliary view is what surface in the right-side view?
17. Surface C in the front view is what surface in the right-side view?
18. What scale was used to draw this drawing?

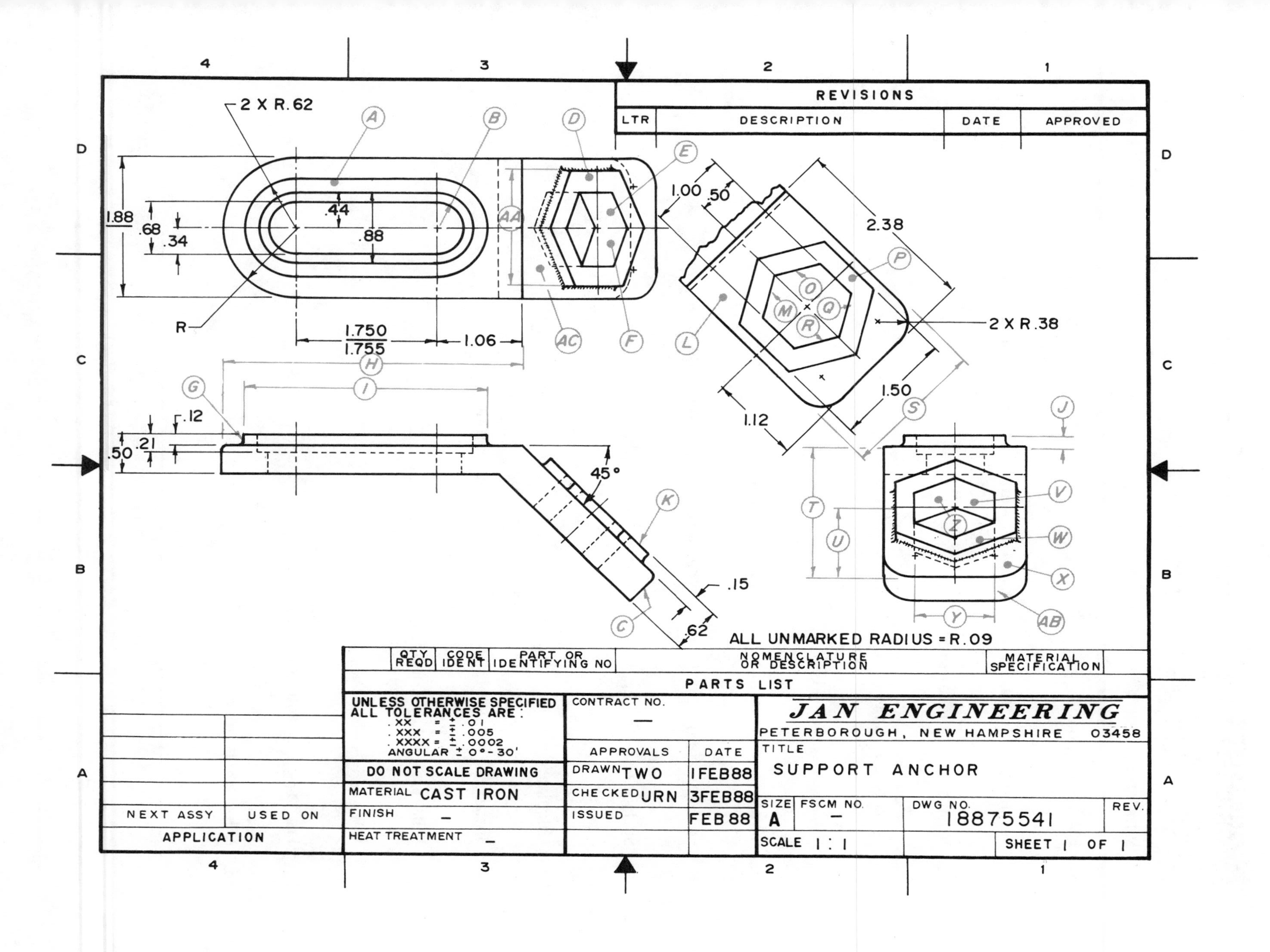
REVISIONS
LTR
DESCRIPTION
DATE
APPROVED
2 X R.62
1.88
.68
.34
.44
.88
R
1.750
1.755
1.06
1.00
.50
2.38
2 X R.38
1.50
1.12
.12
.21
.50
45°
.15
.62
ALL UNMARKED RADIUS = R.09
QTY REQD
CODE IDENT
PART OR IDENTIFYING NO
NOMENCLATURE OR DESCRIPTION
MATERIAL SPECIFICATION
PARTS LIST
UNLESS OTHERWISE SPECIFIED ALL TOLERANCES ARE:
.XX = ± .01
.XXX = ± .005
.XXXX = ± .0002
ANGULAR ± 0°-30'
DO NOT SCALE DRAWING
MATERIAL CAST IRON
FINISH —
HEAT TREATMENT —
CONTRACT NO. —
APPROVALS
DATE
DRAWN TWO 1FEB88
CHECKED URN 3FEB88
ISSUED FEB 88
JAN ENGINEERING
PETERBOROUGH, NEW HAMPSHIRE 03458
TITLE
SUPPORT ANCHOR
SIZE A
FSCM NO. —
DWG NO. 18875541
REV.
SCALE 1 : 1
SHEET 1 OF 1
NEXT ASSY
USED ON
APPLICATION

WORKSHEET 7-3

Instructions: Using drawing A197113 on p. 119, answer the following questions in the spaces provided. (Answers in Appendix A.)

1. What size is radius A?

2. Surface B in the auxiliary view is what surface in the front view?

3. Surface D in the auxiliary view is what surface in the right-side view?

4. What size is dimension Y?

5. What size is dimension E? What size is dimension F?

6. What size is radius H?

7. Surface I in the front view is what surface in the auxiliary view?

8. Surface M in the front view is what surface in the right-side view?

9. Surface O in the right-side view is at what angle to surface D in the top view?

10. Calculate the center to center distance between the .38 diameter holes.

11. Surface X in the right-side view is what surface in the front view?

12. What size is dimension Q?

13. What size is dimension U?

14. List the upper and lower limits of the .38 diameter holes.

15. What size is the radius at J?

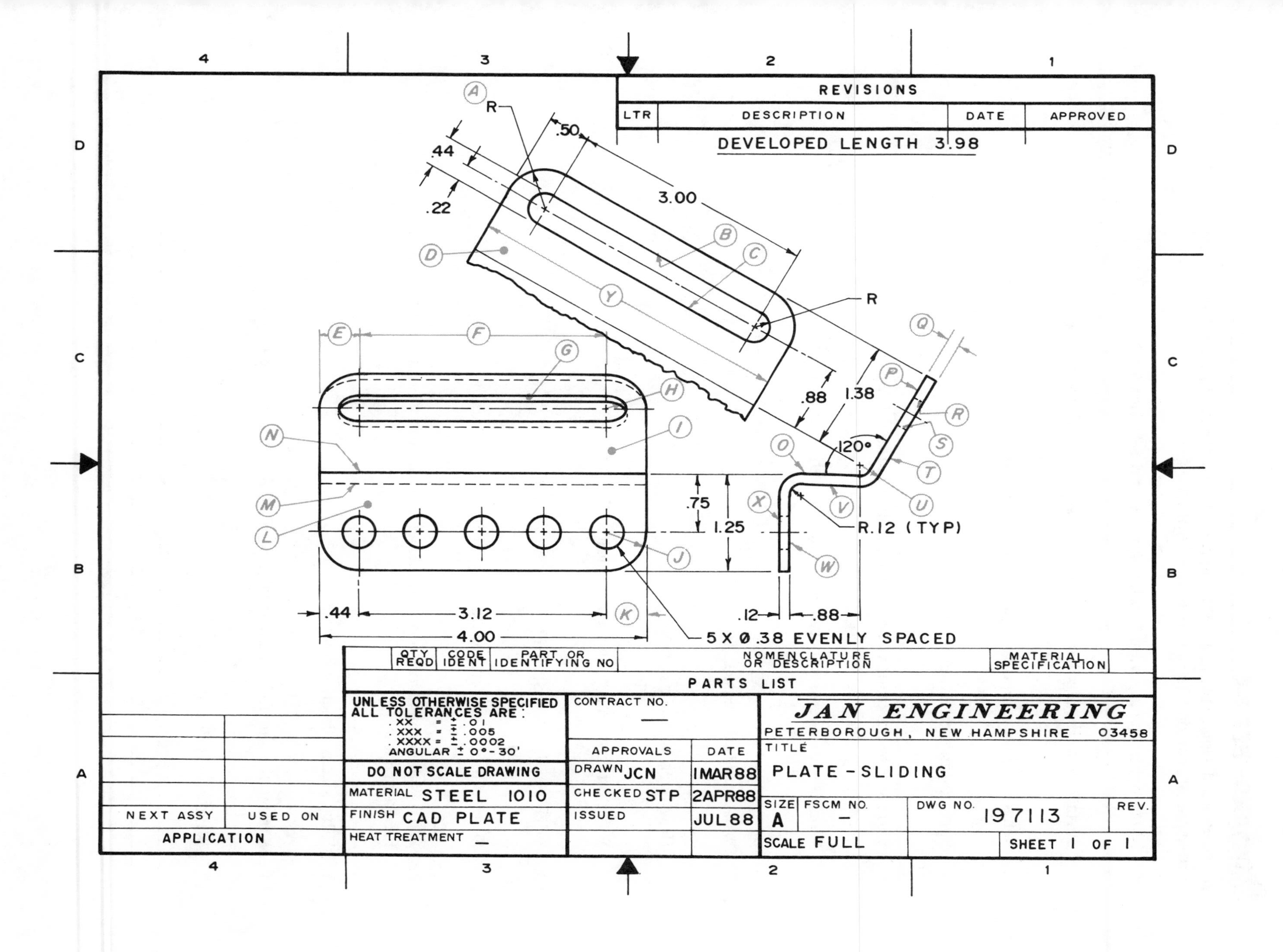
REVISIONS
LTR
DESCRIPTION
DATE
APPROVED
DEVELOPED LENGTH 3.98
R
.50
.44
.22
3.00
R
.88
1.38
120°
.75
1.25
R.12 (TYP)
.44
3.12
4.00
.12
.88
5 X Ø.38 EVENLY SPACED
QTY REQD
CODE IDENT
PART OR IDENTIFYING NO
NOMENCLATURE OR DESCRIPTION
MATERIAL SPECIFICATION
PARTS LIST
UNLESS OTHERWISE SPECIFIED ALL TOLERANCES ARE:
.XX = ± .01
.XXX = ± .005
.XXXX = ± .0002
ANGULAR ± 0°-30'
DO NOT SCALE DRAWING
MATERIAL STEEL 1010
FINISH CAD PLATE
HEAT TREATMENT —
CONTRACT NO. —
APPROVALS
DATE
DRAWN JCN 1MAR88
CHECKED STP 2APR88
ISSUED JUL 88
JAN ENGINEERING
PETERBOROUGH, NEW HAMPSHIRE 03458
TITLE
PLATE - SLIDING
SIZE A
FSCM NO. —
DWG NO. 197113
REV.
SCALE FULL
SHEET 1 OF 1
NEXT ASSY
USED ON
APPLICATION

WORKSHEET 7-4

Instructions: Using drawing A27933 on p. 121, answer the following questions in the spaces provided. (Answers in Appendix A.)

1. Surface A in the top view is what surface in the front view?

2. Surface M in the auxiliary view is what surface in the top view?

3. What kind of an auxiliary view is this drawing?

4. Surface J in the auxiliary view is what surface in the top view?

5. Surface P in the front view is what surface in the top view?

6. What is the dimension between surface O in the front view to surface K in the auxiliary view?

7. What size is dimension I?

8. How deep is dimension N?

9. What dimension is radius Y and radius Z?

10. What size is dimension T?

11. Surface R in the front view is what surface in the auxiliary view?

12. What is the dimension between surface F in the top view and surface Q in the front view?

13. What size is dimension V?

14. Surface S in the front view is what surface in the auxiliary view?

15. What size is dimension U?

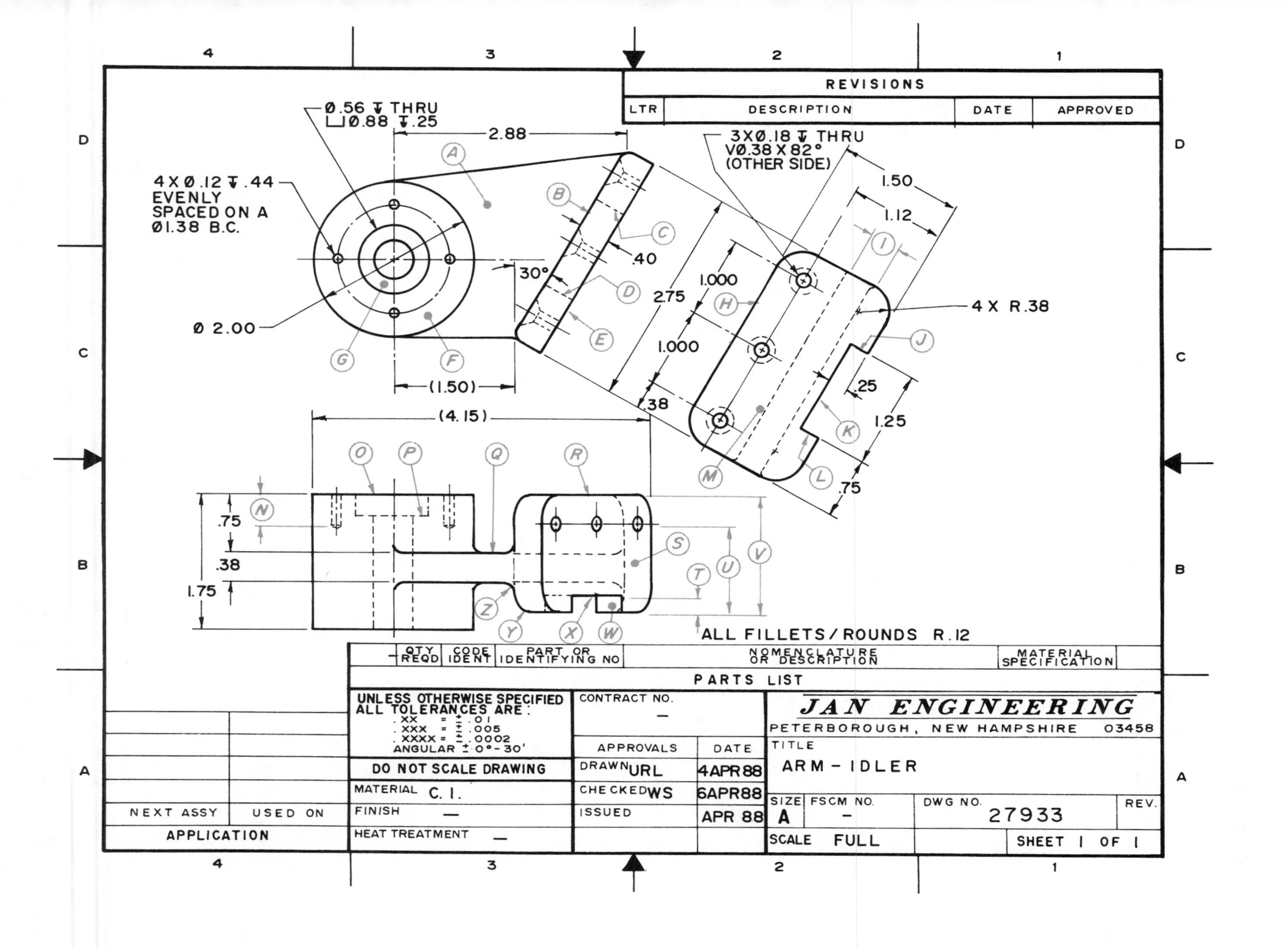
REVISIONS
LTR
DESCRIPTION
DATE
APPROVED
Ø.56 ↧ THRU
⌴Ø.88 ↧.25
2.88
4XØ.12 ↧ .44
EVENLY
SPACED ON A
Ø1.38 B.C.
Ø 2.00
(1.50)
30°
.40
3XØ.18 ↧ THRU
VØ.38 X 82°
(OTHER SIDE)
1.50
1.12
1.000
2.75
1.000
.38
4 X R.38
.25
1.25
.75
(4.15)
.75
.38
1.75
ALL FILLETS/ROUNDS R.12
QTY REQD
CODE IDENT
PART OR IDENTIFYING NO
NOMENCLATURE OR DESCRIPTION
MATERIAL SPECIFICATION
PARTS LIST
UNLESS OTHERWISE SPECIFIED
ALL TOLERANCES ARE:
.XX = ± .01
.XXX = ± .005
.XXXX = ± .0002
ANGULAR ± 0° - 30'
DO NOT SCALE DRAWING
MATERIAL C. I.
FINISH —
HEAT TREATMENT —
CONTRACT NO. —
APPROVALS
DATE
DRAWN URL 4APR88
CHECKED WS 6APR88
ISSUED APR 88
JAN ENGINEERING
PETERBOROUGH, NEW HAMPSHIRE 03458
TITLE
ARM - IDLER
SIZE A
FSCM NO. -
DWG NO. 27933
REV.
SCALE FULL
SHEET 1 OF 1
NEXT ASSY
USED ON
APPLICATION

Threads and Fasteners

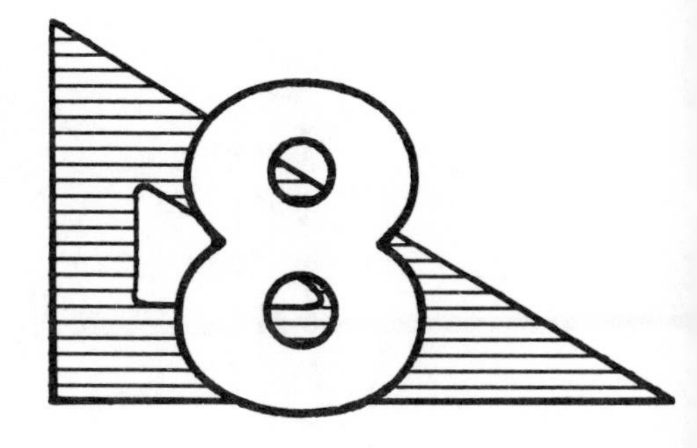

Objective: The reader will identify and understand screw threads, know the various terms associated with screw threads, and know how screw threads are made.

Fastening devices are used in almost all areas of the machine industry. They are used to assemble manufactured goods or products and also in the construction of all kinds of buildings.

CLASSIFICATION

There are two basic classifications of fasteners. The first classification is designed to be *permanently* joined together. The second classification is designed to permit assembly and disassembly. Examples of permanent methods of fastening are welding, brazing, stapling, nailing, gluing, and riveting. Examples of temporary fasteners are screws, bolts, keys, and pins.

A thorough understanding of fasteners, terms associated with them, how to identify each is essential to fully understand and interpret engineering drawings of today.

Threads are used for four purposes: fastening (for instance, with a nut and bolt), adjusting, (say, adjusting feet on a washing machine to level it), transmitting power, (for instance with a car jack), and for measuring.

TAP AND DIE

Various methods are used to produce inside and outside screw threads. The simplest method uses threading tools called *taps* and *dies*. The tap cuts internal threads (see FIG. 8-1), and the die cuts external threads (see FIG. 8-2). To make an internal threaded hole, a tap-drilled hole must be made first. This tap drill is about the same diameter as the minor diameter of the threads. TABLE 8-1 is a drill and tap size chart; it lists the recommended tap size drill for the various threads. Look under "Tap Sizes" for $^{1}/_{4}=20$ ($^{1}/_{4}$ diameter, 20 threads per inch); the tap drill size would be a number 7 or .2010-diameter drill size.

RIGHT- AND LEFT-HAND THREADS

Threads can be either right-handed or left-handed. To distinguish between a right-hand and a left-hand thread, use this simple test: A right-hand thread winding tends to lean toward the *left* (see FIG. 8-3). If the thread leans toward the left, the *right-hand* thumb points in the same direction—to the left. If the threads lean to the *right,* the *left-hand* thumb leans in that direction indicating that it is a left-hand thread.

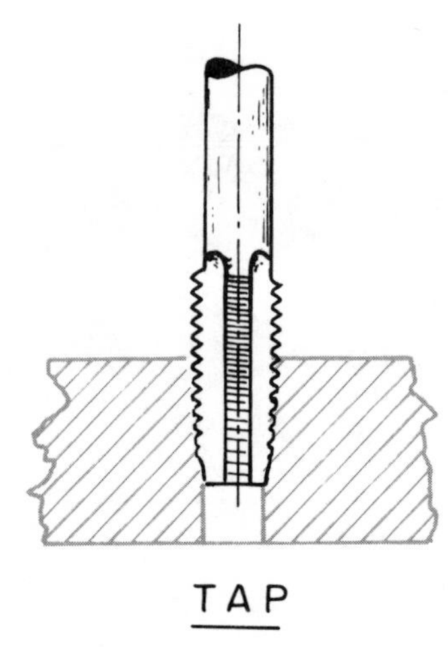

Fig. 8-1. The tap cuts internal threads.

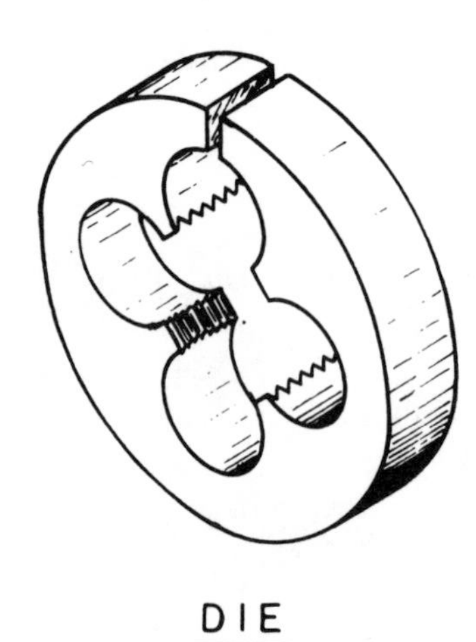

Fig. 8-2. The die cuts external threads.

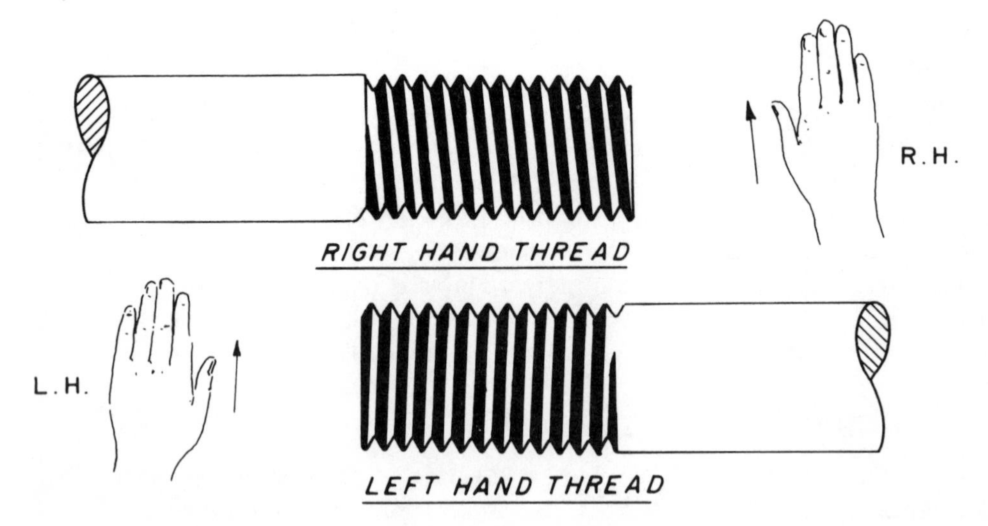

Fig. 8-3. Left- and right-hand threads.

Table 8-1. Drill and Tap Sizes

DECIMAL EQUIVALENTS AND TAP DRILL SIZES OF WIRE GAGE LETTER AND FRACTIONAL SIZE DRILLS (TAP DRILL SIZES BASED ON 75% MAXIMUM THREAD)

Fractional Size Drills	Wire Gage Drills	Decimal Equivalent Inches	Tap Sizes Size of Thread	Tap Sizes Threads Per Inch
	80	.0135		
	79	.0145		
1/64	......	.0156		
	78	.0160		
	77	.0180		
	76	.0200		
	75	.0210		
	74	.0225		
	73	.0240		
	72	.0250		
	71	.0260		
	70	.0280		
	69	.0292		
	68	.0310		
1/32	......	.0312		
	67	.0320		
	66	.0330		
	65	.0350		
	64	.0360		
	63	.0370		
	62	.0380		
	61	.0390		
	60	.0400		
	59	.0410		
	58	.0420		
	57	.0430		
	56	.0465		
3/64	......	.0469	0	80
	55	.0520		
	54	.0550		
	53	.0595	1	64
1/16	......	.0625		72
	52	.0635		
	51	.0670		
	50	.0700	2	56
	49	.0730		64
	48	.0760		
5/64	......	.0781		
	47	.0785	3	48
	46	.0810		
	45	.0820	3	56
	44	.0860		
	43	.0890	4	40
	42	.0935	4	48
3/32	......	.0937		
	41	.0960		
	40	.0980		
	39	.0995		
	38	.1015	5	40
	37	.1040	5	44
	36	.1065	6	32

Fractional Size Inches	Wire Gage Drills	Decimal Equivalent Inches	Tap Sizes Size of Thread	Tap Sizes Threads Per Inch
9/64	......	.1406		
	27	.1440		
	26	.1470		
	25	.1495	10	24
	24	.1520		
	23	.1540		
5/32	......	.1562		
	22	.1570		
	21	.1590	10	32
	20	.1610		
	19	.1660		
	18	.1695		
11/64	......	.1719		
	17	.1730		
	16	.1770	12	24
	15	.1800		
	14	.1820	12	28
	13	.1850		
3/16	......	.1875		
	12	.1890		
	11	.1910		
	10	.1935		
	9	.1960		
	8	.1990		
	7	.2010	1/4	20
13/64	......	.2031		
	6	.2040		
	5	.2055		
	4	.2090		
	3	.2130	1/4	28
7/32	......	.2187		
	2	.2210		
	1	.2280		
	A	.2340		
15/64	......	.2344		
	B	.2380		
	C	.2420		
	D	.2460		
1/4	E	.2500		
	F	.2570	5/16	18
	G	.2610		
17/64	......	.2656		
	H	.2660		
	I	.2720	5/16	24
	J	.2770		
	K	.2810		
9/32	......	.2812		
	L	.2900		
	M	.2950		
19/64	......	.2969		
	N	.3020		

Fractional Size Inches	Wire Gage Drills	Decimal Equivalent Inches	Tap Sizes Size of Thread	Tap Sizes Threads Per Inch
23/64	......	.3594		
	U	.3680	7/16	14
3/8	......	.3750		
	V	.3770		
	W	.3860		
25/64	......	.3906	7/16	20
	X	.3970		
	Y	.4040		
13/32	......	.4062		
	Z	.4130		
27/64	......	.4219	1/2	13
7/16	......	.4375		
29/64	......	.4531		
15/32	......	.4687		
31/64	......	.4844	9/16	12
1/2	......	.5000		
33/64	......	.5156	9/16	18
17/32	......	.5312	5/8	11
35/64	......	.5469		
9/16	......	.5625		
37/64	......	.5781	5/8	18
19/32	......	.5937		
39/64	......	.6094		
5/8	......	.6250		
41/64	......	.6406		
21/32	......	.6562	3/4	10
43/64	......	.6719		
11/16	......	.6875	3/4	16
45/64	......	.7031		
23/32	......	.7187		
47/64	......	.7344		
3/4	......	.7500		
49/64	......	.7656	7/8	9
25/32		......	.7812	
51/64	......	.7969		
13/16	......	.8125	7/8	14
53/64	......	.8281		
27/32	......	.8437		
55/64	......	.8594		
7/8	......	.8750	1	8
57/64	......	.8906		
29/32	......	.9062		
59/64	......	.9219		
15/16	......	.9375	1	14
61/64	......	.9531		
31/32	......	.9687		
63/64	......	.9844	1 1/8	7
1	......	1.000		

Continued

Table 8-1. Continued

Fractional Size Drills	Wire Gage Drills	Decimal Equivalent Inches	Tap Sizes Size of Thread	Tap Sizes Threads Per Inch	Fractional Size Inches	Wire Gage Drills	Decimal Equivalent Inches	Tap Sizes Size of Thread	Tap Sizes Threads Per Inch	Fractional Size Inches	Wire Gage Drills	Decimal Equivalent Inches	Tap Sizes Size of Thread	Tap Sizes Threads Per Inch
7/64	......	.1094			5/16	......	.3125	3/8	16					
	35	.1100				O	.3160							
	34	.1110				P	.3230							
	33	.1130	6	40	21/64	......	.3281							
	32	.1160				Q	.3320	3/8	24					
	31	.1200				R	.3390							
1/8	......	.1250			11/32	......	.3437							
	30	.1285				S	.3480							
	29	.1360	8	32		T	.3580							
	28	.1405		36										

THREAD TERMS

Refer to FIG. 8-4.

external threads—Threads located on the outside of a part such as those found on a bolt.

internal threads—Threads located on the inside of a part such as those found on a nut.

major diameter—The largest diameter of a screw thread, both external and internal.

minor diameter—The smallest diameter of a screw thread, both external and internal.

pitch diameter—The diameter of an imaginary diameter centrally located between the major and minor diameters.

pitch—The distance from a point on a screw thread to a corresponding point on the next thread, as measured parallel to the centerline of the threads.

root—The bottom point joining the sides of a thread.

crest—The top point joining the sides of the thread.

depth of thread—The distance between the crest and the root of the thread, as measured at right angles to the centerline of the threads.

series of threads—A standard number of threads per inch (TPI) for each standard diameter.

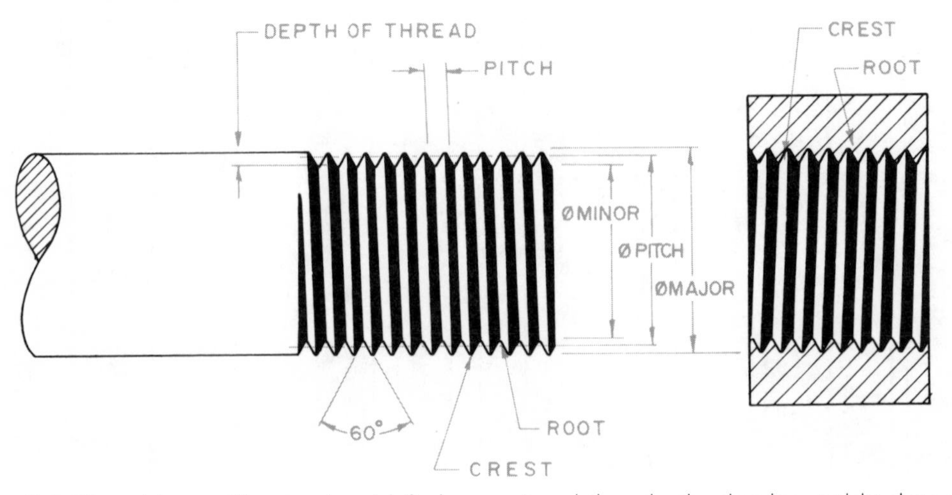

Fig. 8-4. Thread terms. The drawing at left shows external threads, the drawing at right shows internal threads.

Refer to TABLE 8-2 for various information of sizes of threads.

Example: 1/4-inch-diameter-size coarse thread. The major diameter is .250, the minor diameter is .1850, and the tap drill is diameter .2010 and has 20 TPI (threads per inch)—standard for a coarse thread.

Table 8-2. Dimension and Size for Threads

Nominal		Diameter				Tap Drill			Threads		Pitch		T.P.I	
Size		(Major)		(Minor)		(For 75% Th'd.)			Per Inch		(MM)		(Approx.)	
Inch	M.M.	Inch	M.M.	Inch	M.M.	Drill	Inch	M.M.	UNC	UNF	Coarse	Fine	Coarse	Fine
—	M1.4	.055	1.397	—	—	—	—	—	—	—	.3	.2	85	127
0	—	.060	1.524	.0438	1.092	3/64	.0469	1.168	—	80	—	—	—	—
—	M1.6	.063	1.600	—	—	—	—	—	—	—	.35	.2	74	127
1	—	.073	1.845	.0527	1.320	53	.0595	1.499	64	—	—	—	—	—
1	—	.073	1.854	.0550	1.397	53	.0595	1.499	—	72	—	—	—	—
—	M.2	.079	2.006	—	—	—	—	—	—	—	.4	.25	64	101
2	—	.086	2.184	.0628	1.587	50	.0700	1.778	56	—	—	—	—	—
2	—	.086	2.184	.0657	1.651	50	.0700	1.778	—	64	—	—	—	—
—	M2.5	.098	2.489	—	—	—	—	—	—	—	.45	.35	56	74
3	—	.099	2.515	.0179	1.828	47	.0785	1.981	48	—	—	—	—	—
3	—	.099	2.515	.0758	1.905	46	.0810	2.057	—	58	—	—	—	—
4	—	.112	2.845	.0795	2.006	43	.0890	2.261	40	—	—	—	—	—
4	—	.112	2.845	.0849	2.134	42	.0935	2.380	—	48	—	—	—	—
—	M3	.118	2.997	—	—	—	—	—	—	—	.5	.35	51	74
5	—	.125	3.175	.0925	2.336	38	.1015	2.565	40	—	—	—	—	—
5	—	.125	3.175	.9055	2.413	37	.1040	2.641	—	44	—	—	—	—
6	—	.138	3.505	.0975	2.464	36	.1065	2.692	32	—	—	—	—	—
6	—	.138	3.505	.1055	2.667	33	.1130	2.870	—	40	—	—	—	—
—	M4	.157	3.988	—	—	—	—	—	—	—	.7	.35	36	51
8	—	.164	4.166	.1234	3.124	29	.1360	3.454	32	—	—	—	—	—
8	—	.164	4.166	.1279	3.225	29	.1360	3.454	—	36	—	—	—	—
10	—	.190	4.826	.1359	3.429	26	.1470	3.733	24	—	—	—	—	—
10	—	.190	4.826	.1494	3.785	21	.1590	4.038	—	32	—	—	—	—
—	M5	.196	4.978	—	—	—	—	—	—	—	.8	.5	32	51
12	—	.216	5.486	.1619	4.089	16	.1770	4.496	24	—	—	—	—	—
12	—	.216	5.486	.1696	4.293	15	.1800	4.572	—	28	—	—		—
—	M6	.236	5.994	—	—	—	—	—	—	—	1.0	.75	25	34
1/4	—	.250	6.350	.1850	4.699	7	.2010	5.105	20	—	—	—	—	—
1/4	—	.250	6.350	.2036	5.156	3	.2130	5.410	—	28	—	—	—	—
5/16	—	.312	7.938	.2403	6.096	F	.2570	6.527	18	—	—	—	—	—
5/16	—	.312	7.938	.2584	6.553	I	.2720	6.908	—	24	—	—	—	—
—	M8	.315	8.001	—	—	—	—	—	—	—	1.25	1.0	20	25
3/8	—	.375	9.525	.2938	7.442	5/16	.3125	7.937	16	—	—	—	—	—
3/8	—	.375	9.525	.3209	8.153	Q	.3320	8.432	—	24	—	—	—	—
—	M10	.393	9.982	—	—	—	—	—	—	—	1.5	1.25	17	20
7/16	—	.437	11.113	.3447	8.738	U	.3680	9.347	14	—	—	—	—	—
7/16	—	.437	11.113	.3726	9.448	25/64	.3906	9.921	—	20	—	—	—	—
—	M12	.471	11.963	—	—	—	—	—	—	—	1.75	1.25	14.5	20
1/2	—	.500	12.700	.4001	10.162	27/64	.4219	10.715	13	—	—	—	—	—
1/2	—	.500	12.700	.4351	11.049	29/64	.4531	11.509	—	20	—	—	—	—
—	M14	.551	13.995	—	—	—	—	—	—	—	2	1.5	12.5	17
9/16	—	.562	14.288	.4542	11.531	31/64	.4844	12.3031	12	—	—	—	—	—
9/16	—	.562	14.288	.4903	12.446	33/64	.5156	13.096	—	18		—	—	—
5/8	—	.625	15.875	.5069	12.852	17/32	.5312	13.493	11	—	—	—	—	—
5/8	—	.625	15.875	.5528	14.020	37/64	.5781	14.684	—	18	—	—	—	—
—	M16	.630	16.002	—	—	—	—	—	—	—	2	1.5	12.5	17
—	M18	.709	18.008	—	—	—	—	—	—	—	2.5	1.5	10	17
3/4	—	.750	19.050	.6201	15.748	21/32	.6562	16.668	10	—	—	—	—	—
3/4	—	.750	19.050	.6688	16.967	11/16	.6875	17.462	—	16	—	—	—	—
—	M20	.787	19.990	—	—	—	—	—	—	—	2.5	1.5	10	17
—	M22	.866	21.996	—	—	—	—	—	—	—	2.5	1.5	10	17
7/8	—	.875	22.225	.7307	18.542	49/64	.7656	19.446	9	—	—	—	—	—
7/8	—	.875	22.225	.7822	19.863	13/16	.8125	20.637	—	14	—	—	—	—
—	M24	.945	24.003	—	—	—	—	—	—	—	3	2	8.5	12.5
1	—	1.000	25.400	.8376	21.2598	7/8	.8750	22.225	8	—	—	—	—	—
1	—	1.000	25.400	.8917	22.632	59/64	.9219	23.415	—	12	—	—	—	—
—	M27	1.063	27.000	—	—	—	—	—	—	—	3	2	8.5	12.5

SCREW THREAD FORMS

The form of a thread is actually its profile shape. There are many types of screw thread forms. FIGURE 8-5 shows a few major kinds.

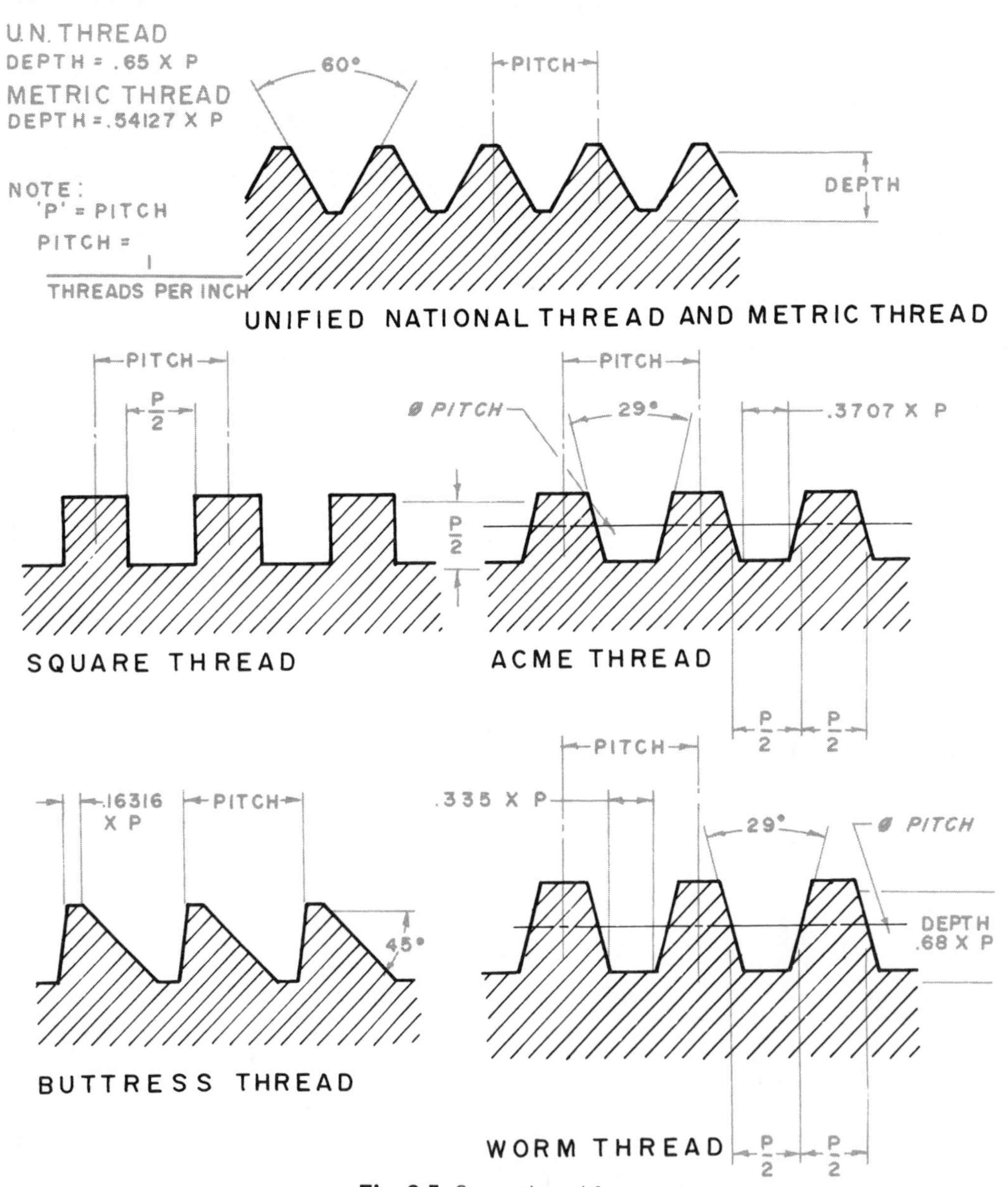

Fig. 8-5. Screw thread form.

unified national thread—The unified national thread form has been the standard thread form used in the United States, Canada, and the United Kingdom since 1948. This is the form most used for fasteners and adjustments.

metric thread—This is the newest standard thread form. It is very similar to the unified national thread form, except its depth is slightly less.

square thread—This profile is square and the teeth are at right angles to the centerline of the threads. It is used to transmit power, but because it is expensive to make, it is being replaced by the acme thread form.

acme thread form—The acme thread is a slight modification of the square thread form. It is stronger and easier to manufacture than the square thread.

It, too, is used to transmit power.

buttress thread form—The buttress thread has certain advantages in application involving exceptionally high stress along its centerline in *one* direction only. It is used for airplane propeller hubs and hydraulic presses.

worm thread form—The worm thread is similar to the acme thread form. It is primarily to transmit power.

THREADS PER INCH

One method to measure threads per inch (TPI) is to place a scale along the crests of a thread and count the number of full threads found in 1 inch (see FIG. 8-6). A better way is to use a screw thread gage (see FIG. 8-7). By trial and error, the various fingers or leaves of the gage are placed over the threads until one is found that exactly fits into the threads. The threads per inch is then read directly off the finger or leaf.

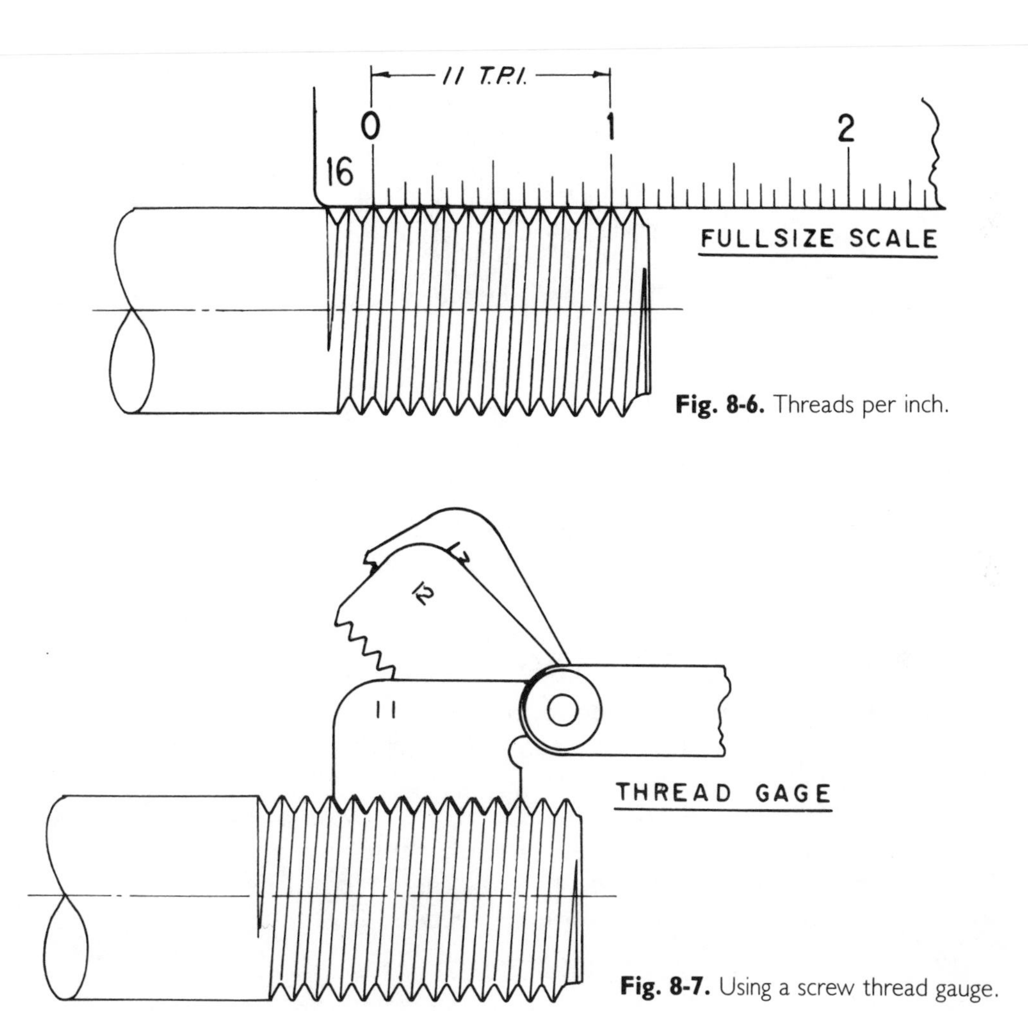

Fig. 8-6. Threads per inch.

Fig. 8-7. Using a screw thread gauge.

PITCH

The *pitch* of any thread, regardless of its thread form or profile, is the distance from one point on a thread to the corresponding point on the adjacent thread (refer again to FIG. 8-4). The actual distance can be found by dividing the TPI into 1 full inch. In the top example in FIG. 8-8, a coarse thread is shown. There

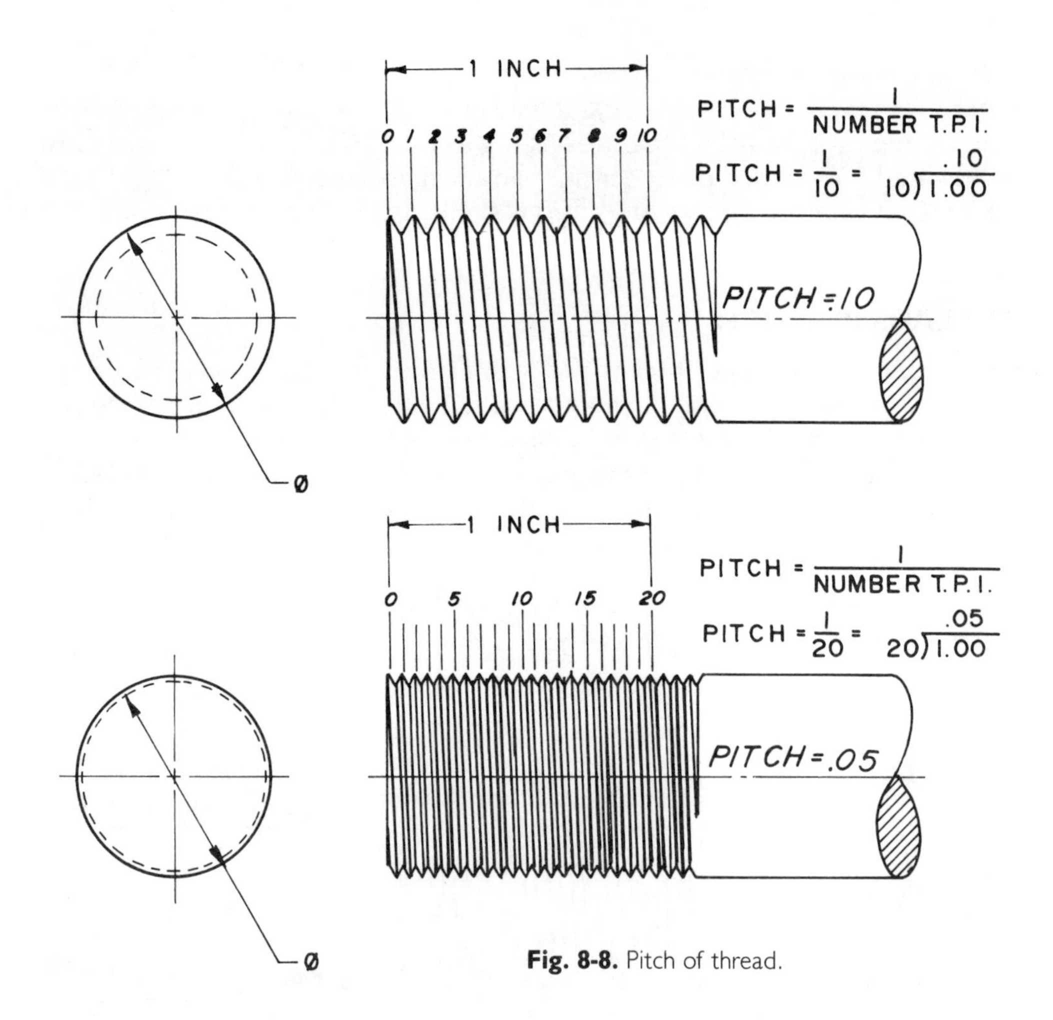

Fig. 8-8. Pitch of thread.

are 10 threads in 1 measured inch: 10 TPI divided into 1 inch equals a pitch of .10. In the fine thread of the same diameter below, there are 20 threads in 1 measured inch: 20 TPI divided into 1 inch equals a pitch of .05.

SINGLE AND MULTIPLE THREADS

A single thread is composed of one continuous ridge (see FIG. 8-9). The *lead* of a single thread is equal to the pitch. Lead is the distance a thread advances along its axis in one full turn. Multiple threads are used when speed or travel distance is an important design consideration. Speed, not power, is a factor in multiple threads. Using a 1/4-20 thread as an example (single thread), one turn of the thread advances one way or the other, a distance equal to the pitch, or 20 divided into 1 full inch = .05. A double-threaded 1/4-20 thread would be 2 × the pitch or .10. Double threads advance a distance equal to twice the pitch or .10.

SCREW REPRESENTATION

FIGURE 8-10 illustrates how external threads are drawn. To draw threads exactly as they look, would take much too long (refer to top illustration). To speed up drawing time, threads are drawn by using either the *schematic representation* method or the *simplified representation* method. The schematic method was developed in 1940 and is still used occasionally. Most drawings, however, use

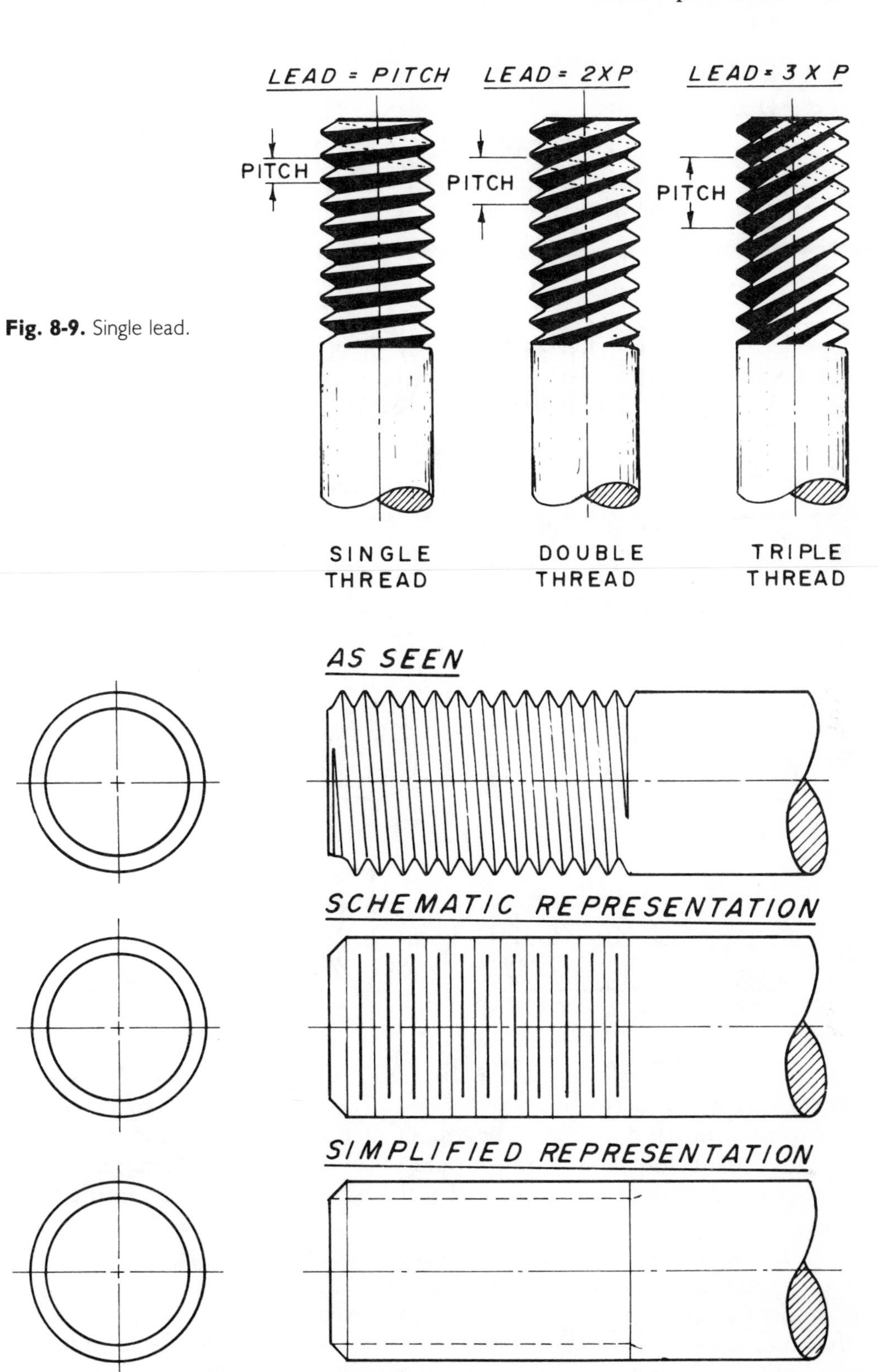

Fig. 8-9. Single lead.

Fig. 8-10. Thread representation.

the simplified method. The dash lines indicate threads. The most recent standard used to draw threads today is illustrated in FIG. 8-11. The schematic system is included only because there are still old drawings in use today. Note the way in which threads in section are illustrated.

There are two kinds of threaded holes: a through hole and a blind hole, FIG. 8-12. A *through hole* is a hole with threads that go all the way through the

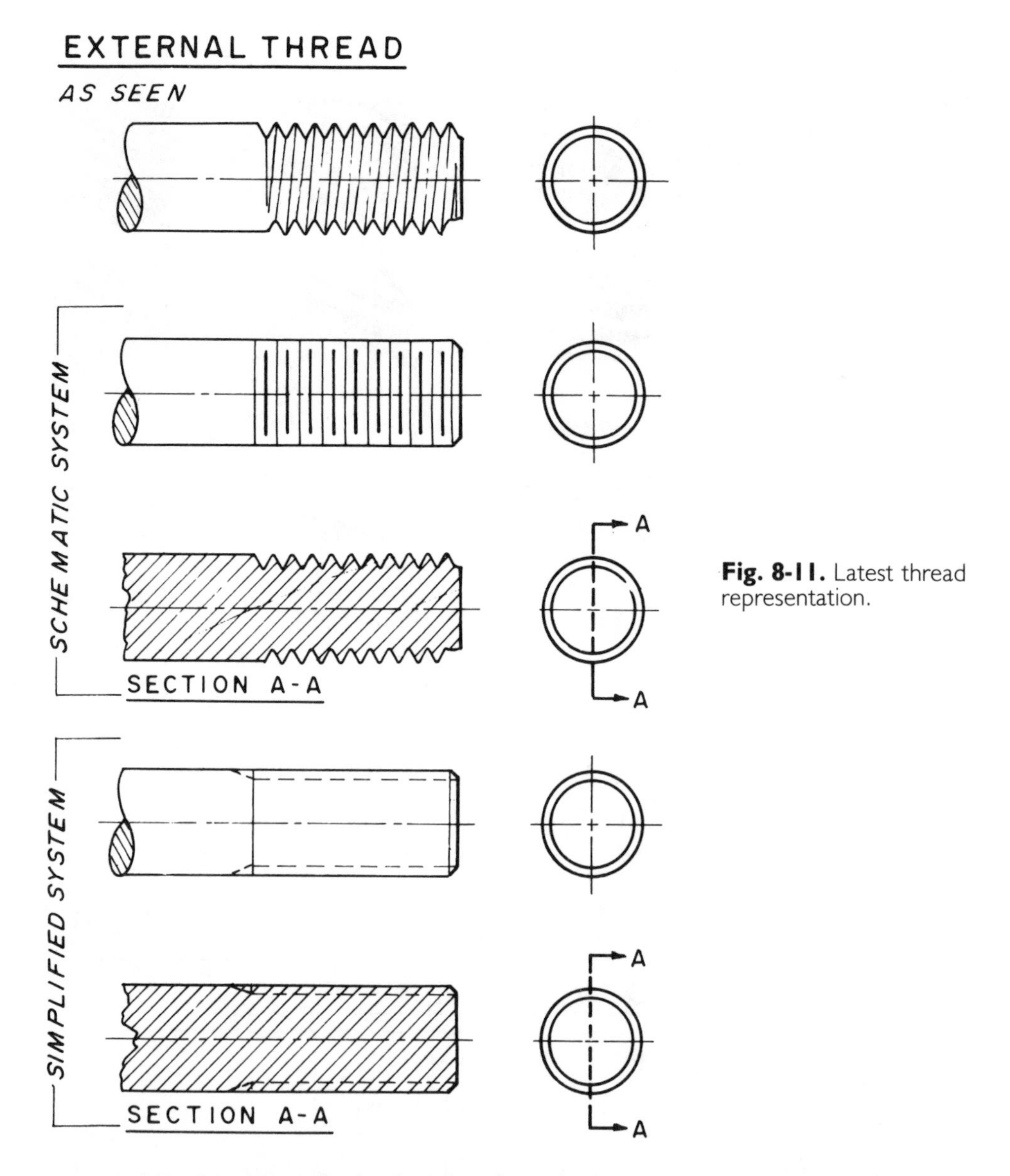

Fig. 8-11. Latest thread representation.

part. A *blind hole* is a hole that the threads do not go completely through the part. In order to make either kind of a hole, a tap drill must be used (refer back to TABLE 8-1 for tap drill sizes). The top illustration shows how the part would actually look; below shows how it would be actually drawn. Threads in section are illustrated at the bottom. Note how the tap drill is illustrated in the blind holes.

FIGURE 8-13 illustrates how the tap drill is drilled first to depth deeper than the required specified, depth of the full threads. This is done because the tap has only partial threads at its tip and cannot make full threads all the way into the hole. Study FIG. 8-13. At the right, note how the tap drill hole is illustrated by the hidden lines and how the threads are also illustrated but not as deep as the tap drill. The cone shape at the end of the hole is the tip of the tap drill.

THREAD RELIEF OR UNDERCUT

On external threads it is impossible to make perfect fully formed threads right up to a shoulder (see FIG. 8-14), the threads tend to runout as shown. To eliminate the problem, a cut is made, usually to the depth of the threads. The illustration in the center shows how such a cut would look; the bottom illustration

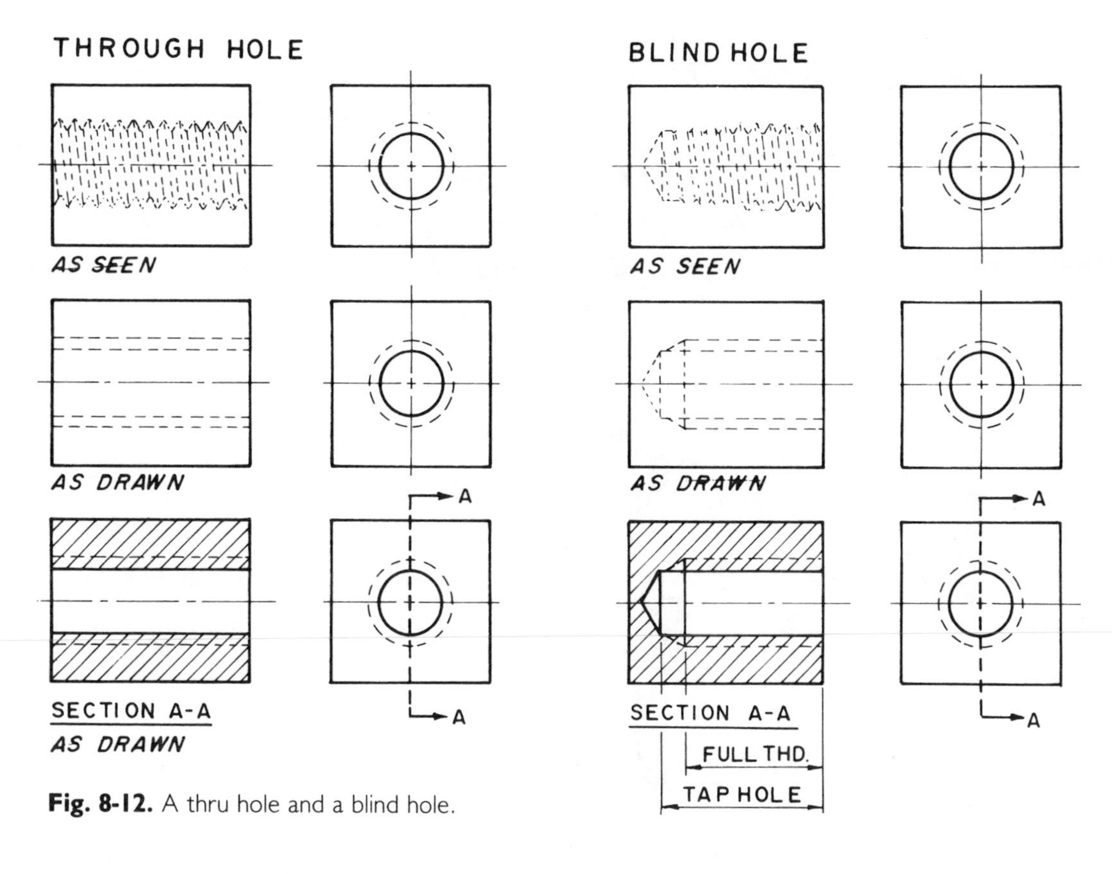

Fig. 8-12. A thru hole and a blind hole.

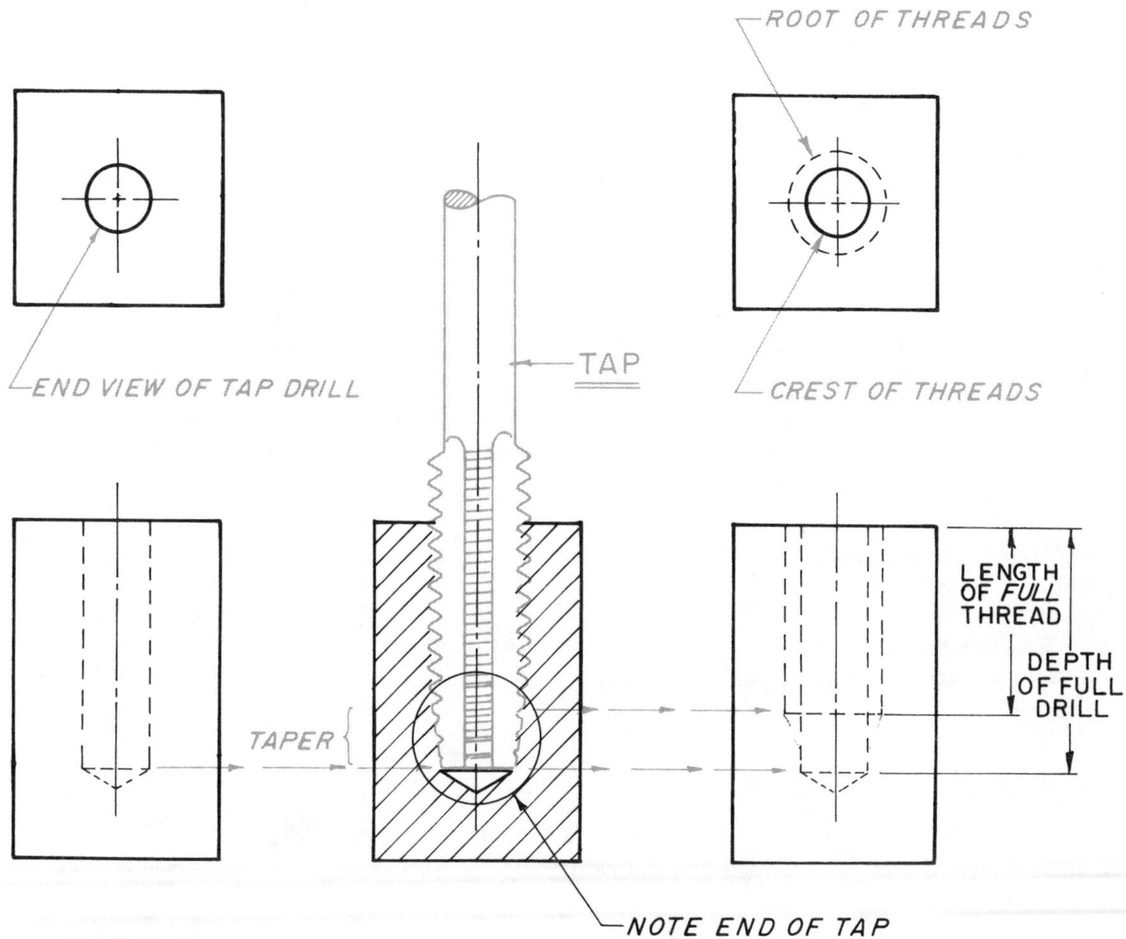

Fig. 8-13. The tap drill hole is drilled first.

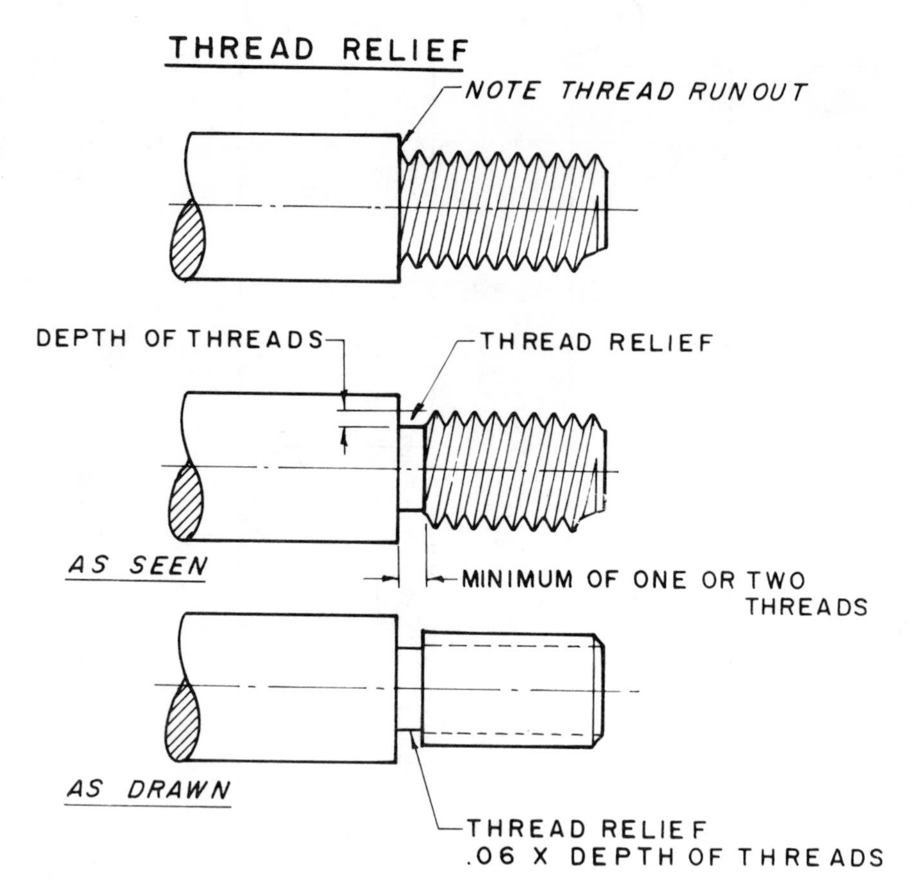

Fig. 8-14. It is impossible to make perfectly formed threads right up to a shoulder.

shows how an undercut or thread relief is drawn. The dimensions give the size of the undercut or relief.

CHAMFER

A chamfer is made to avoid sharp edges at the end of a machined part. Threads usually have chamfers at the end in order to allow the threads to be engaged easier. Chamfers are dimensioned two ways. If the chamfer is at 45 degrees, it is called off by a simple note (see FIG. 8-15). The figure, in this example, .06 indicates how wide the chamfer is to be. Any angle chamfer other than 45 degrees must be dimensioned as illustrated with both the angle given and illustrated and the size of the chamfer.

THREAD CALL-OFFS

FIGURE 8-16 illustrates the standard way a hole is called off on a drawing. The first line notes general identification of the fastener used; the second line specifies detailed requirements. Unless otherwise noted, all holes and fasteners are right-hand.

SCREWS AND RIVETS

There are many kinds of fasteners and only a few will be discussed at this time in order to give you a basic idea about fasteners.

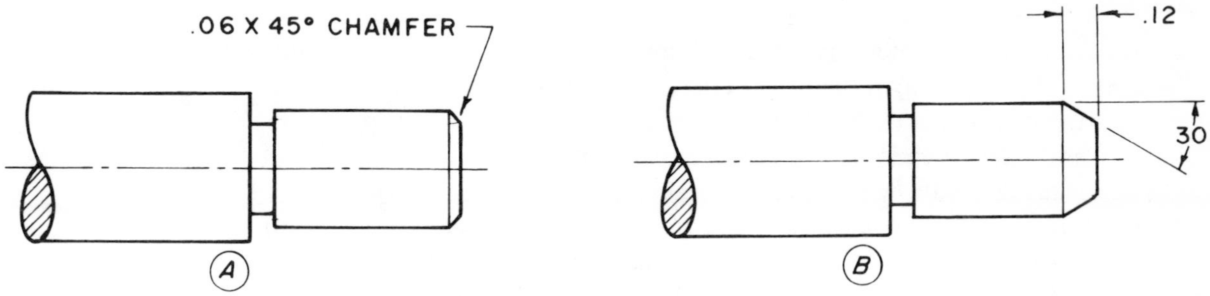

.06 X 45° CHAMFER

(A)

.12

30°

(B)

Fig. 8-15. Chamfers.

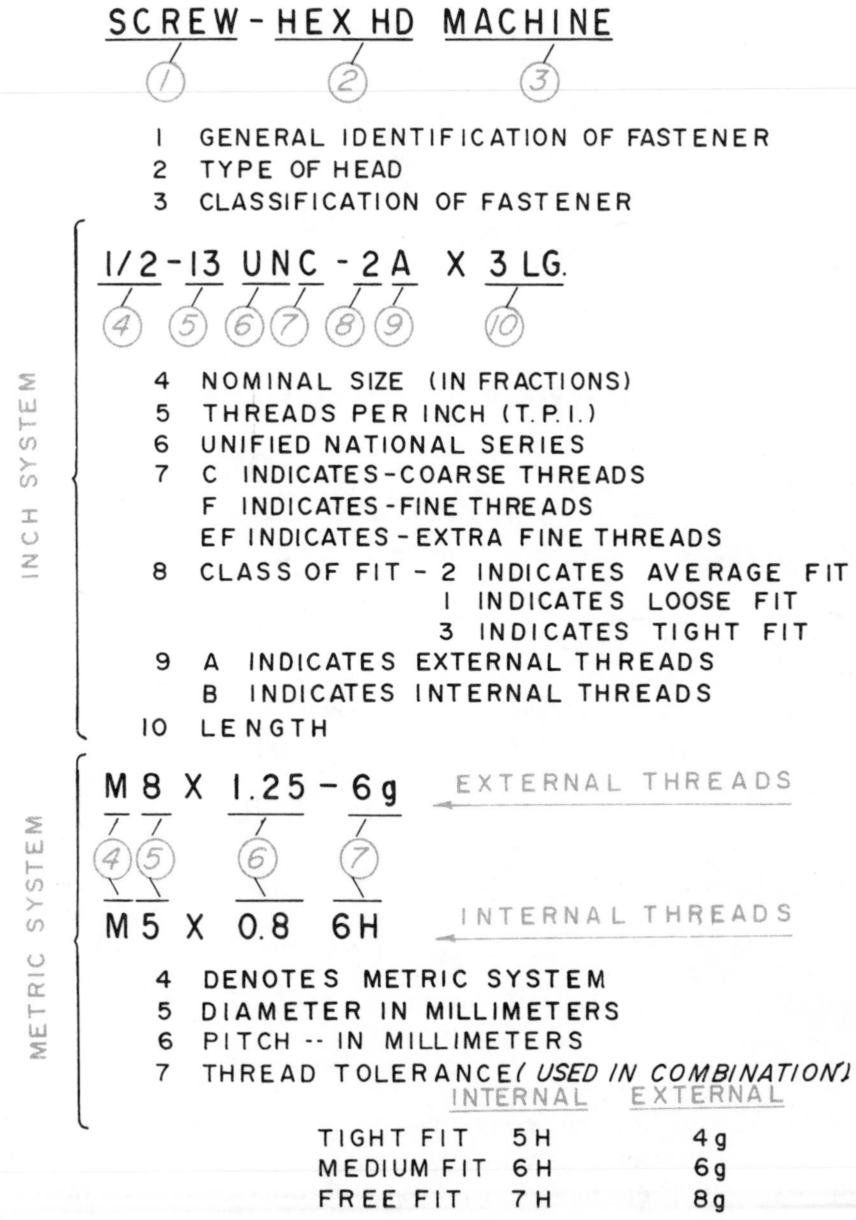

SCREW - HEX HD MACHINE

(1) (2) (3)

1 GENERAL IDENTIFICATION OF FASTENER
2 TYPE OF HEAD
3 CLASSIFICATION OF FASTENER

INCH SYSTEM

1/2-13 UNC - 2A X 3 LG.

(4) (5) (6)(7) (8)(9) (10)

4 NOMINAL SIZE (IN FRACTIONS)
5 THREADS PER INCH (T.P.I.)
6 UNIFIED NATIONAL SERIES
7 C INDICATES-COARSE THREADS
F INDICATES-FINE THREADS
EF INDICATES-EXTRA FINE THREADS
8 CLASS OF FIT - 2 INDICATES AVERAGE FIT
1 INDICATES LOOSE FIT
3 INDICATES TIGHT FIT
9 A INDICATES EXTERNAL THREADS
B INDICATES INTERNAL THREADS
10 LENGTH

METRIC SYSTEM

M 8 X 1.25 - 6g ← EXTERNAL THREADS

(4)(5) (6) (7)

M 5 X 0.8 6H ← INTERNAL THREADS

4 DENOTES METRIC SYSTEM
5 DIAMETER IN MILLIMETERS
6 PITCH -- IN MILLIMETERS
7 THREAD TOLERANCE (*USED IN COMBINATION*)

	INTERNAL	EXTERNAL
TIGHT FIT	5H	4g
MEDIUM FIT	6H	6g
FREE FIT	7H	8g

Fig. 8-16. Thread call-offs.

Screws

Screws are fasteners that are removable. Machine screws run in sizes from .021 (.3 mm) to .750 (20 mm) in diameter. There are eight standard head forms—four are illustrated in FIG. 8-17. Machine screws are usually used to screw into thin material and have threads within a thread or two of the head. The length of any screw, regardless of its head type, is measured from the *top surface* of the part to the bottom end of the screw. Cap screws run in size from .250 (6 mm) and up. There are five kinds of cap screws (see FIG. 8-18). A cap screw is usually used as a true screw; it passes through a clearance hole in one part into a tapped hole in another part. Note how the length is measured.

MACHINE SCREWS

LENGTH

FLAT HEAD
ROUND HEAD
OVAL HEAD
FILLISTER HEAD

Fig. 8-17. Eight standard machine screw head forms.

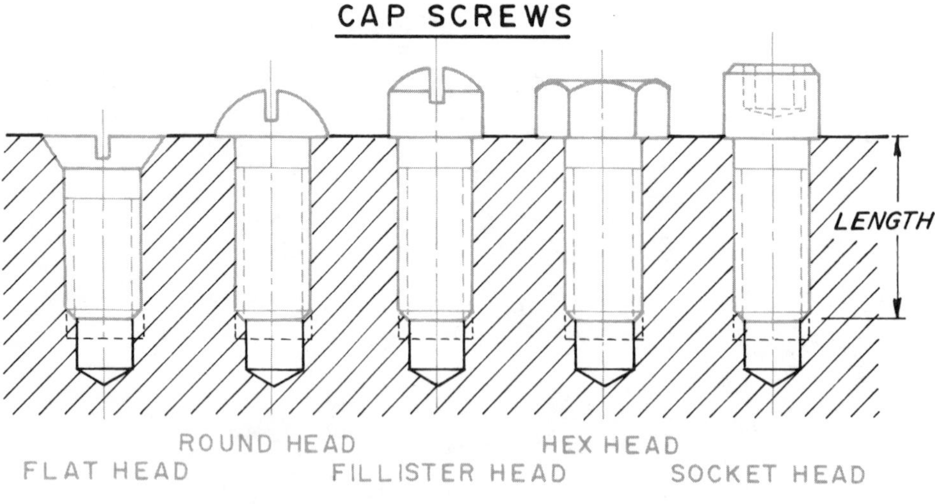

Fig. 8-18. Five standard cap screw head forms.

Rivets

Rivets are fasteners that are permanent and usually used to hold sheet metal together. Most rivets are made of wrought iron, soft steel, copper, aluminum, or other metals. There are five major types of rivets, truss head, button head, pan head, countersunk head, and flat head. A code is used to indicate on which side the head is to be positioned (see FIG. 8-19).

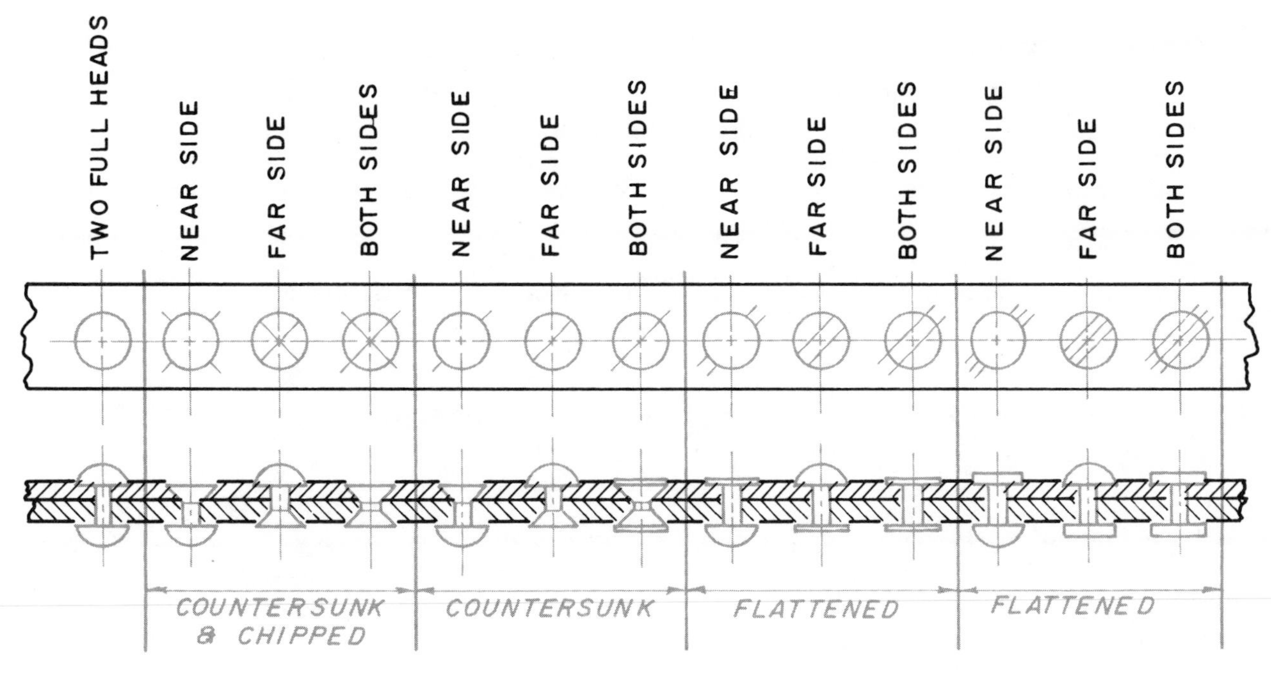

Fig. 8-19. Five major kinds of rivets, along with the code used to identify each.

WORKSHEET 8-1

Instructions: Using drawing number A173334 on p. 139, answer the following questions in the spaces provided. (Answers in Appendix A.)

1. What surface in the right-side view is surface A in the top view?

2. What two surfaces in the top view could be surface P in the right-side view?

3. What surface in the front view is surface F in the top view? What surface in the right-side view is surface F in the top view?

4. Is the 3/8-24-UN thread at E coarse or fine?

5. In the top view, H is what kind of a hole?

6. Give the full specifications for the hole at R in the right-side view?

7. The 1/4 UNF thread in the right-side view at O has how many threads per inch?

8. What is the dimension at W in the front view?

9. What is the specification for the thread at I in the top view?

10. What surface in the right-side view is surface D in the top view?

11. What is the radius at T in the front view and what is it called?

12. What is the maximum distance from surface X in the front view to surface D in the top view?

13. What is the distance from surface V in the front view to surface L in the right-side view?

14. In the right-side view, what is dimension K?

15. In the front view, horizontally, how far is it from the center of the .31 diameter counterbored hole to the .18 diameter countersunk hole?

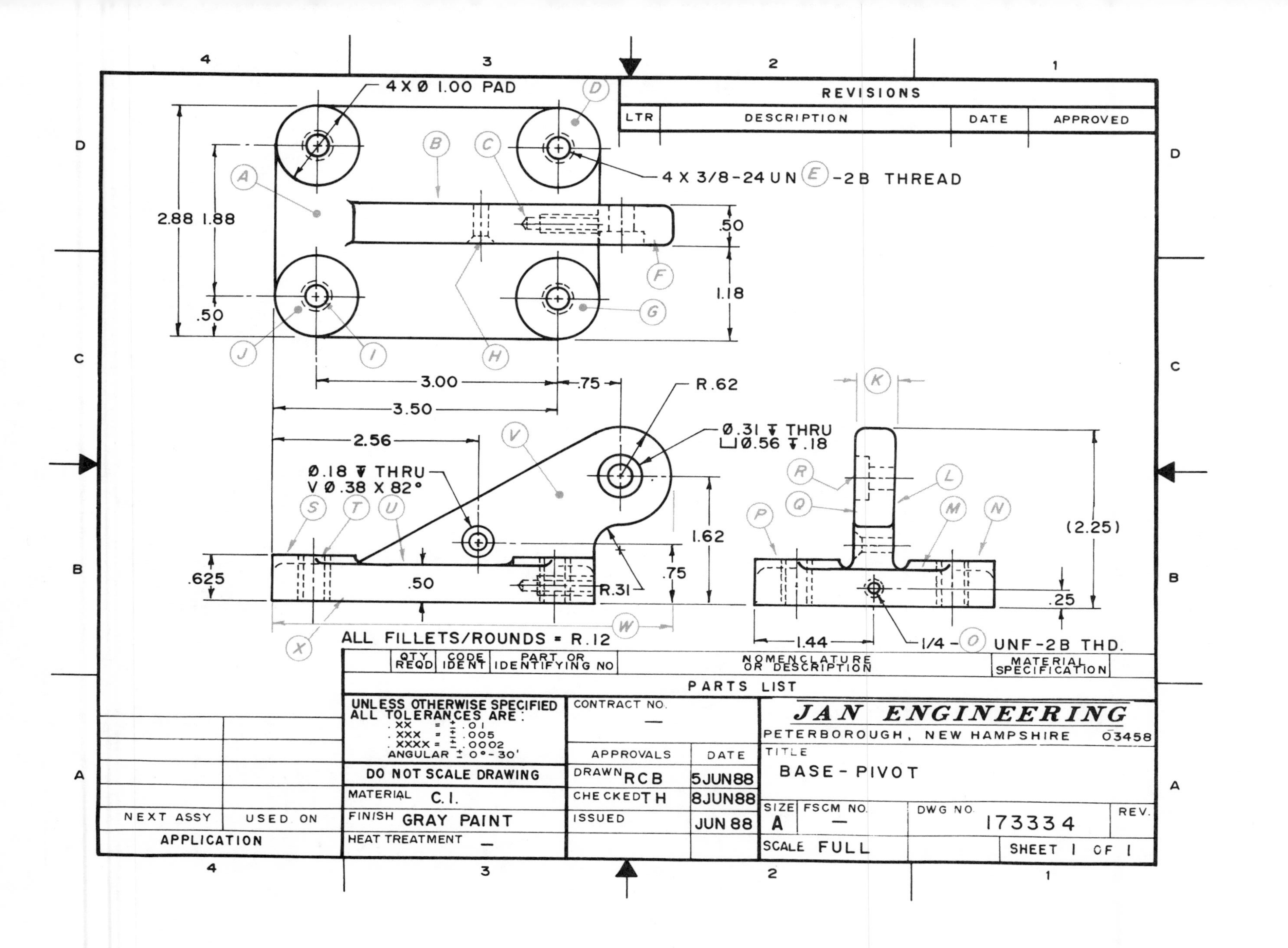
REVISIONS
LTR
DESCRIPTION
DATE
APPROVED
4 X Ø 1.00 PAD
4 X 3/8-24 UN (E) -2B THREAD
2.88 1.88
.50
1.18
3.00
.75
3.50
R.62
Ø.31 ↧ THRU
⌴Ø.56 ↧.18
2.56
Ø.18 ↧ THRU
V Ø.38 X 82°
.625
.50
R.31
.75
1.62
(2.25)
.25
1.44
1/4 - (O) UNF-2B THD.
ALL FILLETS/ROUNDS = R.12
QTY REQD
CODE IDENT
PART OR IDENTIFYING NO
NOMENCLATURE OR DESCRIPTION
MATERIAL SPECIFICATION
PARTS LIST
UNLESS OTHERWISE SPECIFIED ALL TOLERANCES ARE:
.XX = ± .01
.XXX = ± .005
.XXXX = ± .0002
ANGULAR ± 0°-30'
DO NOT SCALE DRAWING
MATERIAL C.I.
FINISH GRAY PAINT
HEAT TREATMENT —
CONTRACT NO. —
APPROVALS
DATE
DRAWN RCB 5JUN88
CHECKED TH 8JUN88
ISSUED JUN 88
JAN ENGINEERING
PETERBOROUGH, NEW HAMPSHIRE 03458
TITLE
BASE - PIVOT
SIZE A
FSCM NO. —
DWG NO. 173334
REV.
SCALE FULL
SHEET 1 OF 1
NEXT ASSY
USED ON
APPLICATION

Technical Information

Objective: The reader will know about keys and keyseats, knurlings, surface finishes, tabulated drawings, dadum planes, and all terms associated with them.

A *key* is a demountable part that fits into a pair of slots called keyseats. One is located on the shaft, and the other is located inside the hub. They provide a means of transferring torque between a shaft and a hub. There are five major kinds of keys used today: square, flat, gib head, Pratt and Whitney, and woodruff (see FIG. 9-1). A *keyseat* is a rectangular groove machined into the shaft and/or hub to receive the key. As a general rule, the key width is about one-fourth the diameter of the shaft.

KNURLING

Knurling is the process of rolling depressions of either a straight or diamond pattern onto a cylindrical surface (see FIG. 9-2). Knurling is either coarse, medium, or fine. A coarse pattern has a pitch of 14 ridges per inch, a medium pattern has a pitch of 21, and a fine pattern has a pitch of 33. Just as in threads, the more grooves or ridges per inch, the finer the knurls. The knurl patterns are used for three purposes: for grip, appearance, and a tighter-pressed fit. Note how knurls are schematically illustrated on a drawing, similar to that of threads—the actual knurls are not drawn (refer back to FIG. 9-2).

SURFACE FINISHES

Surfaces that are machined smooth and to tight tolerances are called *finished surfaces.* These special surfaces are identified by a "V" touching the finished surface with a number inside the "V" (see FIG. 9-3). This number is the distance, in *microinches,* from the highest point to the lowest point on a surface (see FIG. 9-4). One microinch equals .000001. The average machined finish is 125 microinches (i.e., .000001 × 125 = a surface roughness of .000125 of an inch allowed from the highest point to the lowest point on the surface). Each machine process used in industry has its own characteristic (see FIG. 9-5). According to the chart, a drilled hole will create a finish of from 250 (6.3 mm) to 63 (1.6 mm).

To learn how the finish symbol is used, refer back to FIG. 9-3. The tip of the "V" touches the surface that is to be finished and notes what the finish requirement is. The finish symbol is applied to the surface to be finished for *all views in which the surface appears as an edge*—even on hidden lines. In the event there is no room on the actual surface to be finished, the finish mark is applied to the *extension line* of that surface.

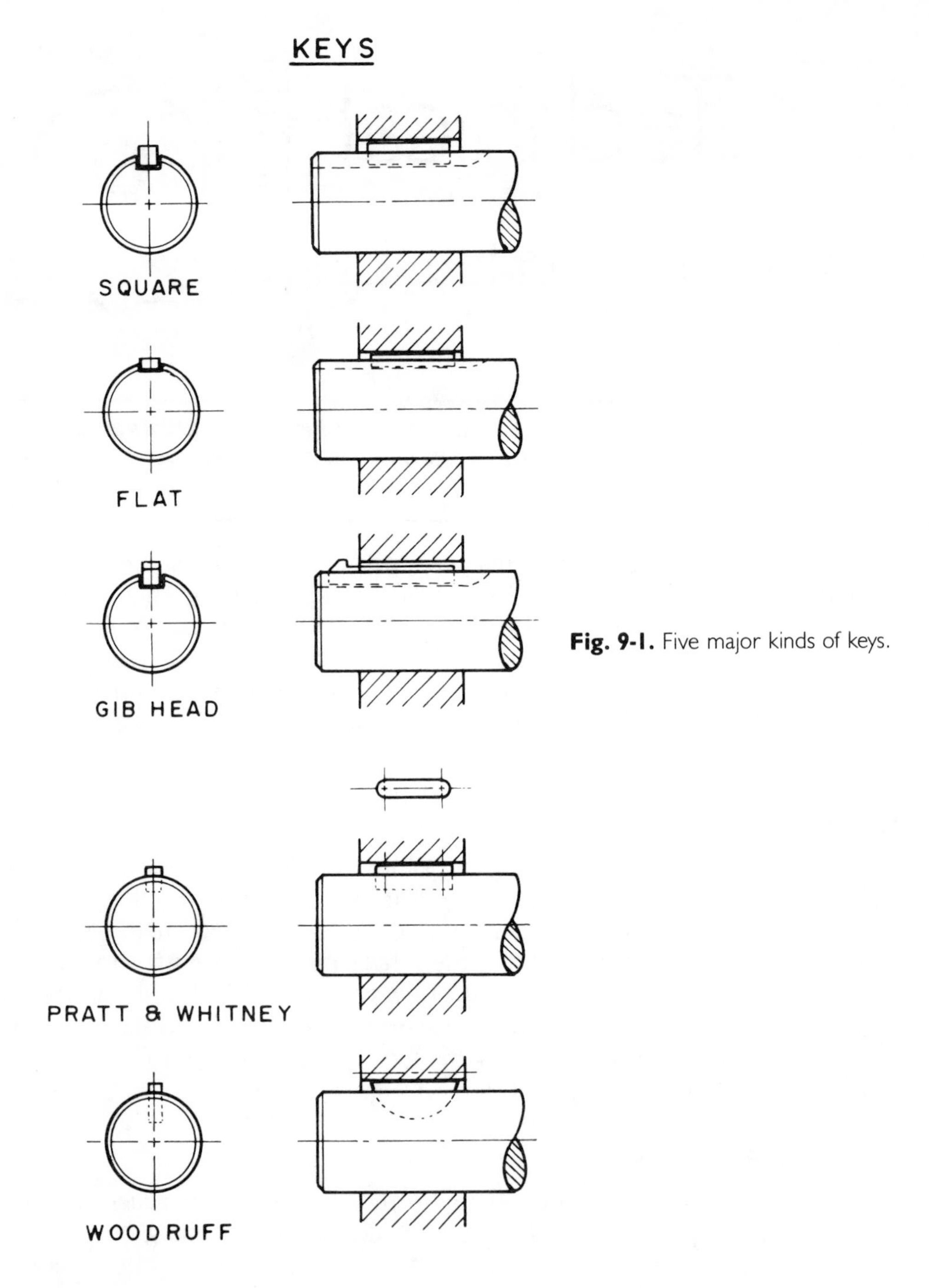

Fig. 9-1. Five major kinds of keys.

Through the years, various finish symbols of finish marks were used to indicate the finished surface. Even today there is a transition to other symbols, yet all are similar to the one illustrated and all mean the same thing and follow all the same rules.

Sometimes a part requires a smooth finish on all surfaces. Rather than applying the surface symbol or mark to all surfaces, the drafter adds a note to the drawing that reads, FINISH ALL OVER, or simply the initials "FAO."

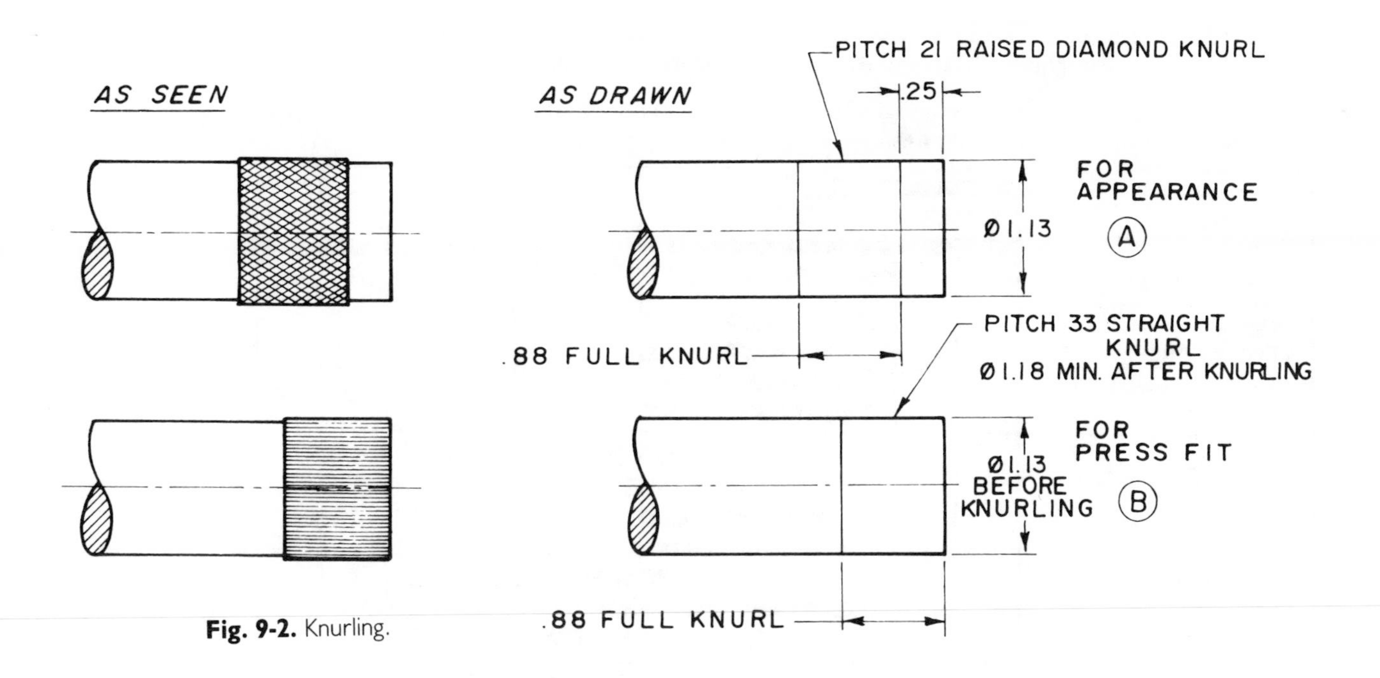

Fig. 9-2. Knurling.

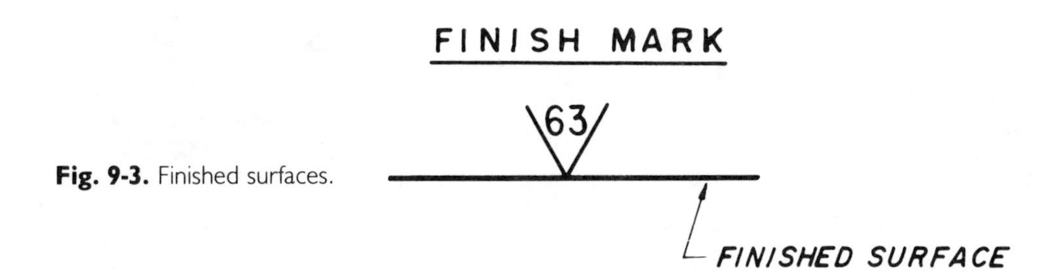

Fig. 9-3. Finished surfaces.

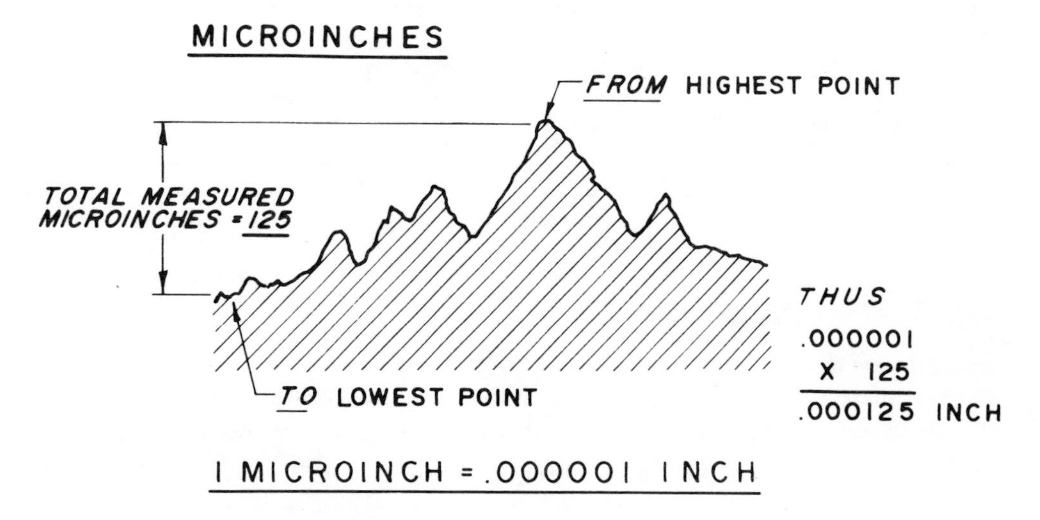

Fig. 9-4. A microinch equals .000001.

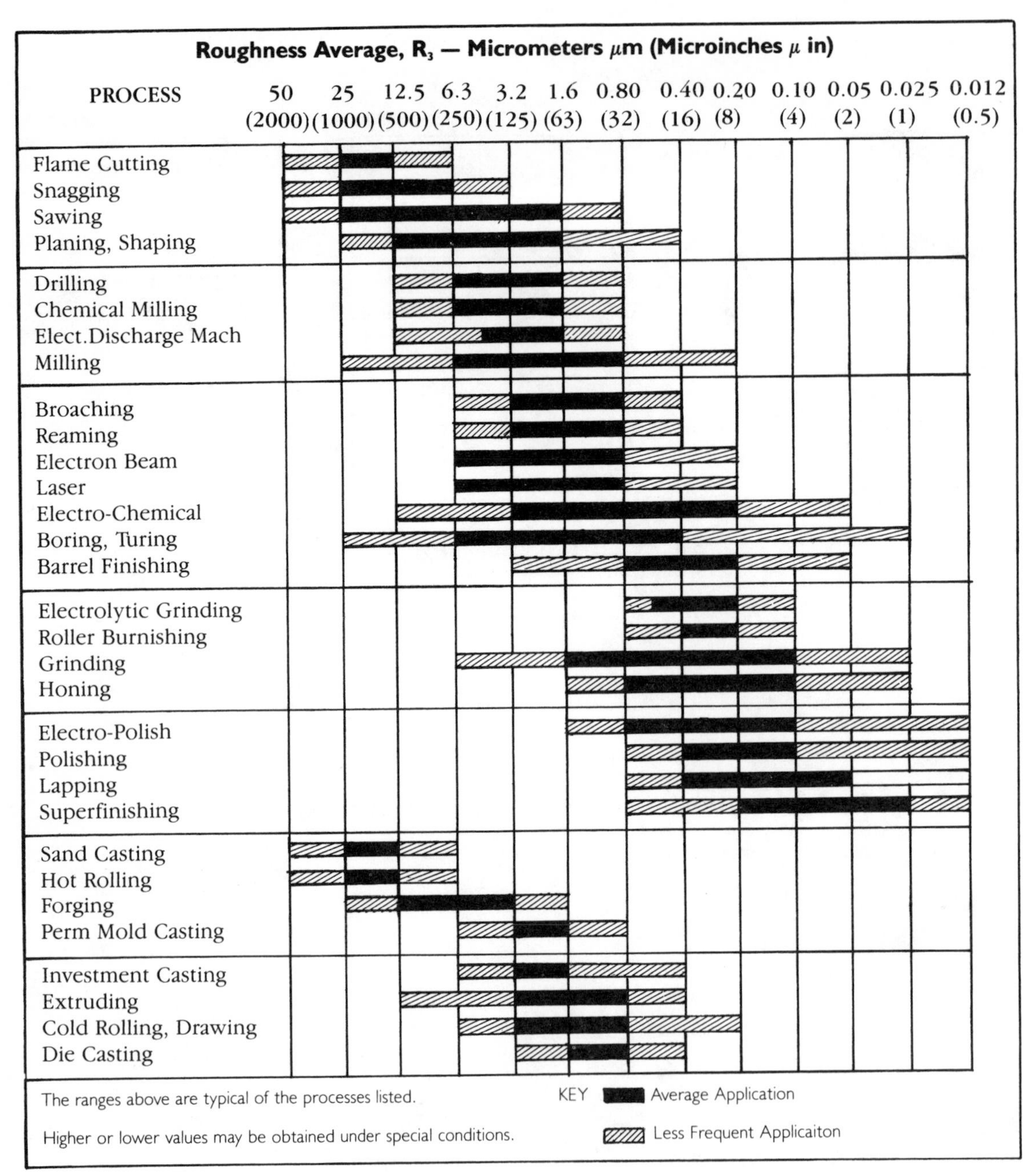

Fig. 9-5. Machine characteristics.

TABULATED DRAWINGS

A *tabulated drawing* is used by many companies in order to save drafting time. One tabulated drawing can be used to illustrate many similar parts. Any parts that are approximately the same size or general shape can be made into a tabular drawing by simply replacing all variable dimensions with a letter and listing these variable letters in a table (see FIG. 9-6). Is this example, three simple U-bolts are illustrated—all with the same overall basic shape. All are made in the shape of a **U**, all have threaded ends with chamfers. The only things that change are the rod diameters, overall lengths, overall widths, thread requirements, and lengths of the threads. These variable dimensions are replaced by letters and read from the table below. (*Note.* Take extreme care when reading a tabular drawing so you don't misread any of the dimensions in the table.)

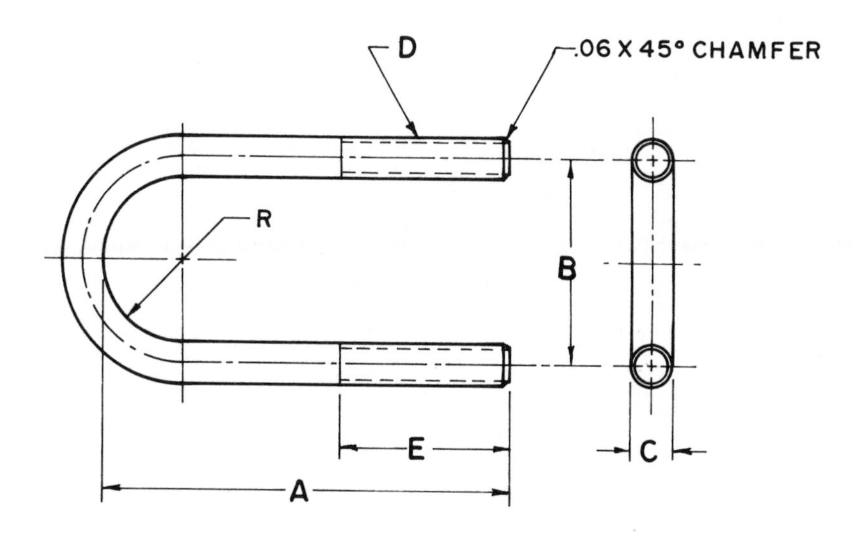

PART NUMBER	A	B	C	D	E
A95254-3	3.62	1.62	Ø.38	3/8-16 UNC-2A	1.50
A95254-2	4.88	2.50	Ø.38	3/8-16 UNC-2A	2.00
A95254-1	2.50	1.50	Ø.25	1/4-20UNC-2A	.88

Fig. 9-6. Tabulated drawing.

PLAN NO A95254

WORKSHEET 9-1

Instructions: Using drawing number A598871 on p. 147, answer the following questions in the spaces provided. (Answers in Appendix A.)

1. What is the radius at A?

2. Dimension O is what?

3. Surface P in the right-side view is what surface in the front view?

4. What is diameter R?

5. The 2.12 diameter, B in the front view is what surface in the right-side view?

6. What is L and what size is it?

7. Dimension H is what?

8. Surface V in the right-side view is what surface in the front view?

9. Give all specifications of the hidden lines at Q in the right-side view.

10. How many finished surfaces are there?

11. How many threads per inch (TPI) are required at K in the front view?

12. What is the maximum and minimum size dimension M can be and still be within limits?

13. What is the surface at G called?

14. The diameter at F is what?

15. What is dimension N?

16. Surface E in the front view is what surface in the right-side view?

17. What is the part made of?

18. Surface C in the front view is what surface in the right-side view?

19. What diameter and width cutter is to cut out the keyway?

20. How many microinches is surface W finished to?

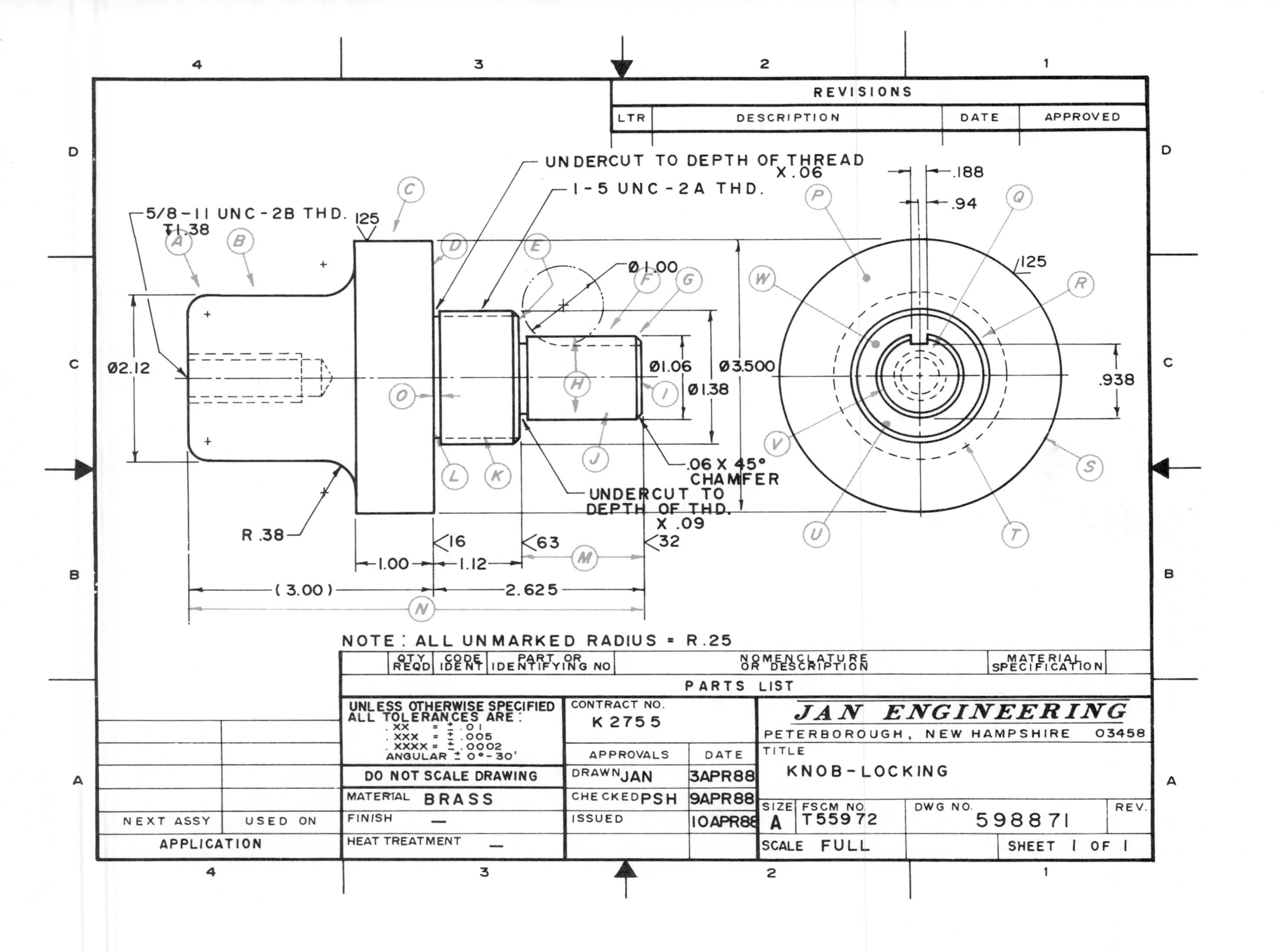
REVISIONS
LTR
DESCRIPTION
DATE
APPROVED
UNDERCUT TO DEPTH OF THREAD X .06
1-5 UNC-2A THD.
5/8-11 UNC-2B THD. ⊽1.38
125
Ø1.00
Ø1.06
Ø1.38
Ø3.500
Ø2.12
.188
.94
.938
125
.06 X 45° CHAMFER
UNDERCUT TO DEPTH OF THD. X .09
R .38
16
63
32
1.00
1.12
(3.00)
2.625
NOTE: ALL UNMARKED RADIUS = R.25
QTY REQD
CODE IDENT
PART OR IDENTIFYING NO
NOMENCLATURE OR DESCRIPTION
MATERIAL SPECIFICATION
PARTS LIST
UNLESS OTHERWISE SPECIFIED ALL TOLERANCES ARE:
.XX = ± .01
.XXX = ± .005
.XXXX = ± .0002
ANGULAR ± 0°-30'
DO NOT SCALE DRAWING
MATERIAL BRASS
FINISH —
HEAT TREATMENT —
CONTRACT NO. K 2755
APPROVALS
DATE
DRAWN JAN 3APR88
CHECKED PSH 9APR88
ISSUED 10APR88
JAN ENGINEERING
PETERBOROUGH, NEW HAMPSHIRE 03458
TITLE
KNOB-LOCKING
SIZE A
FSCM NO. T55972
DWG NO. 598871
REV.
SCALE FULL
SHEET 1 OF 1
NEXT ASSY
USED ON
APPLICATION

WORKSHEET 9-2

Instructions: Using drawing number A22574 on p. 149, answer the following questions in the spaces provided. (Answers in Appendix A.)

1. What scale is used in this drawing? What scale is used in section A-A?

2. What is the I.D. (E) of part number A22574-9?

3. The O.D. (D) of part number A22574-7 is what?

4. Surface A in the front view is what surface in the right-side view?

5. Surface O in section A-A is what surface in the front view?

6. What is dimension L for part number A22574-3?

7. Surface P in section A-A is what surface in the front view?

8. What is dimension C in the front view for part A22574-4?

9. What is the flange collar made of?

10. What is radius I?

11. Surface J in the right-side view is what surface in section A-A?

12. How far is it from surface H in the right-side view to surface N in section A-A for part A22574-10?

13. How far is surface S in section A-A from surface B in the front view for part A22574-5?

14. What is dimension E in the front view for part A22574-8?

15. Which part numbers have tab dimension B and C equal?

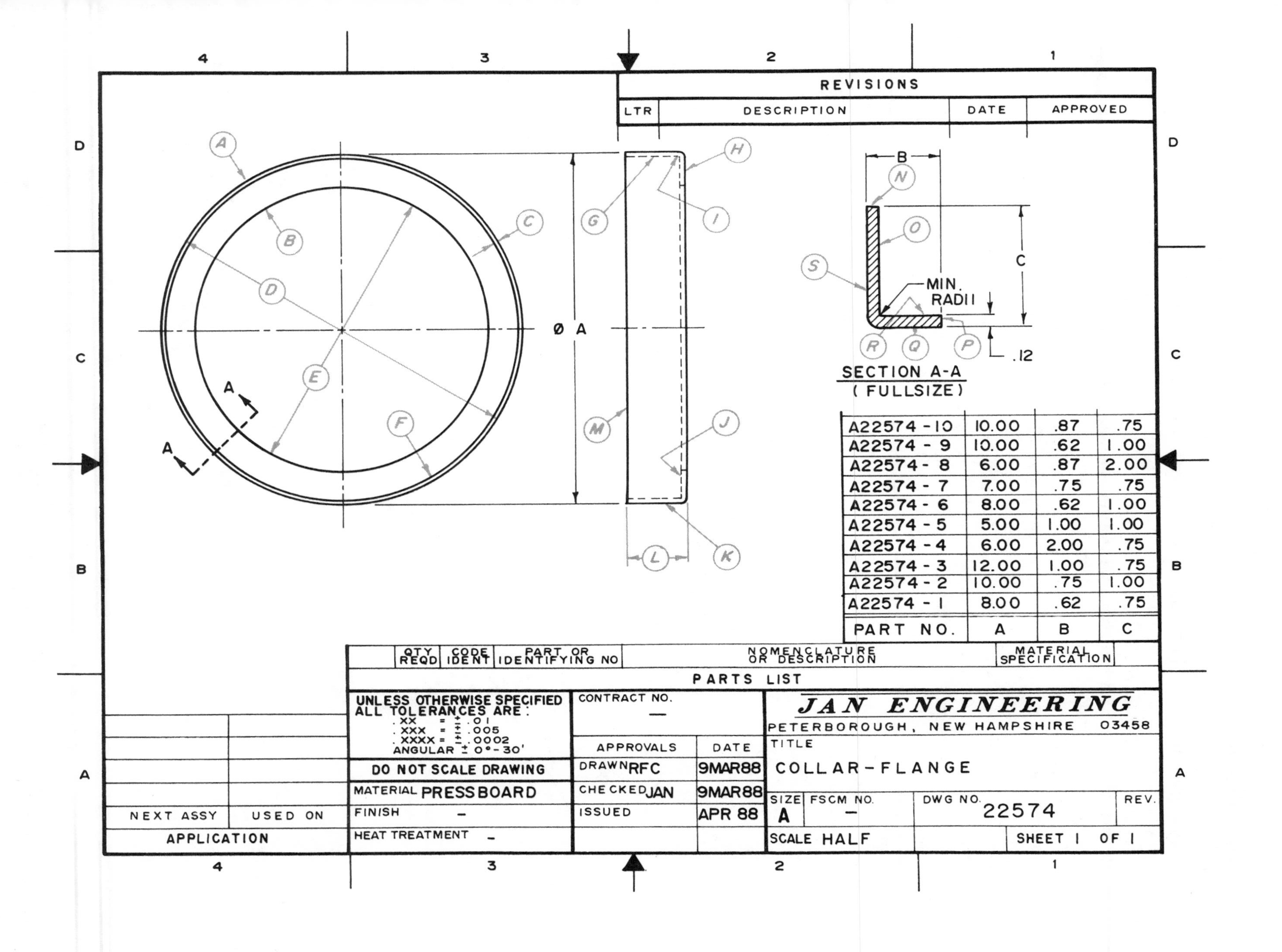
REVISIONS
LTR
DESCRIPTION
DATE
APPROVED
4
3
2
1
D
C
B
A
A
B
C
D
E
F
G
H
I
J
K
L
M
N
O
P
Q
R
S
Ø A
A
A
B
C
MIN. RADII
.12
SECTION A-A
(FULLSIZE)
A22574 - 10 10.00 .87 .75
A22574 - 9 10.00 .62 1.00
A22574 - 8 6.00 .87 2.00
A22574 - 7 7.00 .75 .75
A22574 - 6 8.00 .62 1.00
A22574 - 5 5.00 1.00 1.00
A22574 - 4 6.00 2.00 .75
A22574 - 3 12.00 1.00 .75
A22574 - 2 10.00 .75 1.00
A22574 - 1 8.00 .62 .75
PART NO. A B C
QTY REQD
CODE IDENT
PART OR IDENTIFYING NO
NOMENCLATURE OR DESCRIPTION
MATERIAL SPECIFICATION
PARTS LIST
UNLESS OTHERWISE SPECIFIED ALL TOLERANCES ARE:
.XX = ± .01
.XXX = ± .005
.XXXX = ± .0002
ANGULAR ± 0°-30'
DO NOT SCALE DRAWING
MATERIAL PRESSBOARD
FINISH –
HEAT TREATMENT –
CONTRACT NO. –
APPROVALS
DATE
DRAWN RFC 9MAR88
CHECKED JAN 9MAR88
ISSUED APR 88
JAN ENGINEERING
PETERBOROUGH, NEW HAMPSHIRE 03458
TITLE
COLLAR-FLANGE
SIZE A
FSCM NO. –
DWG NO. 22574
REV.
SCALE HALF
SHEET 1 OF 1
NEXT ASSY
USED ON
APPLICATION

DADUM PLANES

A *dadum plane* can be a reference line, surface, centerline, or important feature of an object from which all dimensions are derived. When dadums are specified or implied, all other features must be located from these dadums. X, Y, and Z dadums or coordinates are located as illustrated in FIG. 9-7. X is always the left-hand surface and is read horizontally from the left surface to the right. Y is always the bottom surface and is read vertically from bottom surface up to the top. Z is always the front surface and is read from the front surface, back into the part. Study FIG. 9-8. The X, Y, and Z coordinates are used exactly as in FIG. 9-7, except now they are used to locate the various hole coordinates. Note the table below the drawing. All X dimensions are from the left-hand surface, all Y dimensions are from the bottom surface, and all Z dimensions are from the front surface. By using this coordinate system of dimensioning, many confusing dimensions can be omitted, thus making the drawing easier to read and understand.

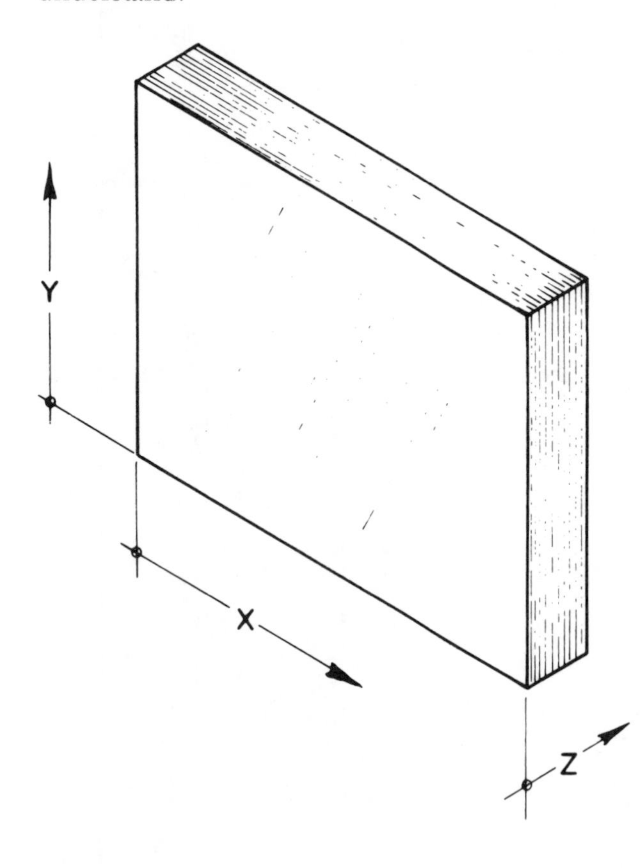

Fig. 9-7. Dadum plane.

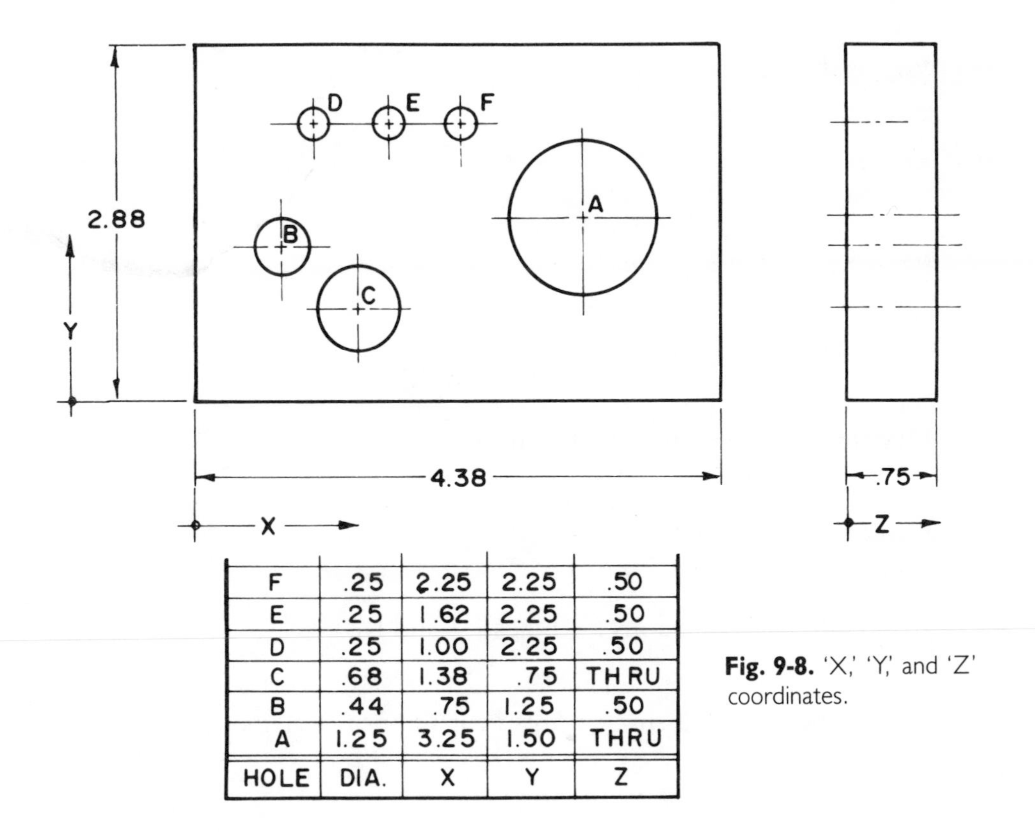

HOLE	DIA.	X	Y	Z
F	.25	2.25	2.25	.50
E	.25	1.62	2.25	.50
D	.25	1.00	2.25	.50
C	.68	1.38	.75	THRU
B	.44	.75	1.25	.50
A	1.25	3.25	1.50	THRU

Fig. 9-8. 'X,' 'Y,' and 'Z' coordinates.

WORKSHEET 9-3

Instructions: Using drawing A37119 on p. 153, answer the following questions in the spaces provided. (Answers in Appendix A.)

1. What is dimension A?

2. What is the maximum distance dimension B could be and still be within limits?

3. Calculate dimension C.

4. What is the minimum size dimension D can be and still be within limits?

5. What is the depth of full thread at E?

6. What is radius F?

7. What is the maximum *and* minimum size dimension G can be and still be within limits?

8. What is the diameter at H?

9. What is the diameter at I?

10. What is the depth at J?

11. How deep is dimension L?

12. What kinds of holes are A1 and A2?

13. What kinds of holes are B1, B2, and B3?

14. What kinds of holes are C1 and C2?

15. Give the full specifications for holes D1 and D2.

16. What is the specified surface finish for the notch at the left side of the plate?

17. What is the tolerance for hole E1?

18. Note the upper/lower limits for holes B1, B2, and B3.

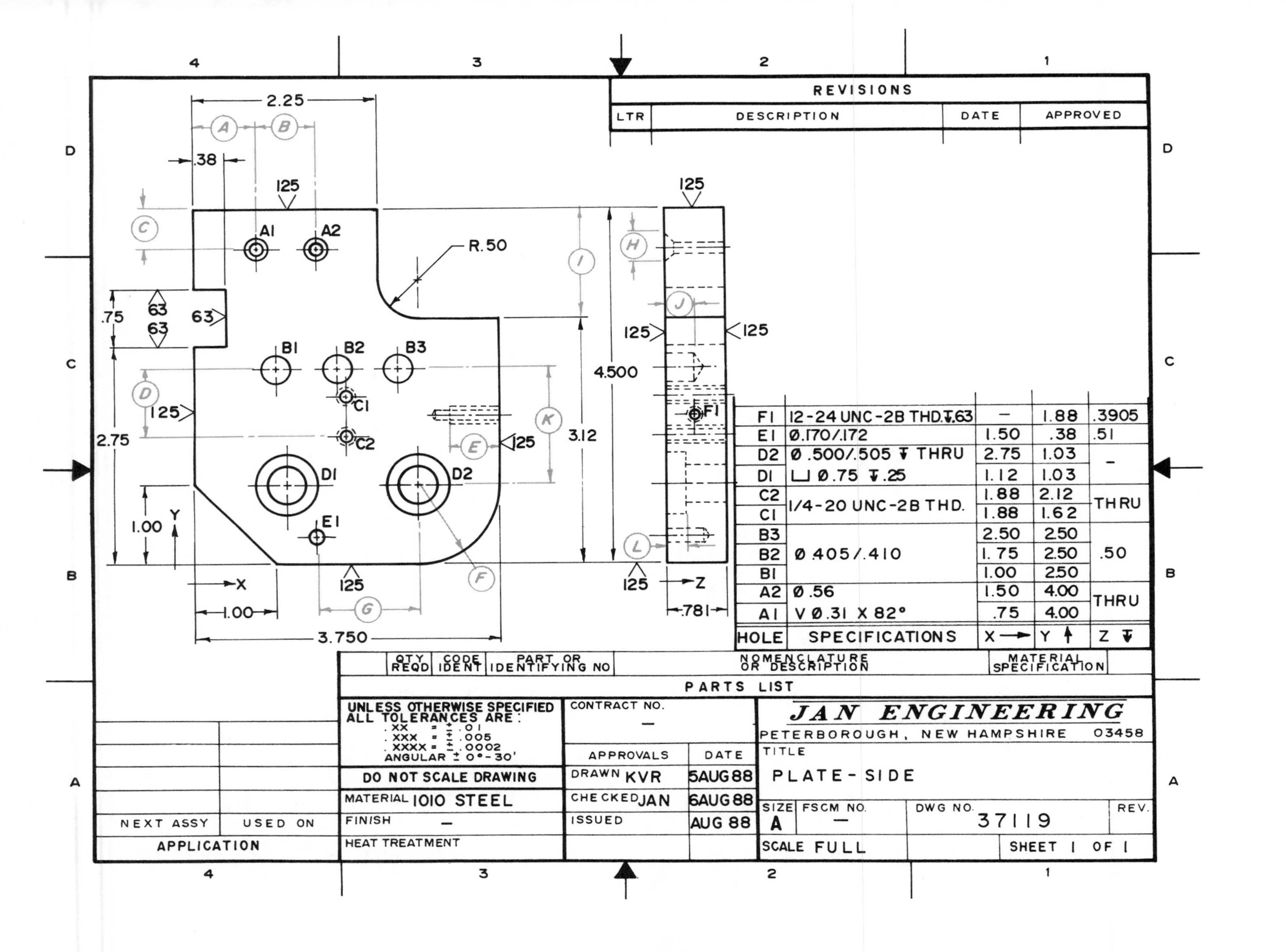
REVISIONS
LTR
DESCRIPTION
DATE
APPROVED
2.25
.38
125
A1
A2
R.50
.75
63
63
63
B1
B2
B3
C1
C2
125
2.75
D1
D2
E1
125
1.00
Y
X
1.00
125
3.750
3.12
4.500
125
125
125
F1
125
Z
.781
HOLE
SPECIFICATIONS
X
Y
Z
F1
12-24 UNC-2B THD. ↧.63
—
1.88
.3905
E1
Ø.170/.172
1.50
.38
.51
D2
Ø .500/.505 ↧ THRU
2.75
1.03
—
D1
⌴ Ø.75 ↧.25
1.12
1.03
C2
1/4-20 UNC-2B THD.
1.88
2.12
THRU
C1
1.88
1.62
B3
2.50
2.50
B2
Ø .405/.410
1.75
2.50
.50
B1
1.00
2.50
A2
Ø .56
1.50
4.00
THRU
A1
V Ø.31 X 82°
.75
4.00
QTY REQD
CODE IDENT
PART OR IDENTIFYING NO
NOMENCLATURE OR DESCRIPTION
MATERIAL SPECIFICATION
PARTS LIST
UNLESS OTHERWISE SPECIFIED ALL TOLERANCES ARE:
.XX = ± .01
.XXX = ± .005
.XXXX = ± .0002
ANGULAR ± 0°-30'
DO NOT SCALE DRAWING
MATERIAL 1010 STEEL
FINISH —
HEAT TREATMENT
CONTRACT NO. —
APPROVALS
DATE
DRAWN KVR 5AUG88
CHECKED JAN 6AUG88
ISSUED AUG 88
JAN ENGINEERING
PETERBOROUGH, NEW HAMPSHIRE 03458
TITLE
PLATE - SIDE
SIZE A
FSCM NO. —
DWG NO. 37119
REV.
SCALE FULL
SHEET 1 OF 1
NEXT ASSY
USED ON
APPLICATION

Metallurgy

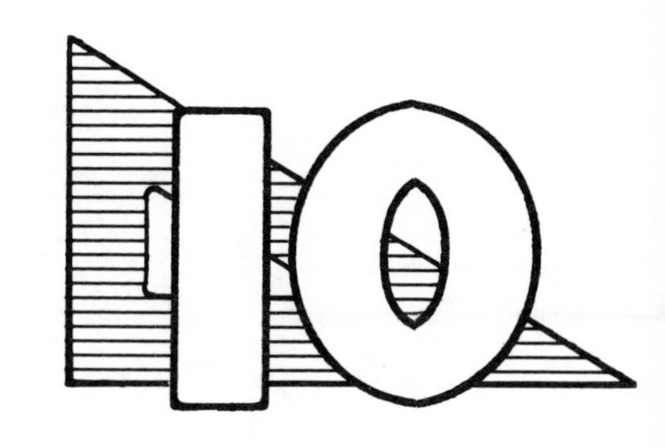

Objective: The reader will know basic metallurgy, all terms associated with metallurgy, and basic characteristics of metals used in industry today. The reader will also know about the heat treatment processes and how metal is tested for hardness.

Anyone using the machine drawings of industry should have basic working knowledge of the behavior, characteristics, and properties of the metals used. The following pages give very general information of these metals. There are two basic classifications of metals: ferrous, or those containing iron, and non-ferrous, or those that do not contain any iron. Each will be discussed in this chapter.

FERROUS METALS

The most common and useful metal used today is iron. Iron is mined from the ground in this country, and it is also imported from foreign countries. Pure iron does *not* rust in water and is seldom used in industry, as it is too soft for most work. Iron rusts only because of the impurities that it contains.

Iron, as used in industry, is first made from *pig iron.* Iron ore becomes pig iron after the impurities are burned out of it in a blast furnace. The pig iron is transferred from the blast furnace and put into large molds called pigs. Pig iron is very hard and brittle and has three uses: for making cast iron, for making wrought iron, and for making steel. Pig iron that has been remelted and poured into sand molds is called *cast iron* (CI). The process is referred to as a casting.

Kinds of Iron

There are five kinds of iron (see FIG. 10-1):

- gray cast iron
- white cast iron
- malleable iron
- ductile iron
- wrought iron

Each vary a large degree from one another by the carbon content. The carbon content in the various irons varies from about 0.4 percent to about 6.0 percent.

Gray cast iron contains from 1.7 to 4.5 percent carbon, melts at 2200 degrees, and is somewhat brittle. It is one of the cheapest kinds of metal used today in industry. *White cast iron* contains from 2.0 to 2.5 percent carbon. Carbon is chemically combined into the iron and forms a very hard iron that cannot be easily machined. White cast iron is mostly used to make malleable iron. *Malleable iron* contains from 2.0 to 2.6 percent carbon. It is produced from white cast iron by heating the iron at a very high temperature for 100 or more

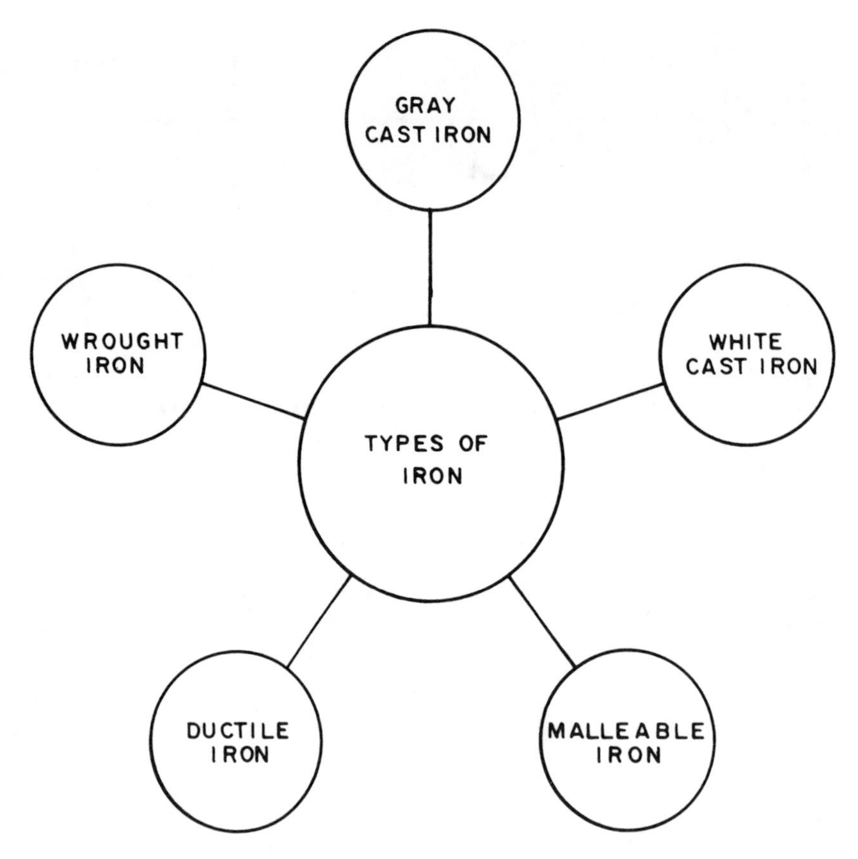

Fig. 10-1. Five kinds of iron.

hours. Malleable castings have many of the tough characteristics of steel. *Ductile cast iron* is produced very much like gray cast iron but is much tougher. It is used for castings that will take a lot of abuse. *Wrought iron* contains almost no carbon, at most 0.4 percent and is the purest form of iron commercially used today. It bends very easily cold or hot but has limited uses today.

Steel

Steel is an alloy made by combining various metals. It is bright and stronger than most forms of iron. There are two major catagories of steel: carbon steel and alloy steel.

Carbon steels. There are three kinds of carbon steel: mild carbon steel, medium carbon steel, and hard carbon steel (see FIG. 10-2). *Mild carbon steel* (10XX) contains from .05 to .30 percent carbon and is used for forged work, rivets, chains, and machine parts that do not need great strength. It can be rolled cold between rollers under great pressure to produce smooth cold-rolled steel sheets. *Medium carbon steel* (10XX) contains from .30 to .60 percent carbon and is used for such things as shafts, screws, nuts, bolts axles, and rails. *High carbon steel* (10XX), also known as *tool steel*, contains from 1.60 to 1.70 percent carbon and is used to make such tools as drills, taps, dies, reamers, files, cold chisels, and hammers.

Alloy steels. An alloy is a mixture of two or more metals melted together to form a totally new metal that is completely different from the original metals. (For example, copper and zinc melted together make brass.) There are five major kinds of alloy steel: nickel-chromium steel, manganese steel, molybdenum steel, vanadium steel, and tungston steel (see FIG. 10-3).

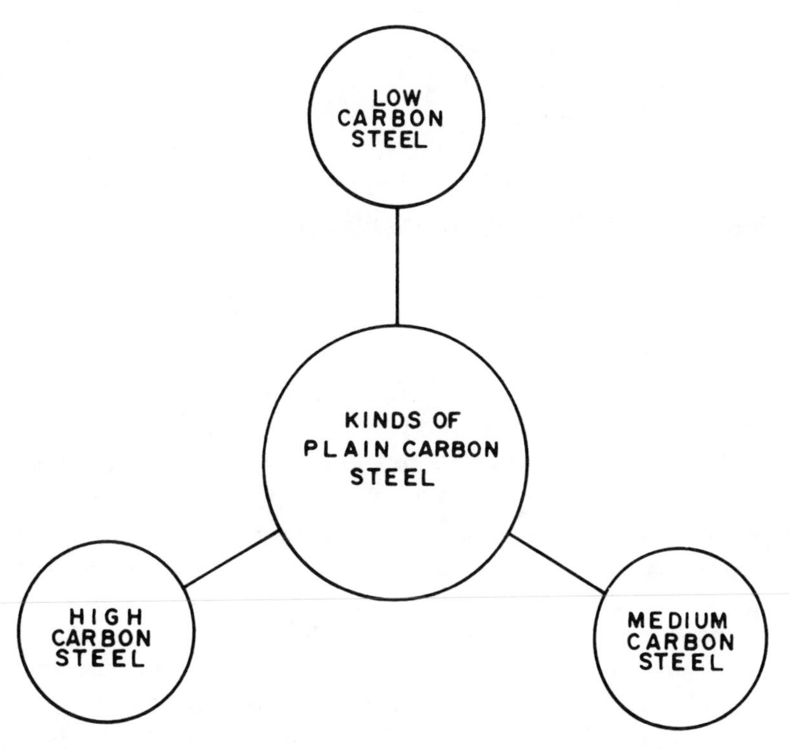

Fig. 10-2. Three kinds of carbon steel.

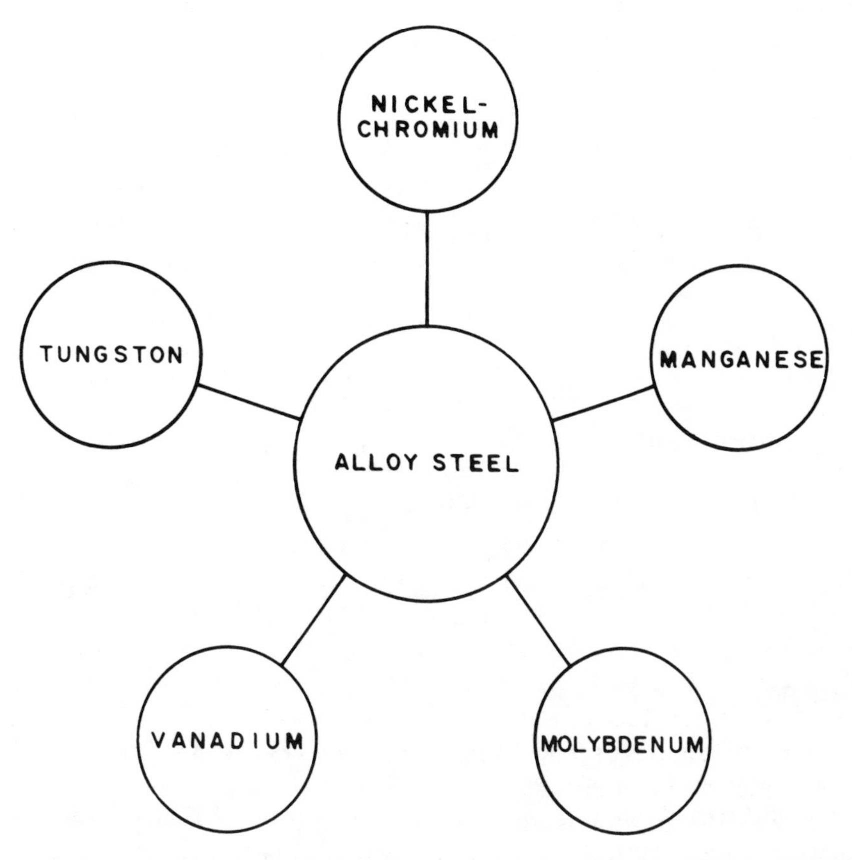

Fig. 10-3. Alloy steel.

Nickel steel (20XX), as its name implies, contains nickel, which adds toughness and strength to the steel. It does not rust easily and is strong and hard. Nickel steel is somewhat elastic and is used for such things as wire cables, car axles, and shafts. *Chromium steel* (50XX) contains 11 to 26 percent chromium. Chromium steel is very tough; it resists rust, stains, and shocks. It is used for such things as sinks, pots and pans, instruments, ball bearings, and valves. *Nickel-chromium* (30XX) is a steel alloy containing both nickel and chromium. It is very hard and strong and is used for such things as gears, cams, shafts, and springs. *Manganese steel* (90XX) contains manganese, which is a very brittle, hard material that adds strength and toughness to the steel. It is used for such things as chains, gears, and safes. *Molybdenum steel* (40XX) is added to steel to produce an alloy that can withstand heat and hard blows. It is used to produce fine wire, ball bearings and roller bearings. *Tungston steel,* made by adding tungston to the steel, has a very high melting point and produces a very hard steel that is used for high-speed tools. *Vanadium steel* gives lightness and toughness to steel. It can resist great shocks and is used for such things as springs, gears, and other parts that must work under heavy vibration.

Steel numbering system. The S.A.E. (Society of Automotive Engineers) developed a numbering system to list chemical composition of various kinds of steels. Each kind of steel has a four- or five-digit number assigned to it (see FIG. 10-4). The first digit of the number indicates the ***kind*** of steel it is. The second digit of the number indicates the approximate percent of the *major alloy* that is added, and the last two digits indicate the average percent of *carbon* the steel contains. Example: A number 3140 means: nickel-chromium steel (3), 1 percent of nickel (1) and it contains .40 percent carbon (40). See TABLE 10-1 for a more complete list of numbers used.

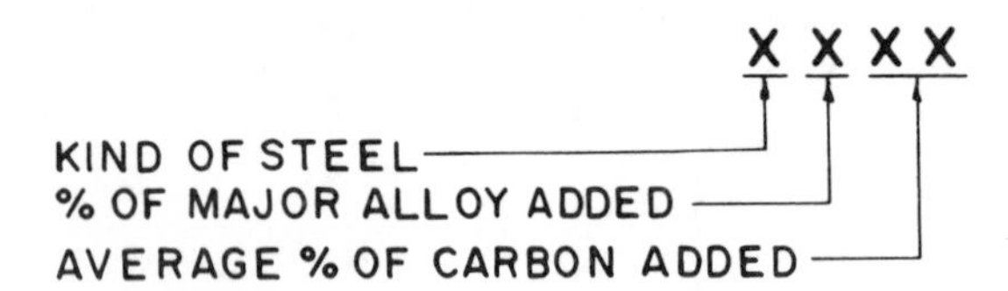

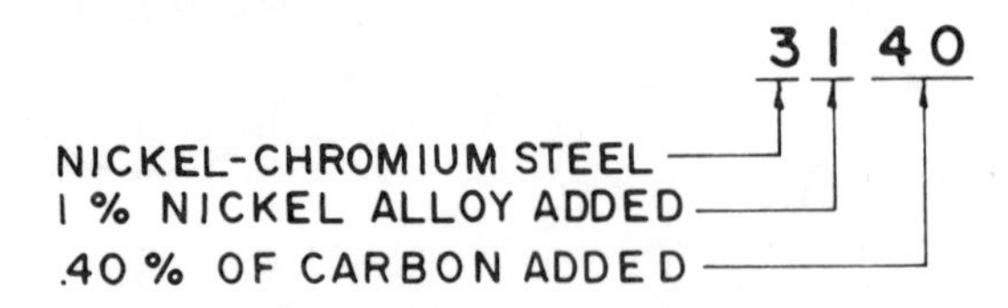

Fig. 10-4. Kinds of steel.

As a general rule, S.A.E. 1020 steel is recommended for parts to be heat treated.

NONFERROUS METALS

Nonferrous metals contain no iron. There are eleven major kinds of nonferrous metals: aluminum, copper, brass, bronze, zinc, magnesium, lead, tin, babbitt, pewter, and nickel (see FIG. 10-5). *Aluminum* is one of the most useful metals used today in industry. It weights about one-third as much as iron or steel and machines much easier. It costs less to transport, does not rust, and is almost maintenance-free. Although it costs more to make, all things considered, it may be more economical than any other metal. Aluminum melts at 1200 degrees

Table 10-1. Properties, Grade Numbers & Usages

Class of Steel	*Grade Number	Properties	Uses
Carbon - Mild 0.3% carbon	10xx	Tough - Less Strength	Rivets - Hooks - Chains Shafts - Pressed Steel Products
Carbon - Medium .03% to 0.6% carbon	10xx	Tough & Strong	Gears - Shafts - Studs- Various Machine Parts
Carbon - Hard 1.6% to 1.7%	10xx	Less Tough - Much Harder	Drills - Knives - Saws
Nickel	20xx	Tough & Strong	Axles - Connecting Rods - Crank Shafts
Nickel Chromium	30xx	Tough & Strong	Rings Gears - Shafts - Piston Pins - Bolts- Studs - Screws
Molybdenum	40xx	Very Strong	Forgings - Shafts - Gears - Cams
Chromium	50xx	Hard W/Strength & Toughness	Ball Bearings - Roller Bearing - Springs - Gears - Shafts
Chromium Vanadium	60xx	Hard & Strong	Shafts - Axles - Gears - Dies - Punches - Drills
Chromium Nickel Stainless	60xx	Rust Resistance	Food Containers - Medical/Dental Surgical Instruments
Silicon Manganese	90xx	Springiness	Large Springs

and can be drawn into a very fine wire. It can be stamped into deep forms and rolled into very thin sheets of foil.

Copper is one of the oldest metals used by man. It is sold in bars, plates, or sheets. Copper is the second-best conductor of electricity (silver is best), and it is used for wire, kettles, window and door screens, and roofing. Copper hardens when hammered but can be easily softened.

Brass is made from copper with zinc added. It is used for inexpensive jewelry, screws, locks, clocks, and so on. *Bronze* is also made from copper but with tin added. It costs more than brass because of the high cost of tin, but it is harder than brass. Bronze is used for such things as boat propellers, bushings, bells, and welding rods.

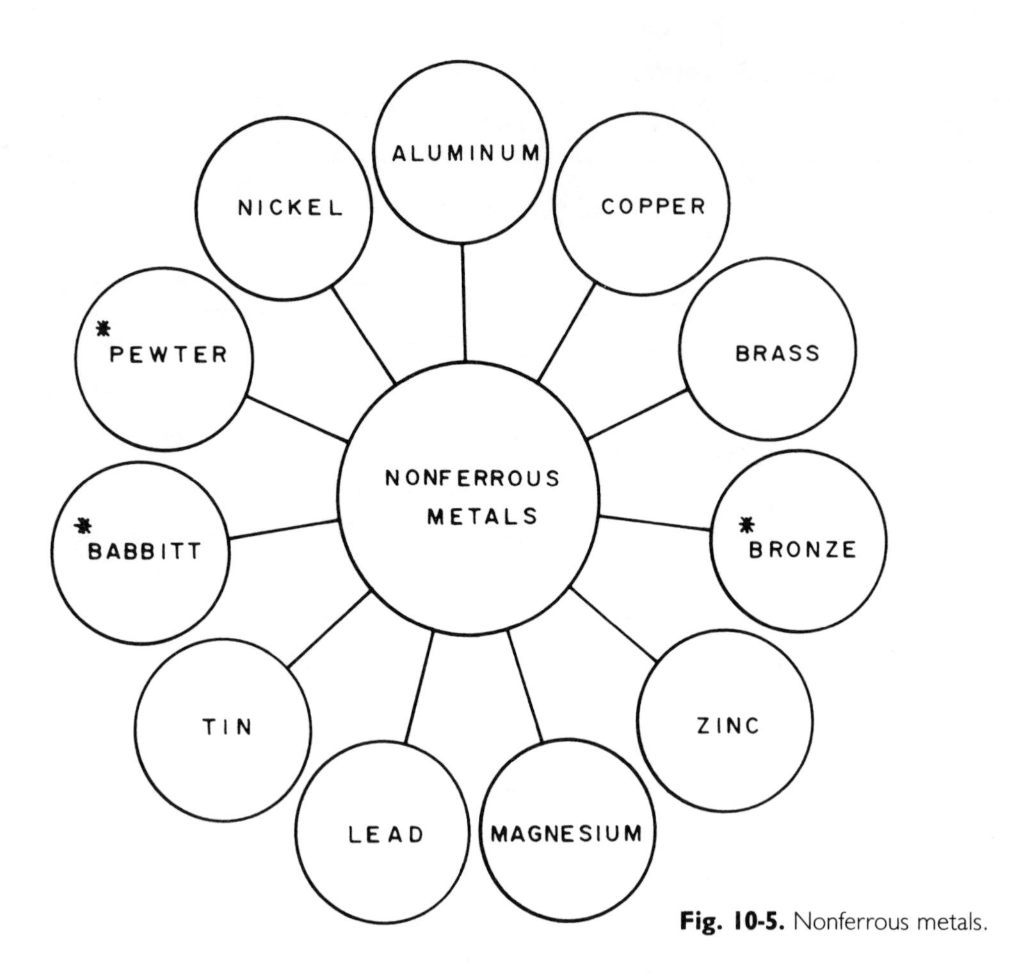

Fig. 10-5. Nonferrous metals.

Zinc is used to coat iron in order to make it rustproof. It is used in the galvanizing process. Man zinc-based alloys are used in making die castings for such things as carburetors, typewriter and computer parts, and small car parts. *Magnesium* is lighter than aluminum but is not used much because of its high cost. In its pure form it burns very easily, so it must be handled with extreme care.

Lead is a heavy metal. It is very soft and is poisonous, therefore it must be worked with care at all times. In years past, it was used in the manufacture of paints. Today it is used in making pewter, solder, and other metals.

Tin does not rust and is very expensive. Very few things are actually made of tin, but it is used in making other metals such as bronze, babbitt metal, pewter, and solder. (*Note.* The "tin can" is actually made of steel and is only *coated* with a very thin layer of tin—less than 1 percent of the can's weight is tin.)

Babbitt comes in two forms: tin-based babbitt or lead-based babbitt. Babbitt does not rust and is used for bearings in machines and engines. *Pewter* is a white metal made up mostly of tin. It is used for tableware or ornamental use. *Nickel* is a very tough, shiny metal that does not rust. It is used in making nickel steel and sometimes for plating.

CHARACTERISTICS OF METALS AND ALLOYS

The composition of metal and the various elements regulate the mechanical, chemical, and electrical properties of that metal. The following terms describe certain characteristics and capabilities associated with metals and alloys.

Strength—the ability to resist deformation

Plasticity—the ability to withstand deformation without breaking. Usually hardened metals have strength but are very low in plasticity. They are rather brittle.

Ductility—refers to how well a material can be drawn out. This is especially important for manufacturing wire and for metal forming.

Malleability—the ability of a metal to be shaped by hammering or rolling.

Elasticity—the ability of metal to be stretched and then returned to its original size and shape.

Brittleness—a characteristic of metal to break with little deformation, such as glass.

Toughness—describes a metal that has high strength and malleability.

Fatigue limit—the stress, measured in pounds per square inch, at which a metal will break after a certain number of repeated applications of a load.

Conductivity—describes how well a metal transmits electricity or heat.

Corrosion resistance—describes how well a metal resists rust. (*Note.* Rust *adds* weight, reduces strength, and ruins the overall appearance of the metal.)

HEAT TREATMENT

Heat treatment is a controlled heating and cooling process of metals in their solid state, in order to change their internal properties. Steel can be made harder, tougher, softer, or stronger by various kinds of heat treatment processes. All metals have an internal grain structure, and by heat treating, the grain structures can be changed in order to bring about desired qualities in the metal.

There are five major heat treatment processes: hardening, tempering, annealing, normalizing, and case hardening, see FIG. 10-6. *Hardening* is the

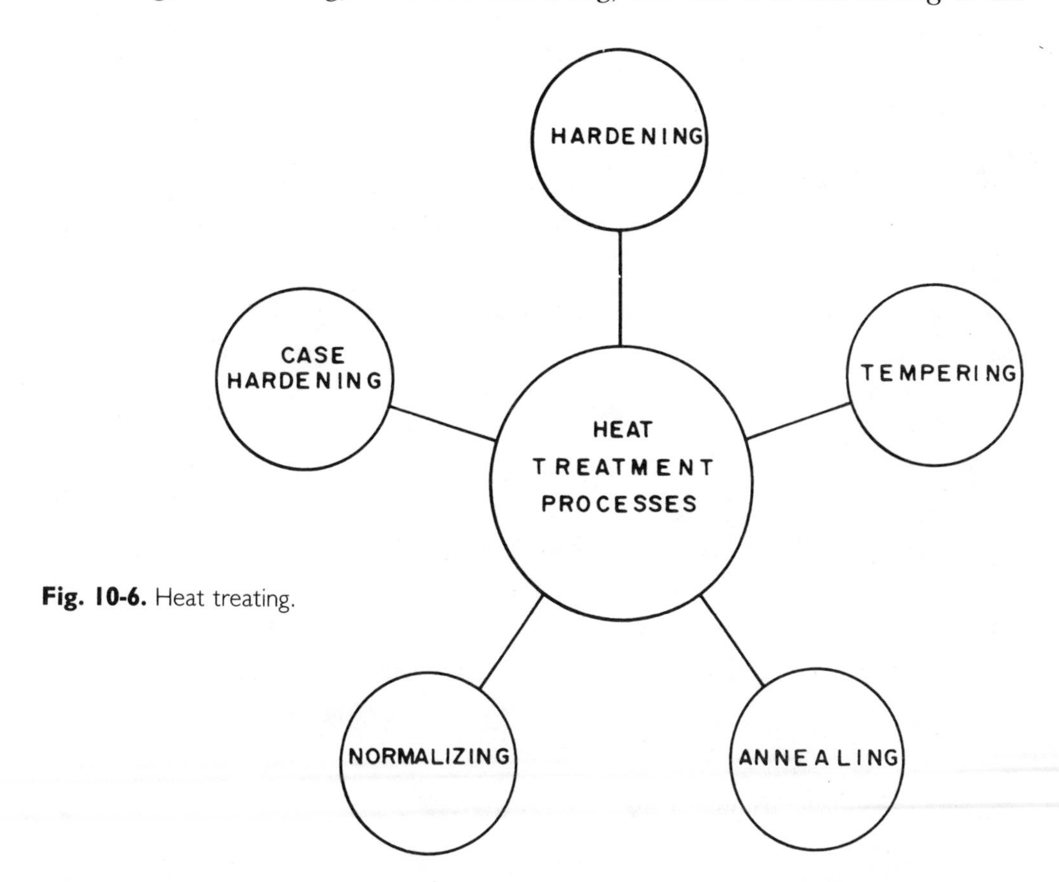

Fig. 10-6. Heat treating.

process that makes the steel harder. Files, drills, taps, and dies are examples of steel that has been hardened after the part was formed. Medium-carbon steel and high-carbon steel are hardened by heating slowly to a determined temperature and cooling very rapidly in water or oil. *Tempering* is the process that relieves internal stresses created within the part during the hardening process, thus creating toughness. Tempering is done by slowly reheating the part and allowing it to cool slowly in the air. *Annealing* is used to soften the part in order that it can be machined. This process is just the opposite of hardening. The part is heated for about one hour per inch of thickness of the part and allowed to cool very slowly. *Normalizing* is similar to annealing and it too relieves internal stresses due to forging, machining or cold-working. *Case Hardening* involves hardening only a small area and only to a thin depth on the part where a hard surface is needed on the part. Usually this process is applied only to low-carbon steels. Case hardening is a two-step process. Carbon is added to the determined area of surfaces, referred to as *carburizing*. It causes carbon to be absorbed directly into the steel, thus changing that particular area into a high carbon steel area. The second step is to apply the hardening process to the area of the part to be case hardened. Examples of parts that use case hardening are hammers, wrenches, gears, screws, and bolts. The depth of case hardening can vary from .03 to .12 of an inch or more deep.

HARDNESS TESTING

There are several methods of testing the hardness of metals. The two most common methods are the Brinell hardness test and the Rockwell hardness. The *Brinell method* uses a hardened steel ball that is pressed into the surface of the metal under a given pressure or load. The diameter of the hole made by the ball is measured by a microscope and converted into a hardness reading. *The Rockwell method* is very much the same except the hardness value is read directly from a scale built into the tester.

WORKSHEET 10-1

Instructions: Answer the following questions in the spaces provided. (Answers in Appendix A.)

1. Explain the difference between ferrous and nonferrous metals.

2. What is an alloy?

3. List five kinds of iron and their basic differences.

4. List the two major catagories of steels used today in industry.

5. What are the five major kinds of alloy steels?

6. What does S.A.E. stand for?

7. List the eleven kinds of nonferrous metals.

8. Explain the term *plasticity*.

9. Note two reasons why magnesium is not used much in industry.

10. How much tin is used in a "tin can"?

11. What does the term *brittleness* refer to?

12. Explain briefly what the process of *heat treatment* is.

13. What are the two hardness test methods used to test the hardness of a surface and explain the difference between the two?

14. *Tempering* a part does what?

15. Approximately, what percentage of carbon would be expected in white cast iron?

16. What kind of iron contains almost *no* carbon?

17. A rivet is usually made of what kind of carbon steel?

18. A steel with a number of 4010 would be what kind of steel and how much carbon would it contain?

19. What nonferrous metal is used to make clocks?

20. Babbitt is made up of two bases; list the two major kinds of metal used to manufacture babbitt. Also, what is babbitt usually used for?

WORKSHEET 10-2

Instructions: Using drawing A371162 on p. 165, answer the following questions in the spaces provided. (Answers in Appendix A.)

1. What is the microinch value or number at surface D?
2. Surface C in view I is what surface in view II?
3. Name the three views.
4. How many T.P.I. are there at A?
5. What is dimension L?
6. Surface N is finished to what?
7. Assuming upper limits on all given dimensions, what is dimension J?
8. Is the 3/4-10 UN thread at P coarse or fine?
9. Surface R in view III is what surface in view II?
10. What is dimension K in view II?
11. What manufactured process is done at M?
12. What manufactured process is done at G?
13. Dimension O in view OO is what?
14. What is the carbon content of this part?
15. What is the *maximum* distance from surface E in view I to surface I in view II?
16. *Clockwise,* what is the angle between hole "a" in view I and hole "b" in view III?
17. Surface R in view III is how far from surface C in view I?
18. Surface F in view I is which surface in view III?

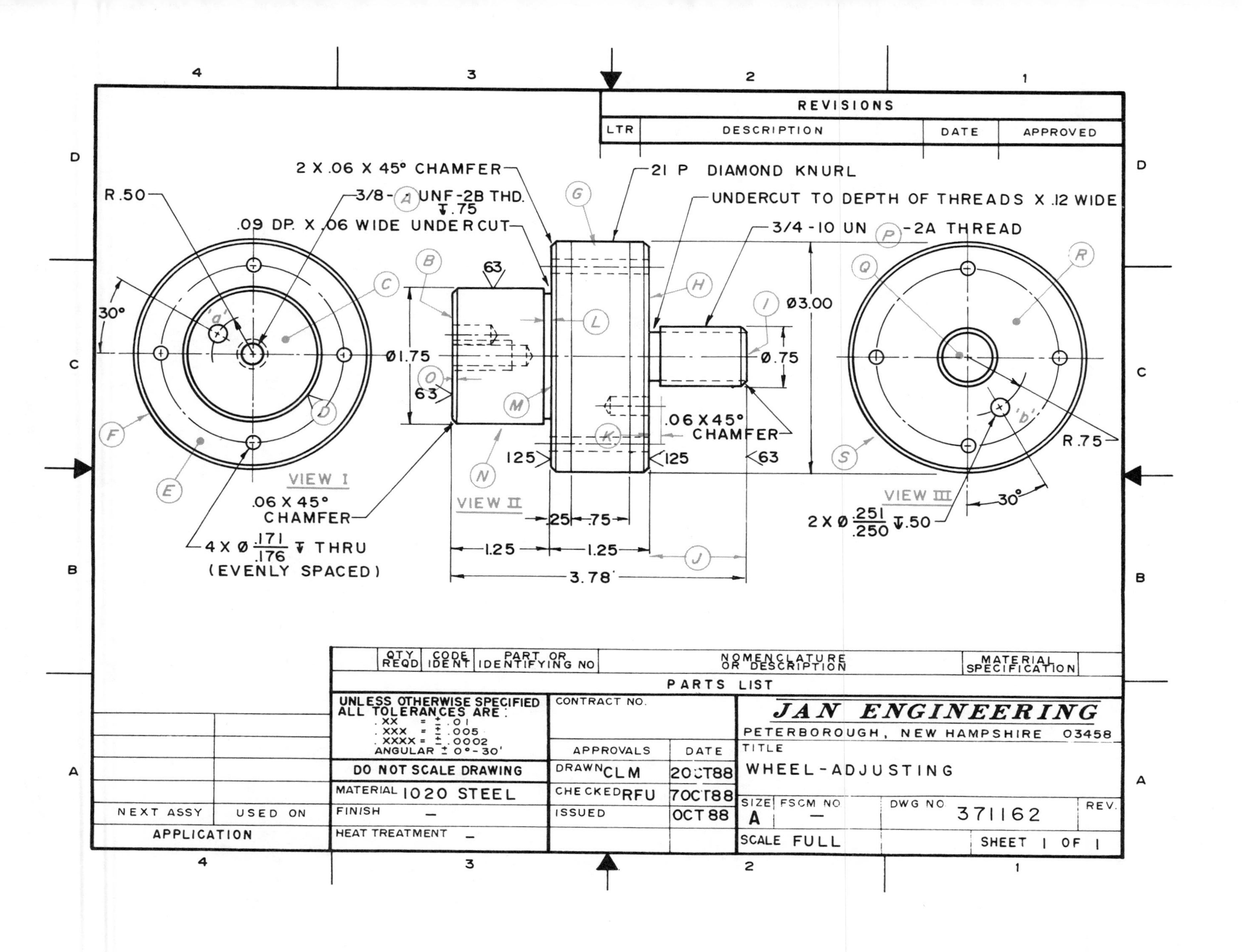
REVISIONS
LTR
DESCRIPTION
DATE
APPROVED
2 X .06 X 45° CHAMFER
3/8- A UNF-2B THD. ↧.75
.09 DP. X .06 WIDE UNDERCUT
R.50
30°
21 P DIAMOND KNURL
UNDERCUT TO DEPTH OF THREADS X .12 WIDE
3/4-10 UN P -2A THREAD
Ø3.00
Ø1.75
Ø.75
63
125
.06 X 45° CHAMFER
R.75
VIEW I
VIEW II
VIEW III
.06 X 45° CHAMFER
4 X Ø .171/.176 ↧ THRU
(EVENLY SPACED)
.25
.75
1.25
1.25
3.78
2 X Ø .251/.250 ↧.50
30°
QTY REQD
CODE IDENT
PART OR IDENTIFYING NO
NOMENCLATURE OR DESCRIPTION
MATERIAL SPECIFICATION
PARTS LIST
UNLESS OTHERWISE SPECIFIED ALL TOLERANCES ARE:
.XX = ± .01
.XXX = ± .005
.XXXX = ± .0002
ANGULAR ± 0°-30'
DO NOT SCALE DRAWING
MATERIAL 1020 STEEL
FINISH –
HEAT TREATMENT –
CONTRACT NO.
APPROVALS
DATE
DRAWN CLM 20OCT88
CHECKED RFU 7OCT88
ISSUED OCT 88
JAN ENGINEERING
PETERBOROUGH, NEW HAMPSHIRE 03458
TITLE
WHEEL-ADJUSTING
SIZE A
FSCM NO –
DWG NO 371162
REV.
SCALE FULL
SHEET 1 OF 1
NEXT ASSY
USED ON
APPLICATION

Basic Welding

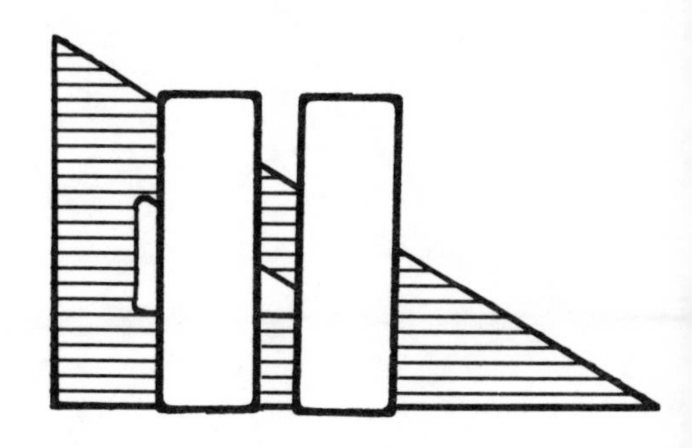

Objective: The reader will know about the various kinds of welds used in industry and the terms associated with welding.

As discussed in Chapter 8, pieces of metal can be fastened together with mechanical fasteners. These kind of fasteners are temporarily fastened together. Parts can be fastened together permanently by means of *welding*.

In years past, parts were welded together by pressure. In this process, parts to be welded together are heated to a plastic state and forced together by pressure or by hammering. This was done in the local blacksmith shop and was called *forge welding*.

WELDING METHODS

The two major methods used today to weld parts together are *fusion welding* and *resistance welding*. *Fusion welding* is usually used for larger parts. In this process, the edges of the parts to be joined are heated until they melt and fuse into each other as one piece. A filler rod is sometimes used to mix with the molten metal to make the joint even stronger. Heat for this process comes from a torch, burning gases, or a very high electric current. It should be noted, parts are sometimes distorted in using this process because of the extreme heat involved.

Resistance welding is the process of passing an electric current through the exact location where the parts are to be joined. The heat caused by the resistance of the part to the flow of electric current and pressure applied by the electrode forms the weld. Resistance welding is usually done on thinner parts such as sheet metal.

Many other welding processes are used today. This chapter will cover only kinds of fusion and resistance welding. The five types of welds to be covered in this chapter are: fillet welds, groove welds, plug or slot welds, flange welds, and spot welds (see FIG. 11-1).

BASIC WELDING SYMBOL

The basic welding symbol consists of a reference line, a leader and arrow, and, if needed, a tail (see FIG. 11-2). The tail is added only if specific information or notes are needed in regard to the weld. The reference line is usually drawn horizontally and the welding symbol is added to it (see FIG. 11-3). If the welding symbol is added *above* the reference line it means *weld opposite side*. If the welding symbol is added *below* the reference line it means *weld arrow side*. The direction of the leader line and arrow has no significance whatsoever to the

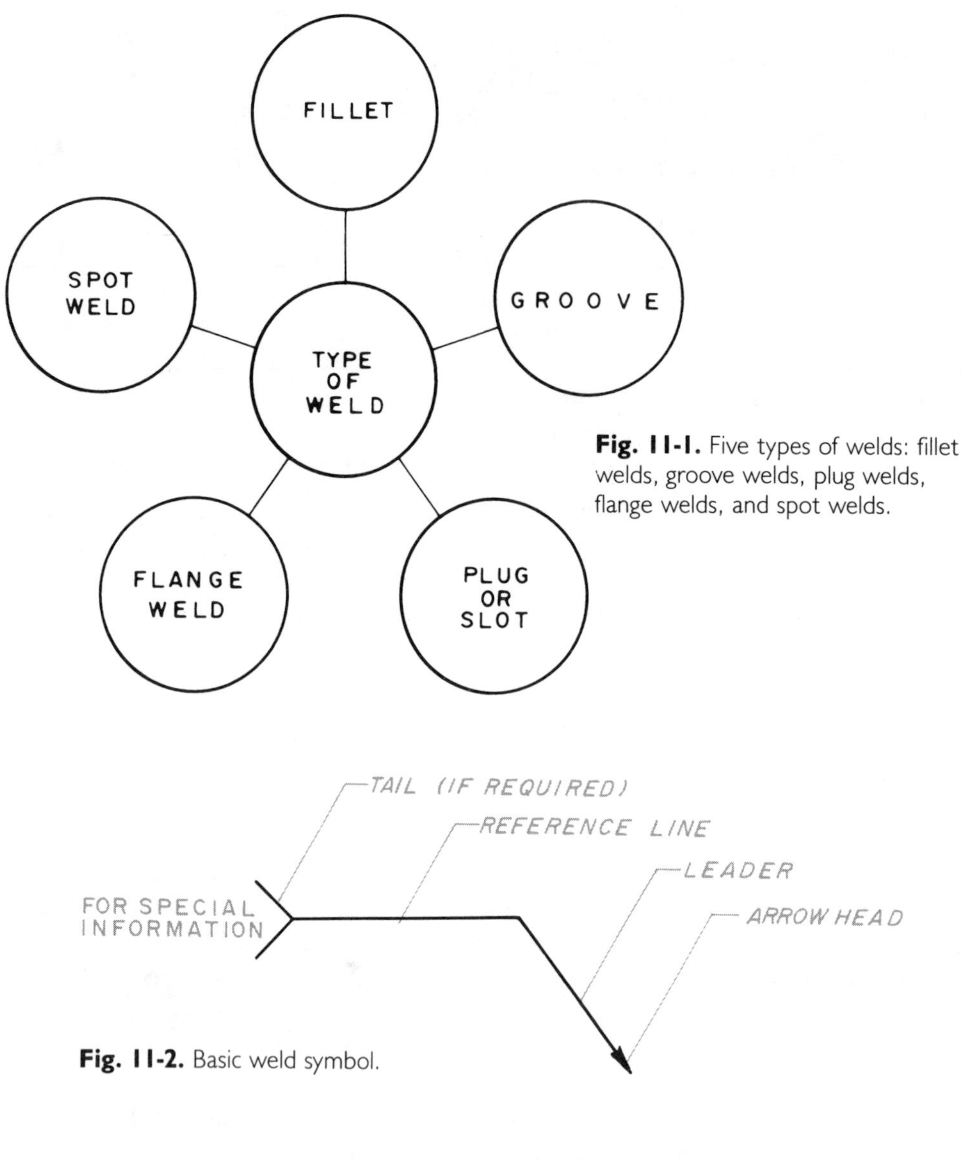

Fig. 11-1. Five types of welds: fillet welds, groove welds, plug welds, flange welds, and spot welds.

Fig. 11-2. Basic weld symbol.

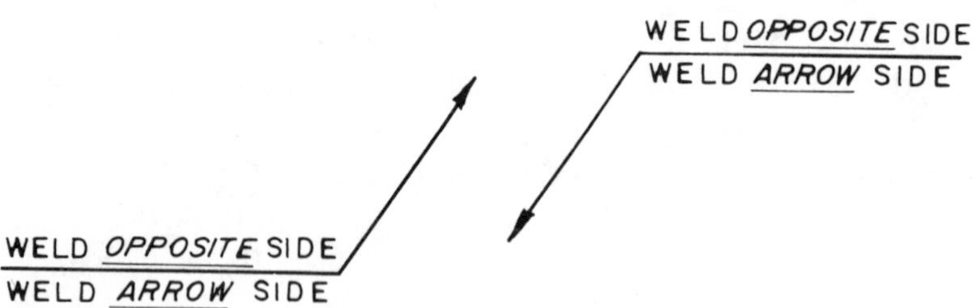

Fig. 11-3. A reference line with a welding symbol added to it.

reference line. The left leg of the welding symbol is always drawn vertically (see FIG. 11-4). Refer back to FIG. 11-3. Again, if the welding symbol is added *above* the reference line it means weld *opposite* side—if *below* the reference line, it means weld *arrow* side (see FIG. 11-5).

FILLET WELD

A *fillet weld* is a weld that fills in between parts (see FIG. 11-6). Given is the fillet weld symbol and examples of how the symbol is used. FIGURE 11-7 indicates a

Fig. 11-4. The left leg of the welding symbol is always drawn vertical.

LEFT LEG OF SYMBOL ALWAYS VERTICAL

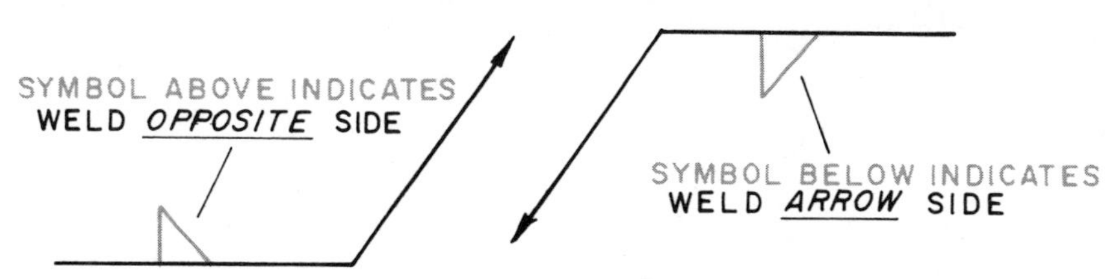

Fig. 11-5. If the welding symbol is *below* the reference line, weld *arrow* side.

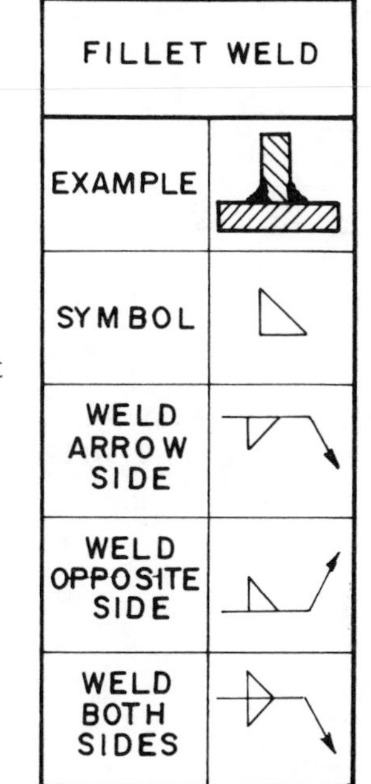

Fig. 11-6. Fillet welds are welds that *fill* in between parts.

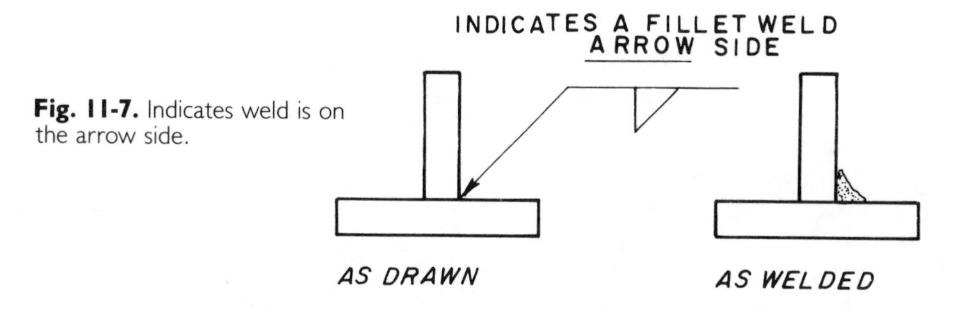

Fig. 11-7. Indicates weld is on the arrow side.

fillet weld on the arrow side; FIG. 11-8 indicates a fillet weld on the opposite side of the arrow. In the event a fillet weld is required on both sides, it is indicated as shown in FIG. 11-9.

Many times welds are dimensioned so as to give its overall size. FIGURE 11-10 indicates a fillet weld, welded on the opposite side with each side of the weld .25 inch in size. If the weld is to be on both sides, it would be dimensioned as in FIG. 11-11. If the weld length is not given it is assumed to be full-length. When a weld is not continuous, extra dimensions are added to the

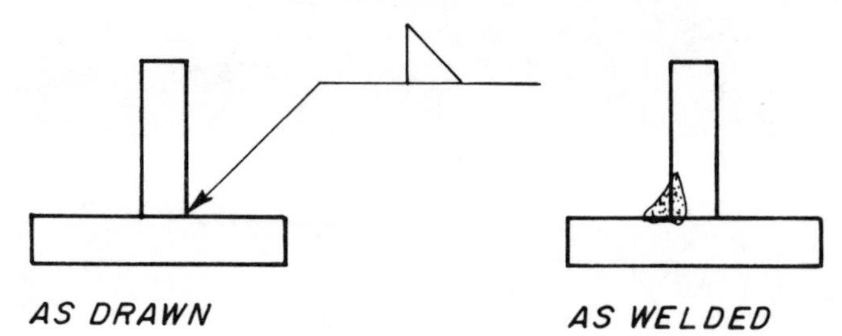

Fig. 11-8. Indicates weld is on opposite side of the arrow.

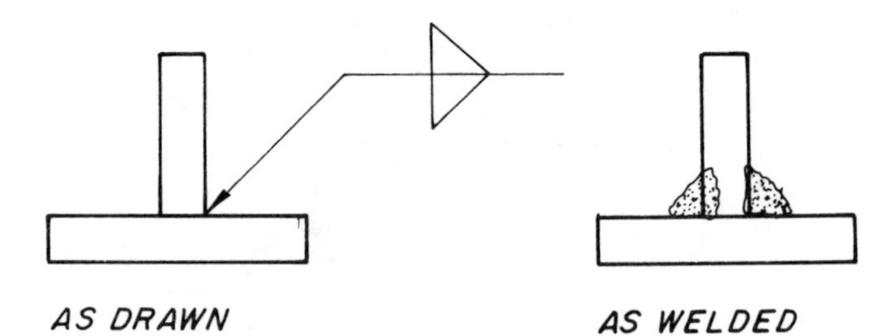

Fig. 11-9. Indicates weld on both sides.

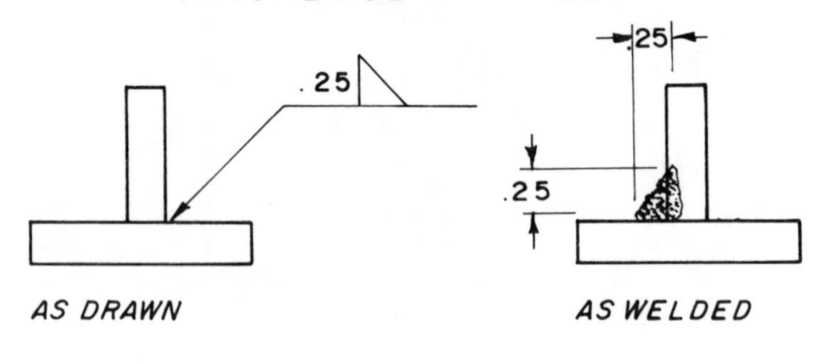

Fig. 11-10. Size of weld is given.

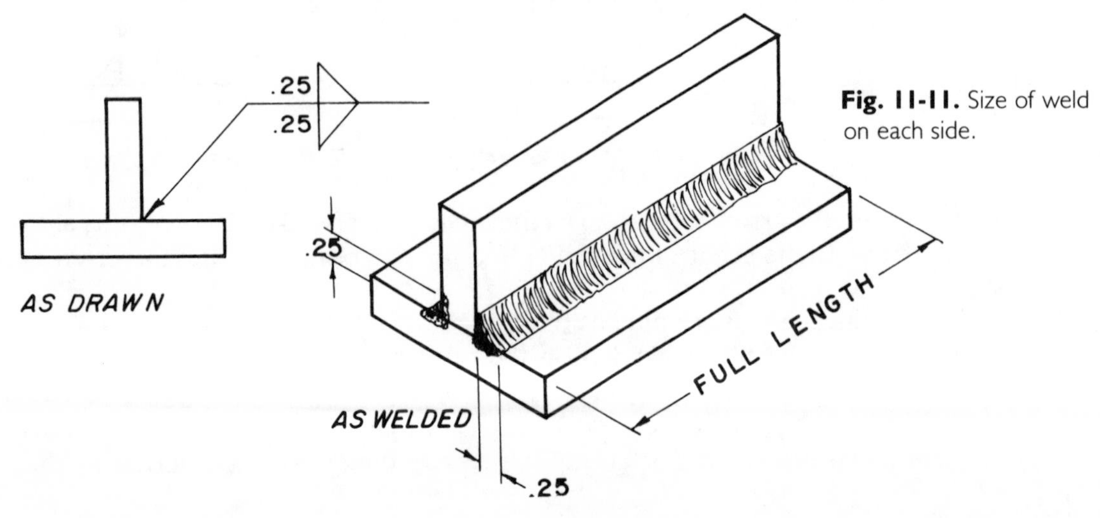

Fig. 11-11. Size of weld on each side.

welding symbol (see FIG. 11-12). In the illustration, the size of weld is given, the type of weld is given, the length of weld is given, and the pitch is given. On both threads and knurls the pitch is measured from one point on the thread or knurl to the *same* point on the next thread or knurl. On welds it is the *center* of one weld to the *center* of the next weld. FIGURE 11-12 shows the required dimensions for the illustration in FIG. 11-13.

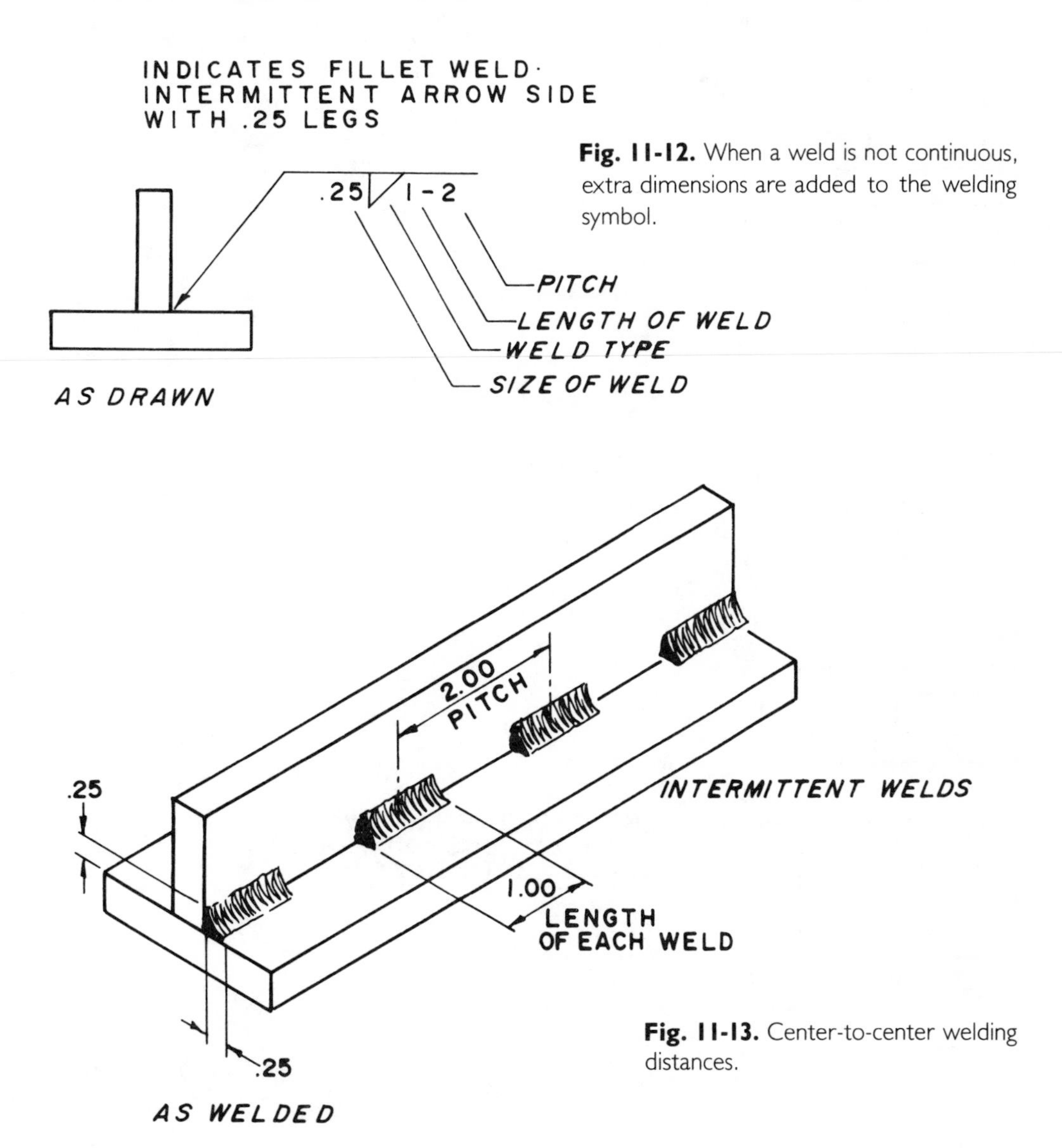

Fig. 11-12. When a weld is not continuous, extra dimensions are added to the welding symbol.

Fig. 11-13. Center-to-center welding distances.

When a weld is to be made continuously all *around* an object, it is indicated on the welding symbol by a small circle between the reference line and the leader line (see FIG. 11-14). If a weld is to be placed only in a few places, it is indicated by the use of multiple leader lines and arrows (see FIG. 11-15).

PROCESS REFERENCE

There are many welding processes developed by the American Welding Society. Each process has a standard abbreviation designation (see TABLE 11-1). Letter designations are used to specify the *method* to be used to obtain a particular

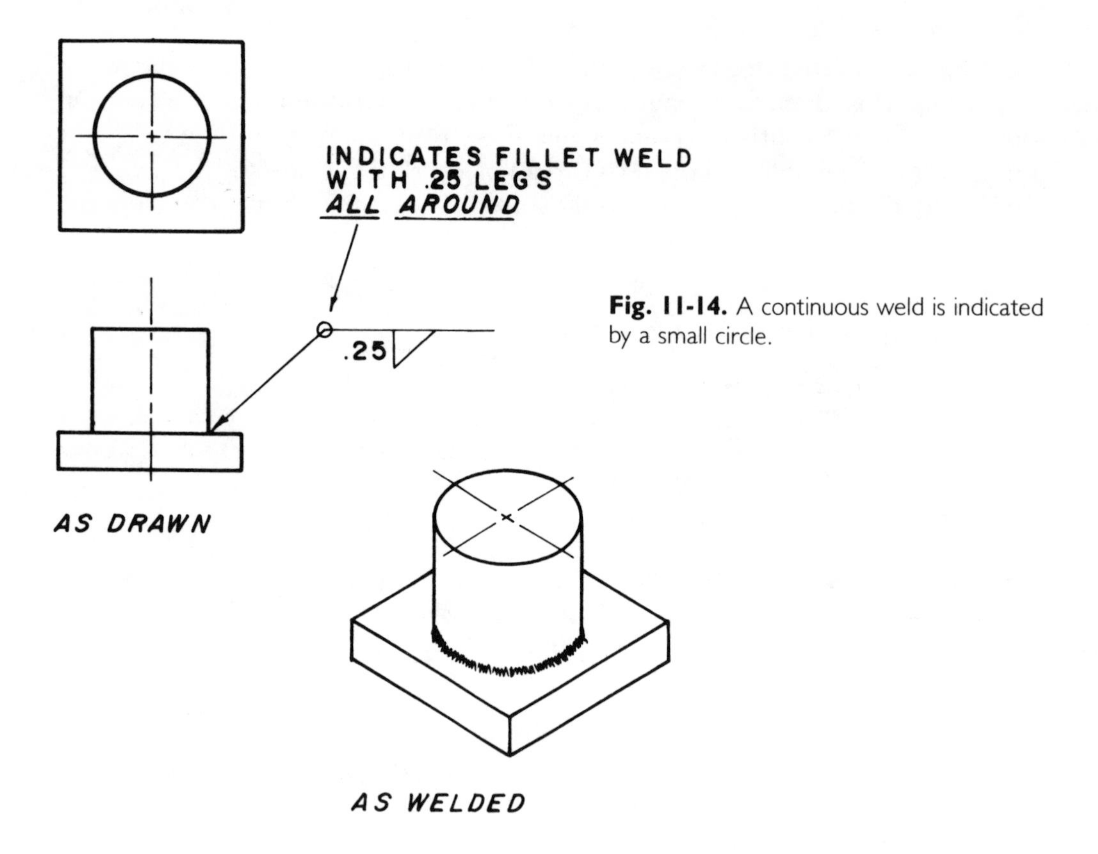

Fig. 11-14. A continuous weld is indicated by a small circle.

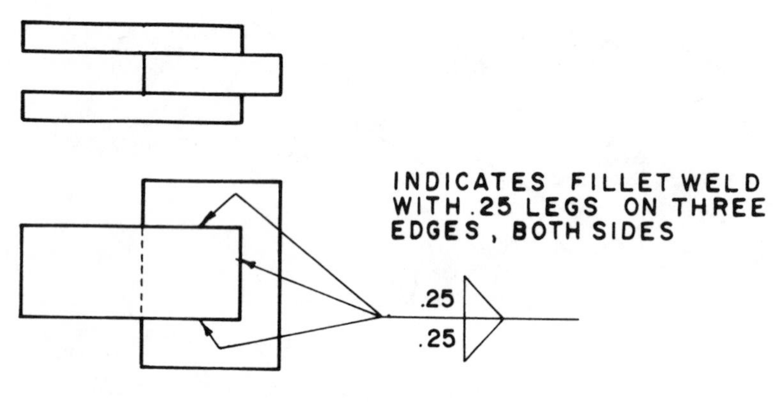

Fig. 11-15. A weld is to be placed only in a few places.

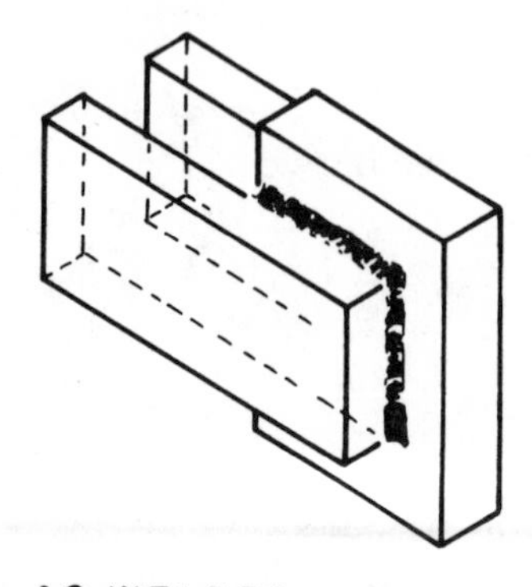

Table 11-1. Standard Welding Abbreviations.

Letter designation	Welding and allied processes	Letter designation	Welding and allied processes
AAC	air carbon arc cutting	GTAW	gas tungsten arc welding
AAW	air acetylene welding	GTAW-P	gas tungsten arc welding—pulsed arc
ABD	adhesive bonding	HFRW	high frequency resistance welding
AB	arc brazing	HPW	hot pressure welding
AC	arc cutting	IB	induction brazing
AHW	atomic hydrogen welding	INS	iron soldering
AOC	oxygen arc cutting	IRB	infrared brazing
AW	arc welding	IRS	infrared soldering
B	brazing	IS	induction soldering
BB	block brazing	IW	induction welding
BMAW	bare metal arc welding	LBC	laser beam cutting
CAC	carbon arc cutting	LBW	laser beam welding
CAW	carbon arc welding	LOC	oxygen lance cutting
CAW-G	gas carbon arc welding	MAC	metal arc cutting
CAW-S	shielded carbon arc welding	OAW	oxyacetylene welding
CAW-T	twin carbon arc welding	OC	oxygen cutting
CW	cold welding	OFC	oxyfuel gas cutting
DB	dip brazing	OFC-A	oxyacetylene cutting
DFB	diffusion brazing	OFC-H	oxyhydrogen cutting
DFW	diffusion welding	OFC-N	oxynatural gas cutting
DS	dip soldering	OFC-P	oxypropane cutting
EASP	electric arc spraying	OFW	oxyfuel gas welding
EBC	electron beam cutting	OHW	oxyhydrogen welding
EBW	electron beam welding	PAC	plasma arc cutting
ESW	electroslag welding	PAW	plasma arc welding
EXW	explosion welding	PEW	percussion welding
FB	furnace brazing	PGW	pressure gas welding
FCAW	flux cored arc welding	POC	metal powder cutting
FCAW-EG	flux cored arc welding—electrogas	PSP	plasma spraying
FLB	flow brazing	RB	resistance brazing
FLOW	flow welding	RPW	projection welding
FLSP	flame spraying	RS	resistance soldering
FOC	chemical flux cutting	RSEW	resistance seam welding
FOW	forge welding	RSW	resistance spot welding
FRW	friction welding	ROW	roll welding
FS	furnace soldering	RW	resistance welding
FW	flash welding	S	soldering
GMAC	gas metal arc cutting	SAW	submerged arc welding
GMAW	gas metal arc welding	SAW-S	series submerged arc welding
GMAW-EG	gas metal arc welding—electrogas	SMAC	shielded metal arc cutting
GMAW-P	gas metal arc welding-pulsed arc	SMAW	shielded metal arc welding
GMAW-S	gas metal arc welding—short circuiting arc	SSW	solid state welding
		SW	stud arc welding
GTAC	gas tungsten arc cutting	TB	torch brazing
Automatic	AU	Manual	MA
Machine	ME	Semiautomatic	SA

contour of a weld (see FIG. 11-16). These abbreviations and letters are added to the welding symbol in order to specify all required information (see FIGS. 11-17 and 11-18). (*Note.* The *process* is added inside the *tail* of the basic welding symbol, and the *method* is added *below* the welding symbol.)

FIELD WELDS

Any weld that is not made in the factory and is to be made at a later date, perhaps at the final assembly of the product, is called a *field weld*. A field weld is indicated by a filled-in flag, added to the basic weld symbol and located between the reference line and the leader line (see FIG. 11-19).

FINISHING METHOD SYMBOLS

LETTER	METHOD
C	CHIPPING
G	GRINDING
H	HAMMERING
R	ROLLING
M	MACHINING

Fig. 11-16. Method to be used to obtain a particular contour of a weld.

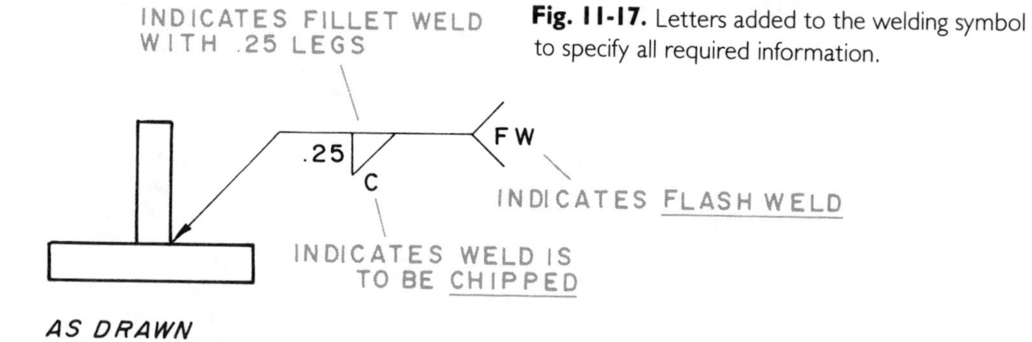

Fig. 11-17. Letters added to the welding symbol to specify all required information.

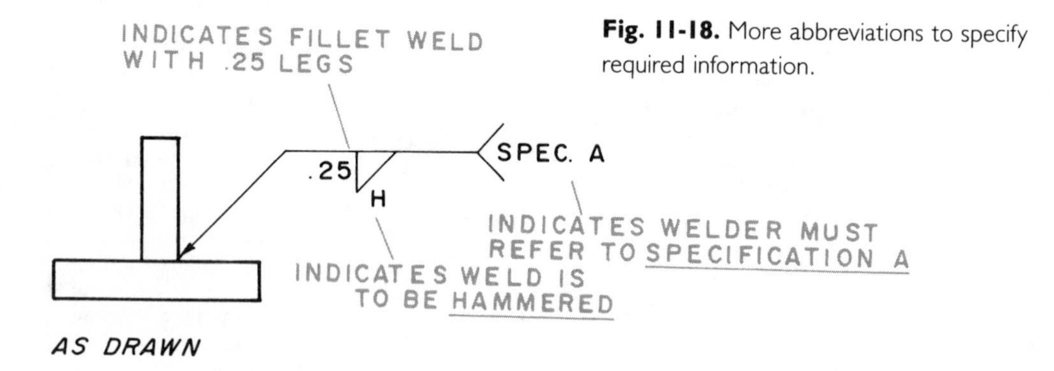

Fig. 11-18. More abbreviations to specify required information.

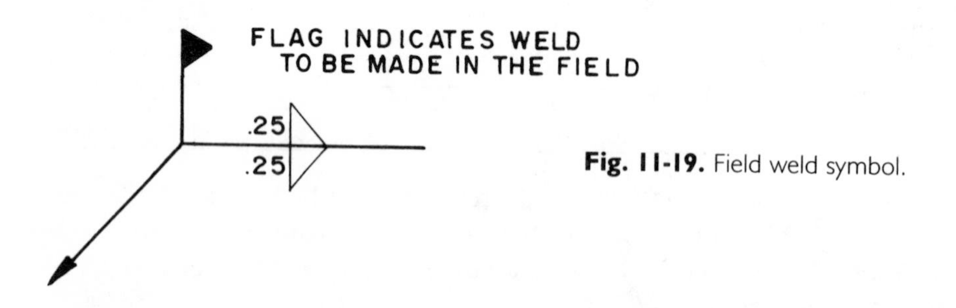

Fig. 11-19. Field weld symbol.

WELDING JOINTS

There are five kinds of welded joints, classified according to the position of the parts that are being joined: *Butt joint, corner joint,* T-*joint, lap joint, and edge joint* (see FIGS. 11-20 through 11-24).

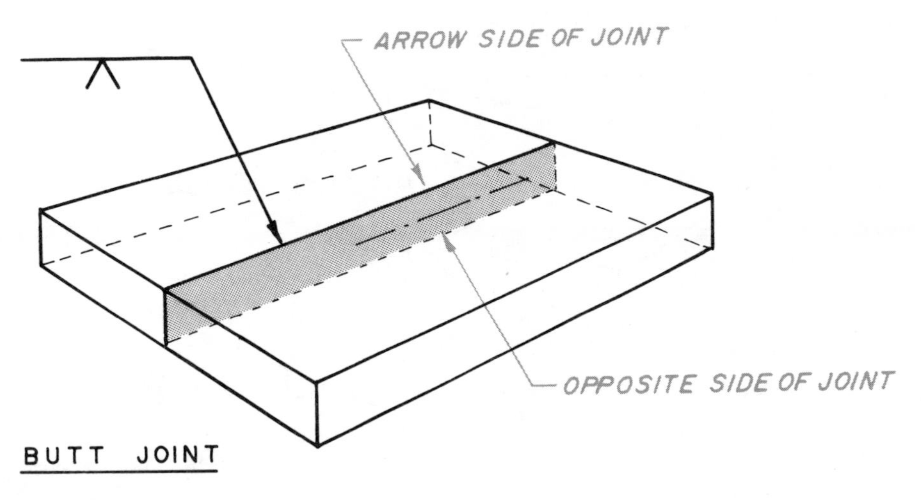

Fig. 11-20. Butt joint.

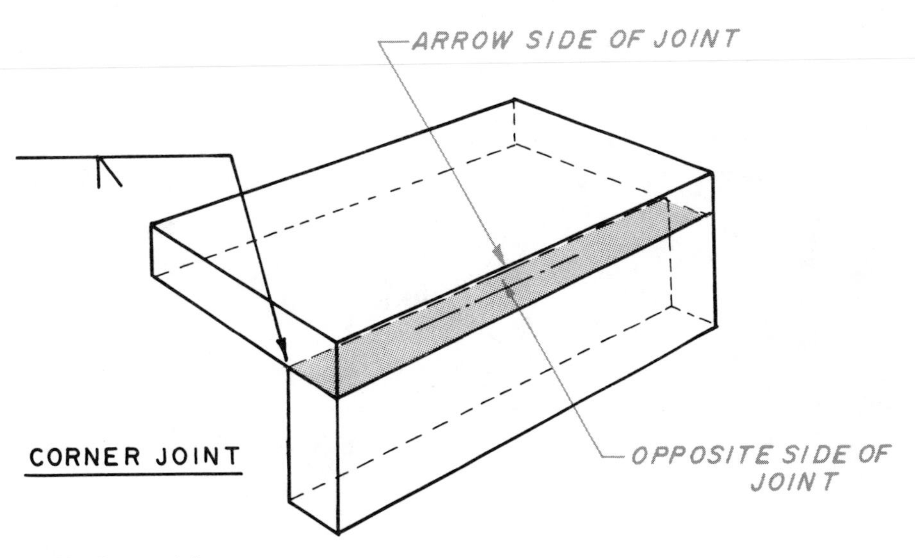

Fig. 11-21. Corner joint.

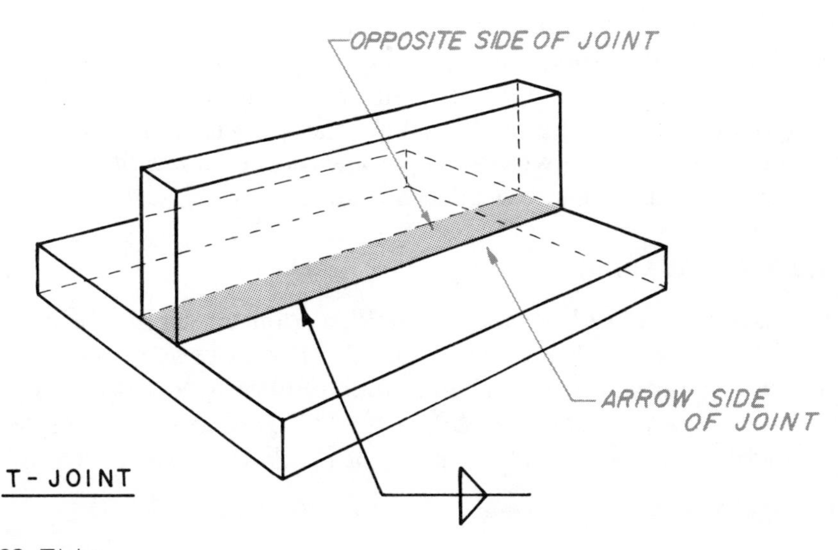

Fig. 11-22. T-joint.

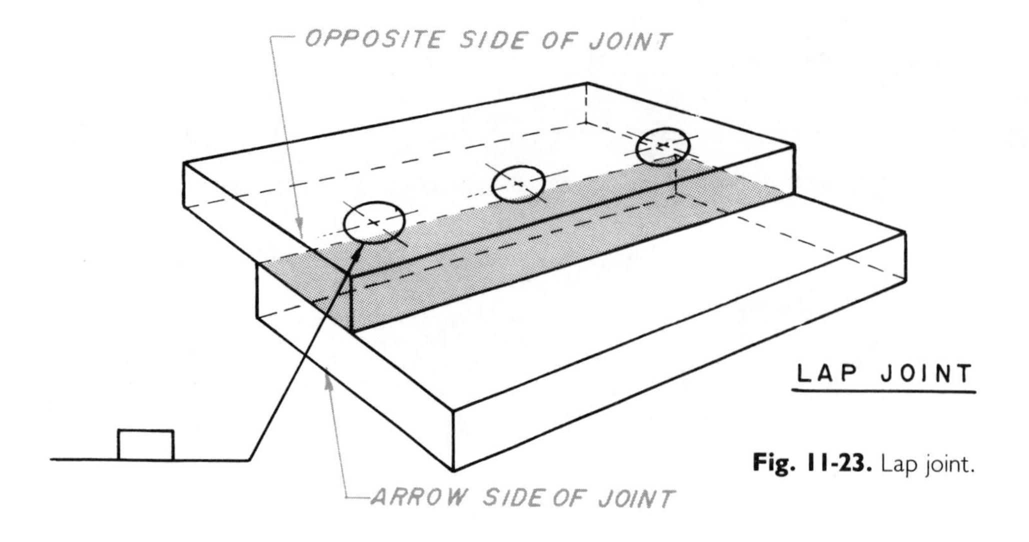

Fig. 11-23. Lap joint.

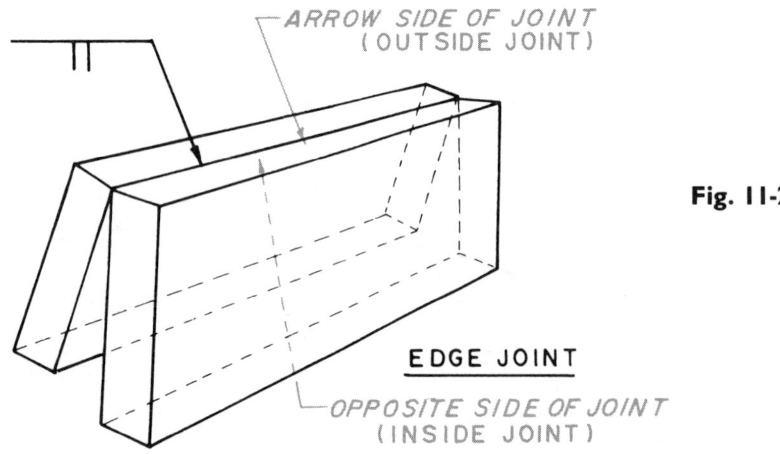

Fig. 11-24. Edge joint.

TYPES OF GROOVE WELD JOINTS

There are six major kinds of groove weld joints: *square groove, V-groove, bevel groove, U-groove, J-groove,* and *flare bevel groove welds* (see FIG. 11-25). Refer to the illustration, study each example, noting its symbol, and how each is added to the welding symbol. Sometimes the *same* type of weld is used on the opposite side of a single weld. Such a joint is called a *double weld*—for example, a double square-groove weld or a double V-groove weld.

CONTOUR SYMBOL

If a special finish must be made to a weld, a contour symbol must be added to the welding symbol. There are three kinds of contour symbols: *flush, convex,* and *concave* (see FIG. 11-26). The contour symbol is added directly above or below the welding symbol (see FIG. 11-27). If the contour must be *finished*, the microinch requirement is added above or below the contour symbol (see FIG. 11-28).

PLUG OR SLOT WELD

Plug or *slot welds* use the same welding symbol. The only difference between them is the shape of the hole through which the actual weld is applied. A plug

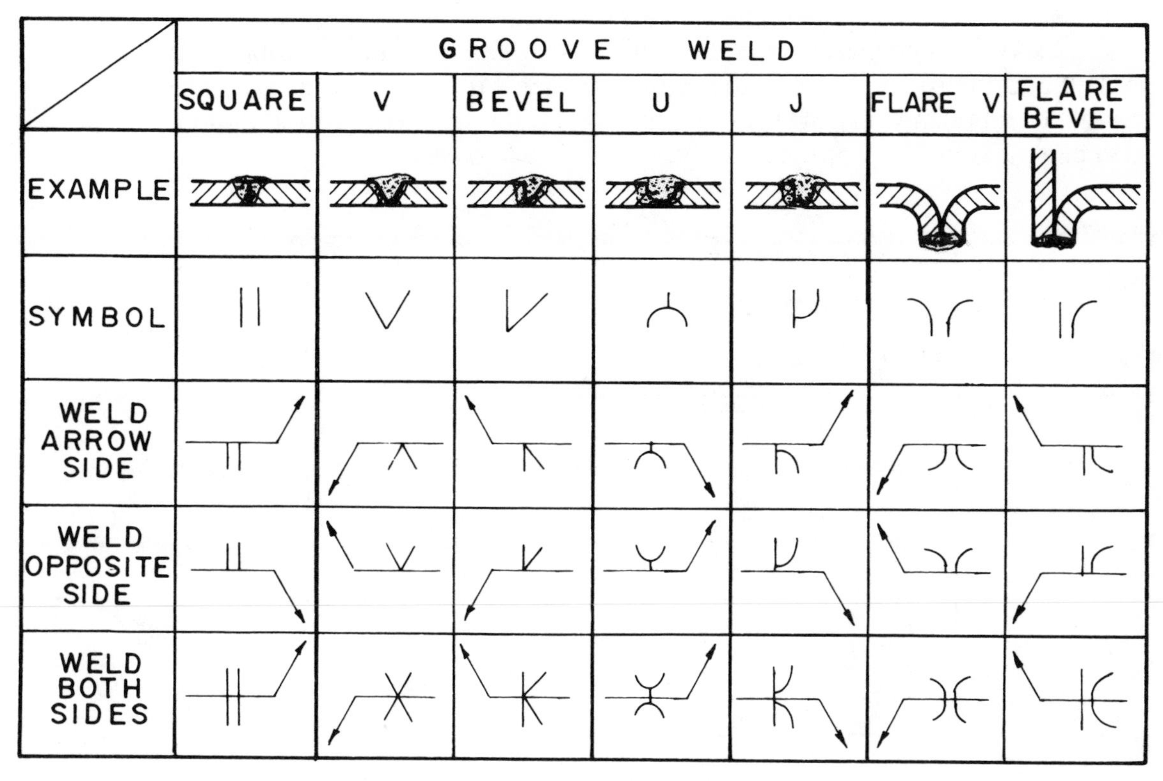

Fig. 11-25. Six major kinds of groove weld joints.

Fig. 11-26. Contour symbols: flush, convex, concave.

FLUSH CONVEX CONCAVE

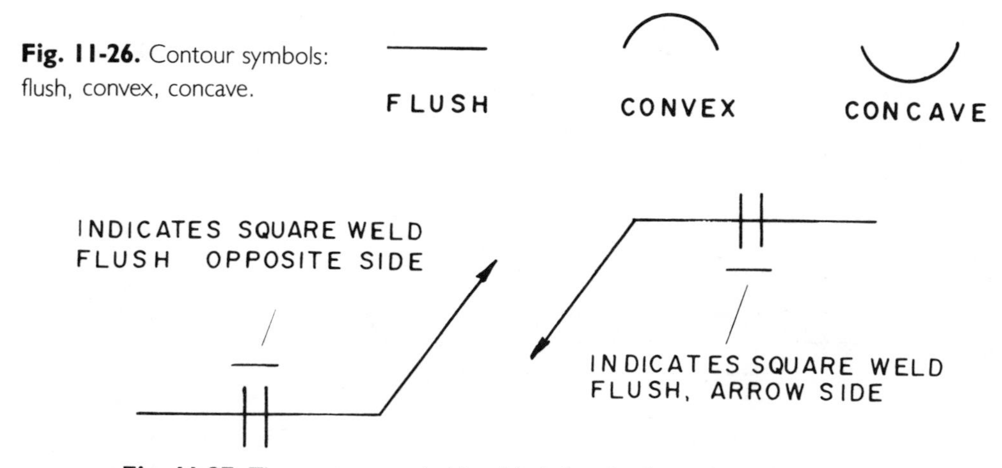

Fig. 11-27. The contour symbol is added directly above the welding symbol.

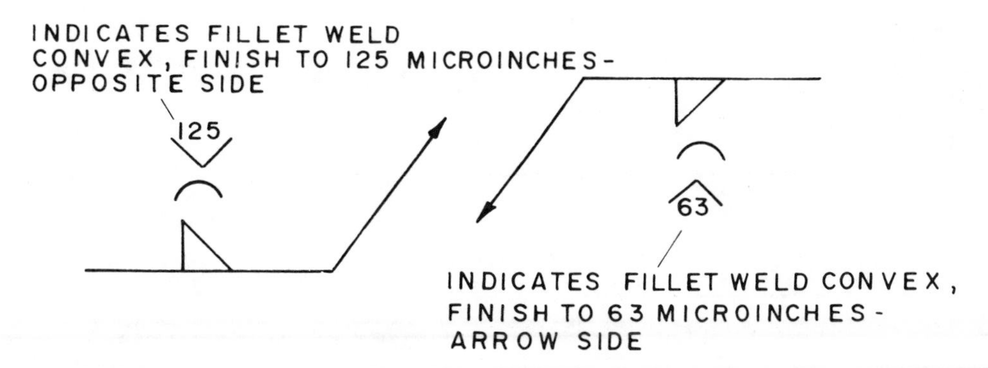

Fig. 11-28. If the contour must be finished, the microinch requirement is added above or below the contour symbol.

weld uses a round hole, the slot weld uses an elongated hole. Both symbols are applied and interpreted exactly the same way (see FIG. 11-29). Again, note the example symbol and how the symbol is applied to the welding symbol. FIGURE 11-30 illustrates how the plug or slot weld is drawn and welded.

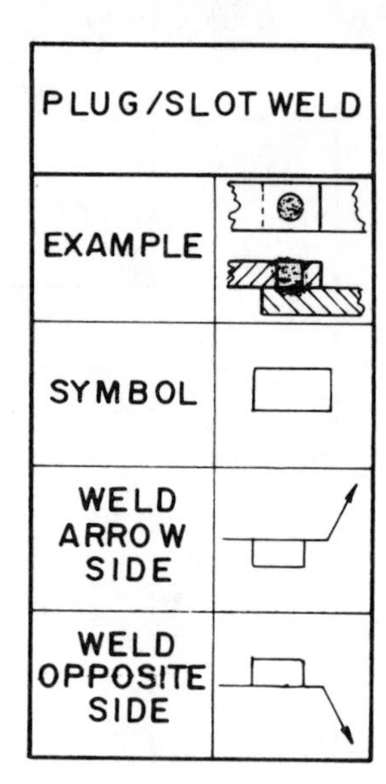

Fig. 11-29. The plug weld uses a round hole, the slot weld uses an elongated hole.

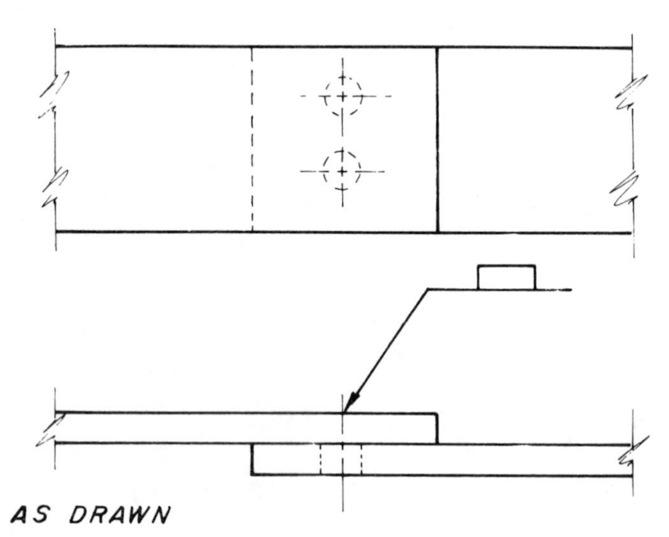

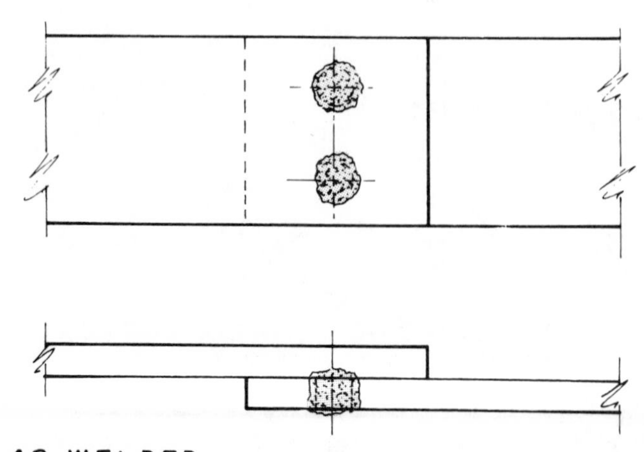

Fig. 11-30. How plug or slot welds are drawn.

FLANGE WELD

Flange welds are used to join thin parts, such as sheet metal parts, together. Instead of applying the weld to the surface of the parts, the weld is applied to the edges of the parts, as a weld to thin parts could cause a burn-through. FIGURE 11-31 illustrates the flange weld, an example of its use, and its symbol. There are two kinds of flange welds: a flange weld and a corner flange weld (see FIG. 11-32).

	FLANGE WELD	
	EDGE	CORNER
EXAMPLE		
SYMBOL		
WELD ARROW SIDE		
WELD OPPOSITE SIDE		

Fig. 11-31. Flange weld and symbol.

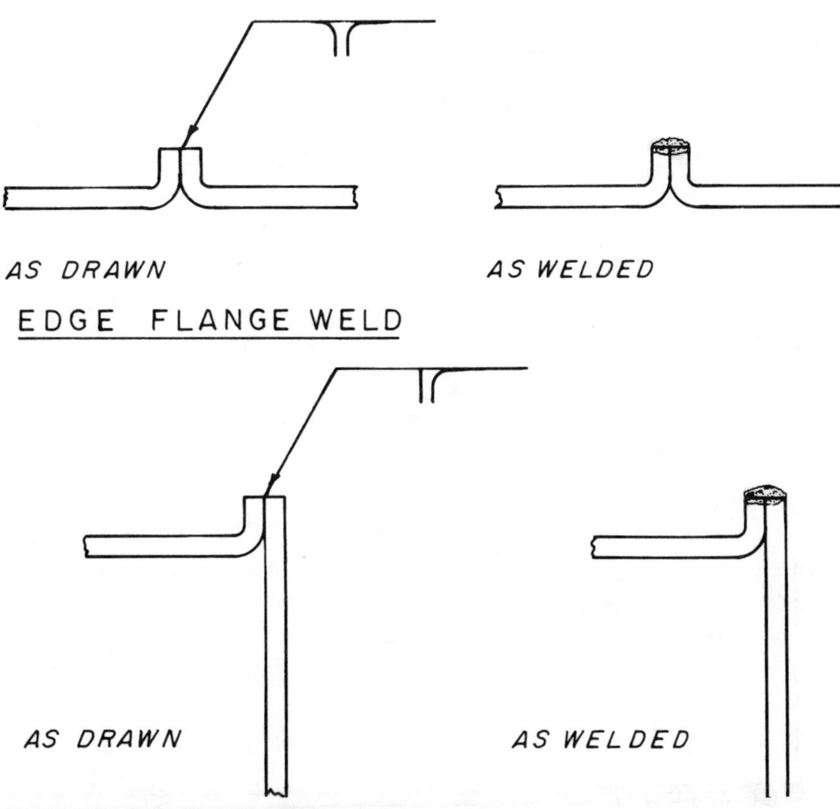

Fig. 11-32. Two kinds of flange welds.

MULTIPLE REFERENCE LINES

If there is more than one operation on a particular joint, each operation is noted on its own reference line (see FIG. 11-33). The first reference line *closest* to the arrow is for the *first* operation, the second reference line closest to the arrow is for the second operation or for supplementary information, and the third reference line is for the third operation or for test information.

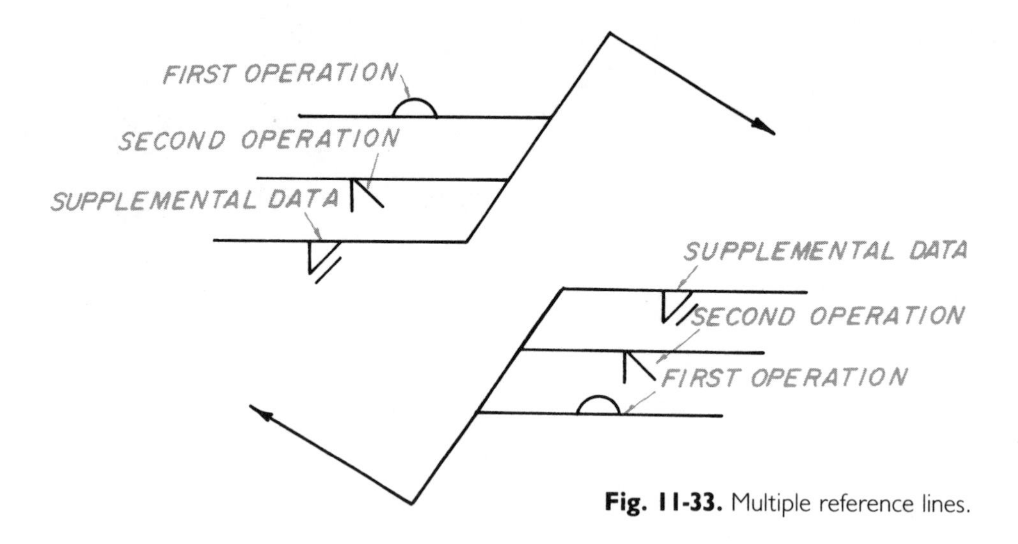

Fig. 11-33. Multiple reference lines.

SPOT WELD

Spot welding is a resistance welding process done by passing an electric current through the exact location where the parts are to be joined. Spot welding is usually done on thin pieces such as sheet metal. Two kinds of resistance welding processes are illustrated in this text: the spot weld and the seam weld (see FIG. 11-34). Note the given examples, the weld symbols, and how each is used.

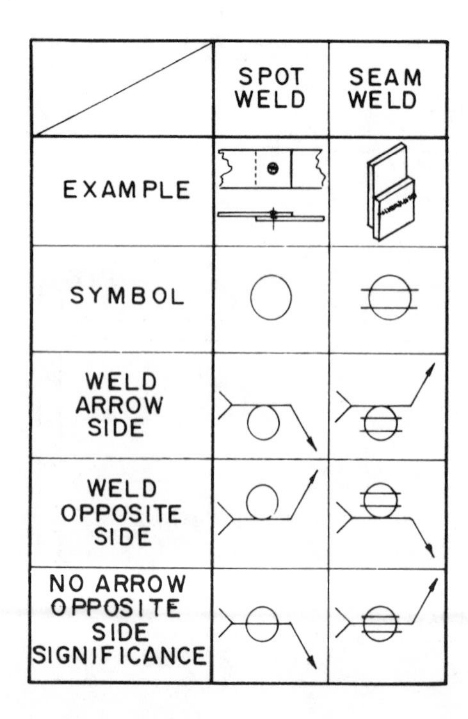

Fig. 11-34. Spot welds.

The symbol of the spot weld is illustrated in FIG. 11-35. It is placed above, below, or on the horizontal reference line as any welding symbol (see FIG. 11-36). Note that the *tail* is always added to a spot weld symbol.

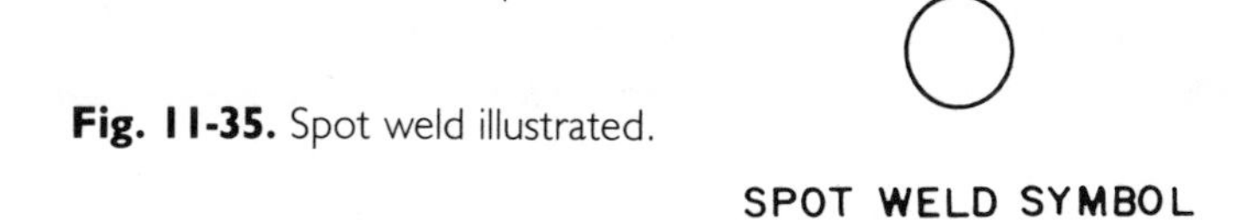

Fig. 11-35. Spot weld illustrated.

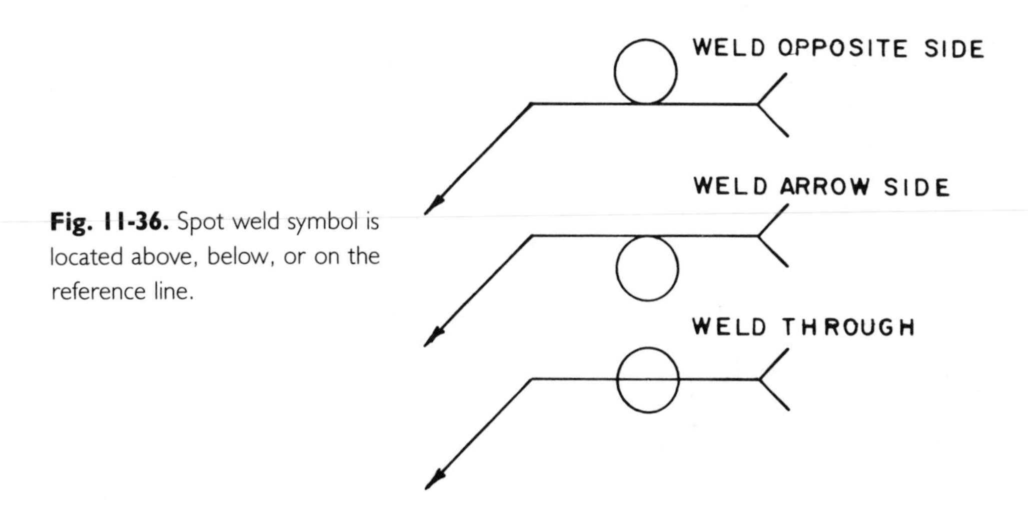

Fig. 11-36. Spot weld symbol is located above, below, or on the reference line.

A spot weld is dimensioned one of two ways: by diameter or by minimum acceptable strength *per spot*. If it is to be dimensioned by the diameter of each spot weld, it is dimensioned per FIG. 11-37. If it is to be dimensioned by

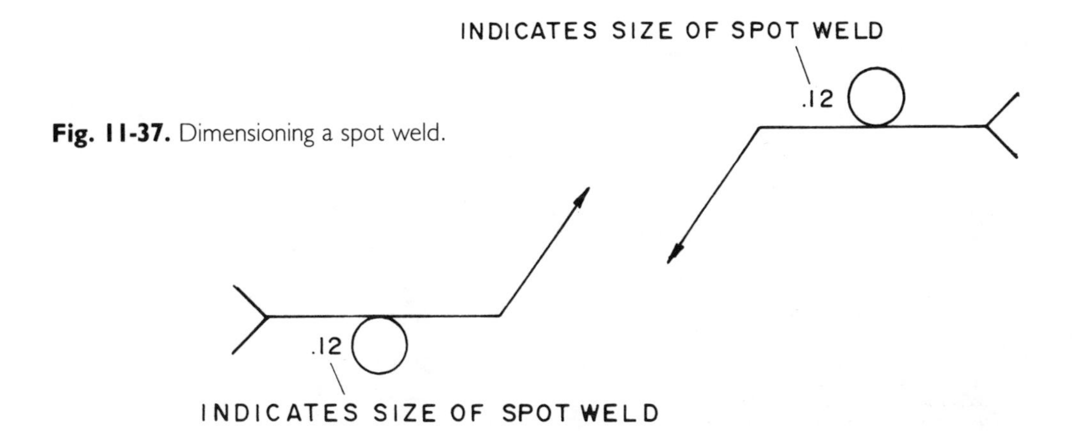

Fig. 11-37. Dimensioning a spot weld.

strength, it is dimensioned per FIG. 11-38. If there are to be a string of spot welds, they are dimensioned per FIG. 11-39. The dimension directly to the *left* of the weld symbol indicates the diameter, the dimension directly to the *right* of the weld symbol indicates the center-to-center distance of the welds. The reference number *below* the welding symbol notes how many spot welds are required.

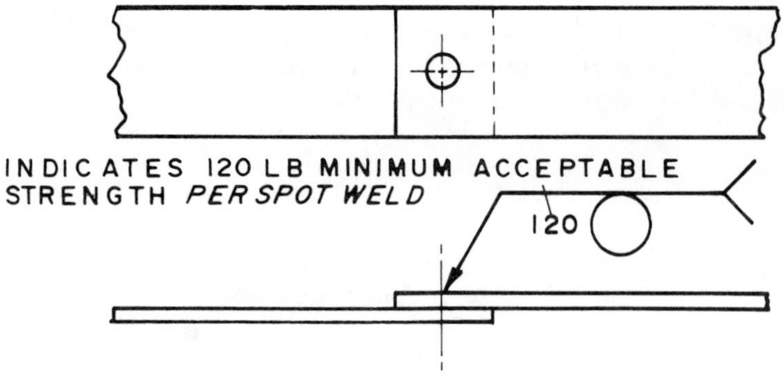

INDICATES 120 LB MINIMUM ACCEPTABLE STRENGTH *PER SPOT WELD*

120

Fig. 11-38. Strength of a spot weld is noted.

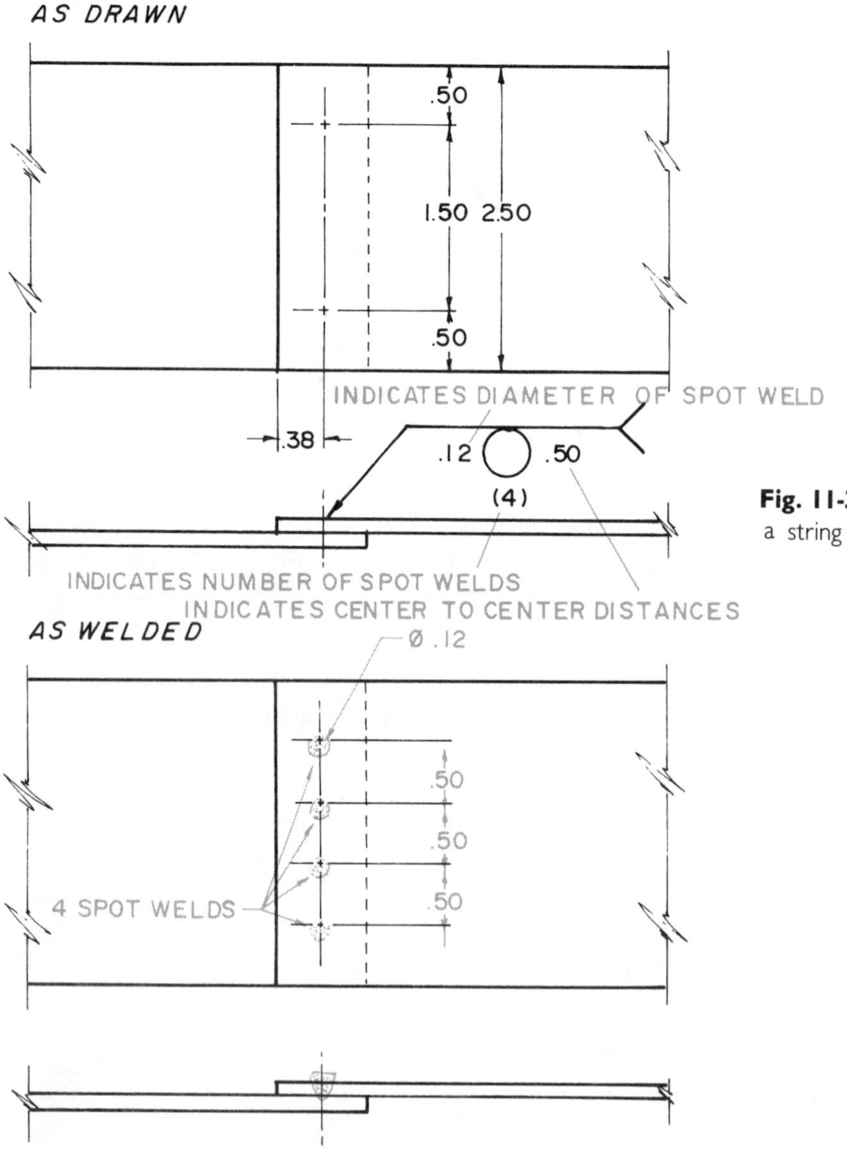

AS DRAWN

.50

1.50 2.50

.50

INDICATES DIAMETER OF SPOT WELD

.38

.12 .50

(4)

INDICATES NUMBER OF SPOT WELDS

INDICATES CENTER TO CENTER DISTANCES

AS WELDED

Ø .12

.50

.50

.50

4 SPOT WELDS

Fig. 11-39. Dimensioning a string of spot welds.

CONTOUR AND FINISH SYMBOLS

Contour and finish symbols are added to the welding symbol exactly as they are added to the fusion welding symbol (see FIG. 11-40). Note the information inside the tail: RSW. Refer back to FIG. 11-16 for its meaning: resistance spot weld.

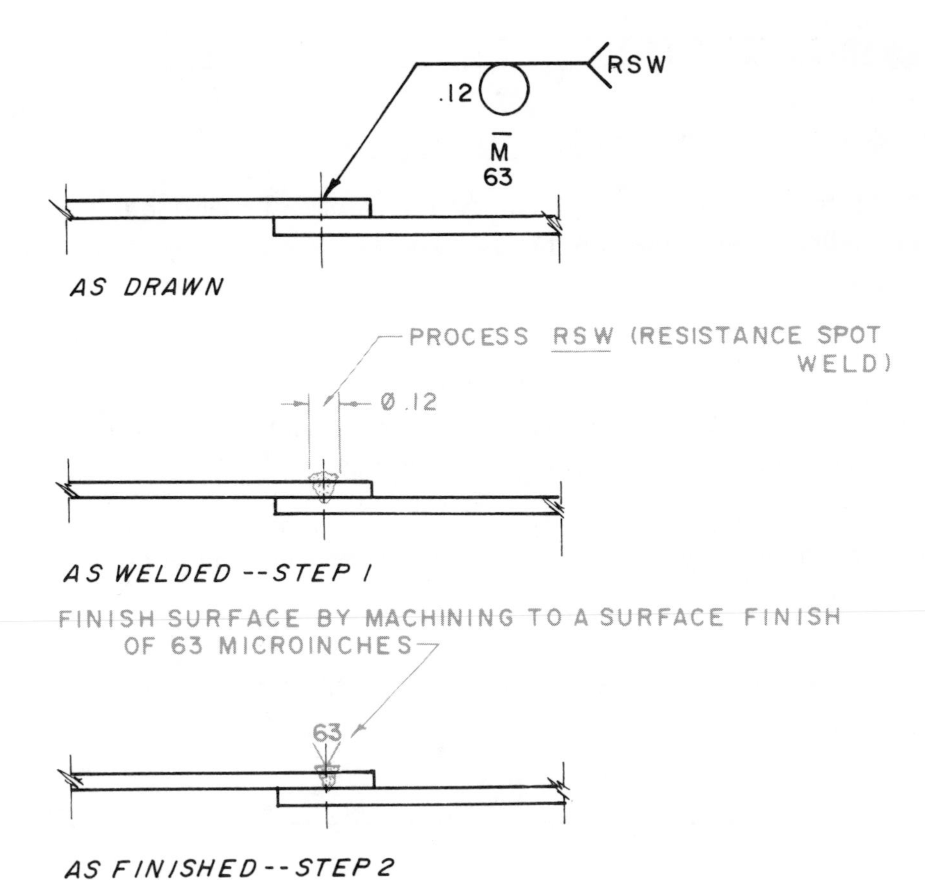

Fig. 11-40. Contour and finish symbols.

SEAM WELD

A *seam weld* is very much like a spot weld except that the weld is continuous from start to finish. The seam weld symbol is a circle with two horizontal lines through it (see FIG. 11-41). FIGURE 11-42 illustrates how the weld symbol is added to the reference line of the welding symbol.

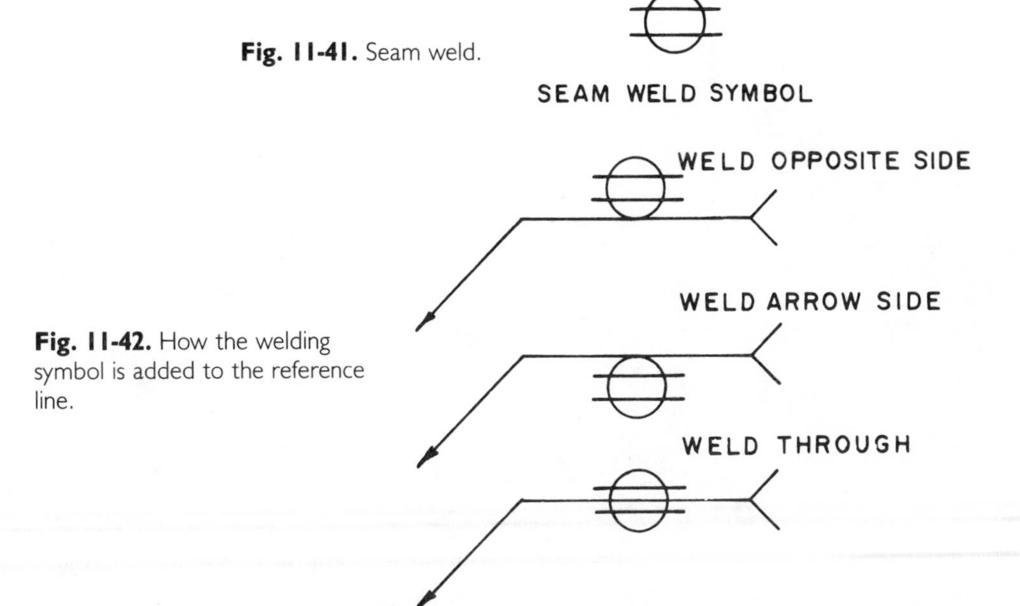

Fig. 11-41. Seam weld.

Fig. 11-42. How the welding symbol is added to the reference line.

WORKSHEET 11-1

Using the illustrations on p. 185, answer the following questions in the spaces provided. (Answers in Appendix A.)

1. Add the required *fillet* weld symbol and required dimension to the reference line. Make it a field weld.

2. Add the *fillet* weld symbol and required information to the reference line.

3. Completely show what the welding symbol should be to indicate an intermittent fillet weld. Add all other required notes and dimensions.

4–7. Add required *groove* weld symbols to the reference line for each kind of joint illustrated.

8. Add the required *spot* weld symbol to the reference line. Be sure to add weld diameter and pitch.

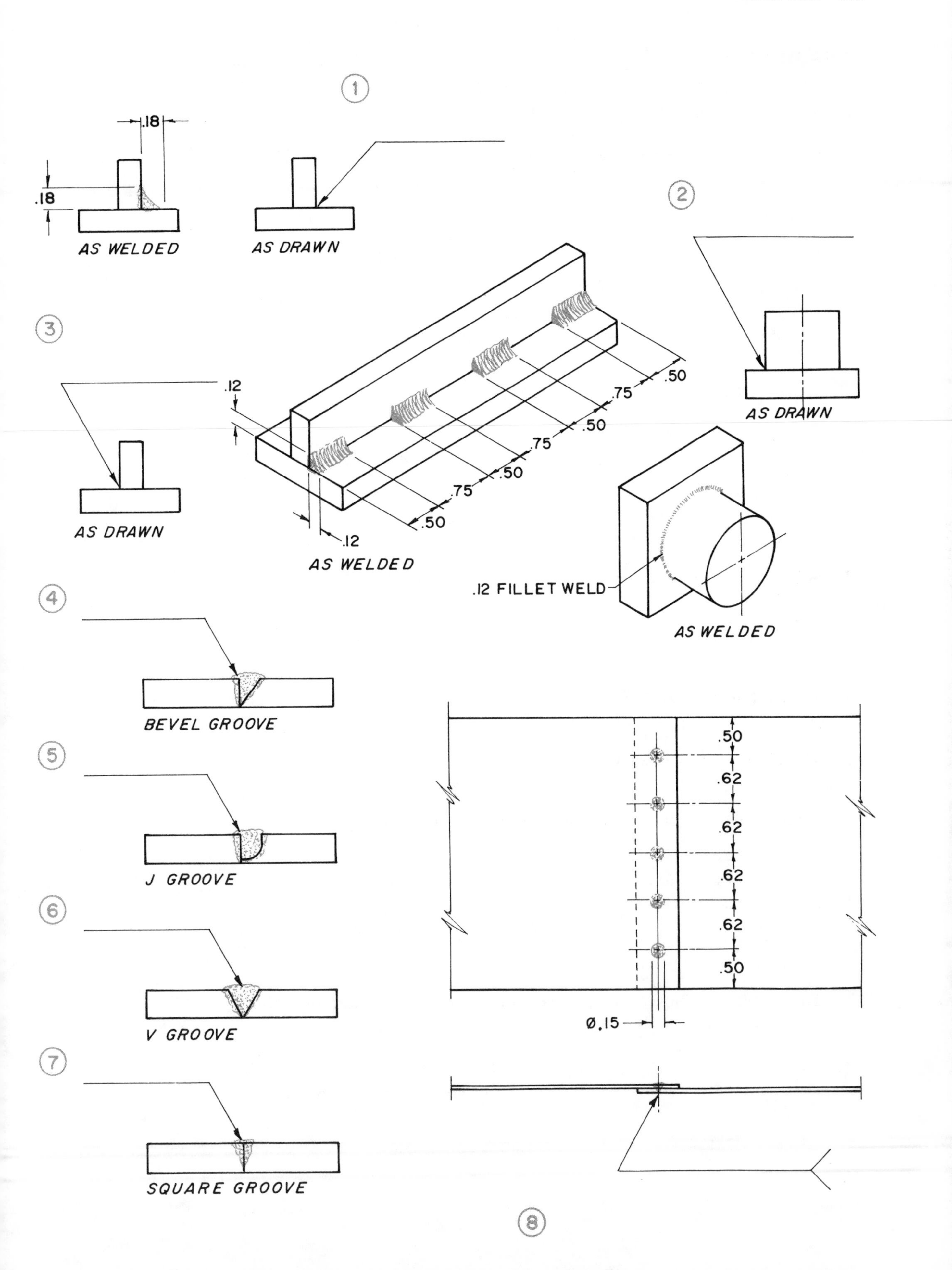
1
.18
.18
AS WELDED
AS DRAWN
2
AS DRAWN
3
.12
.50
.75
.50
.75
.50
.75
.50
.12
AS DRAWN
AS WELDED
.12 FILLET WELD
AS WELDED
4
BEVEL GROOVE
5
J GROOVE
6
V GROOVE
7
SQUARE GROOVE
.50
.62
.62
.62
.62
.50
Ø.15
8

WORKSHEET 11-2

Using drawing A296988 on p. 187, answer the following questions in the spaces provided (Answers in appendix A).

1. What surface in the right-side view is surface A in the top view?
2. Surface W in the right-side view is what surface in the top view?
3. Explain in full what all the information given at L means.
4. Surface K in the front view is what surface in the right-side view?
5. What does the small circle at N indicate?
6. Surface I in the top view is what surface in the front view?
7. Surface R in the front view is what surface in the top view?
8. Surface P in the front view is what surface in the top view?
9. Surface Y in the right-side view is what surface in the top view?
10. Surface E in the top view is what surface in the front view?
11. What kind of a weld is a Q?
12. Surface S in the front view is what surface in the top view?
13. What kind of a weld is at Z?
14. What is the carbon content of the steel used?
15. The welding symbols are *below* the reference lines in most of the welding symbols. What does this indicate?

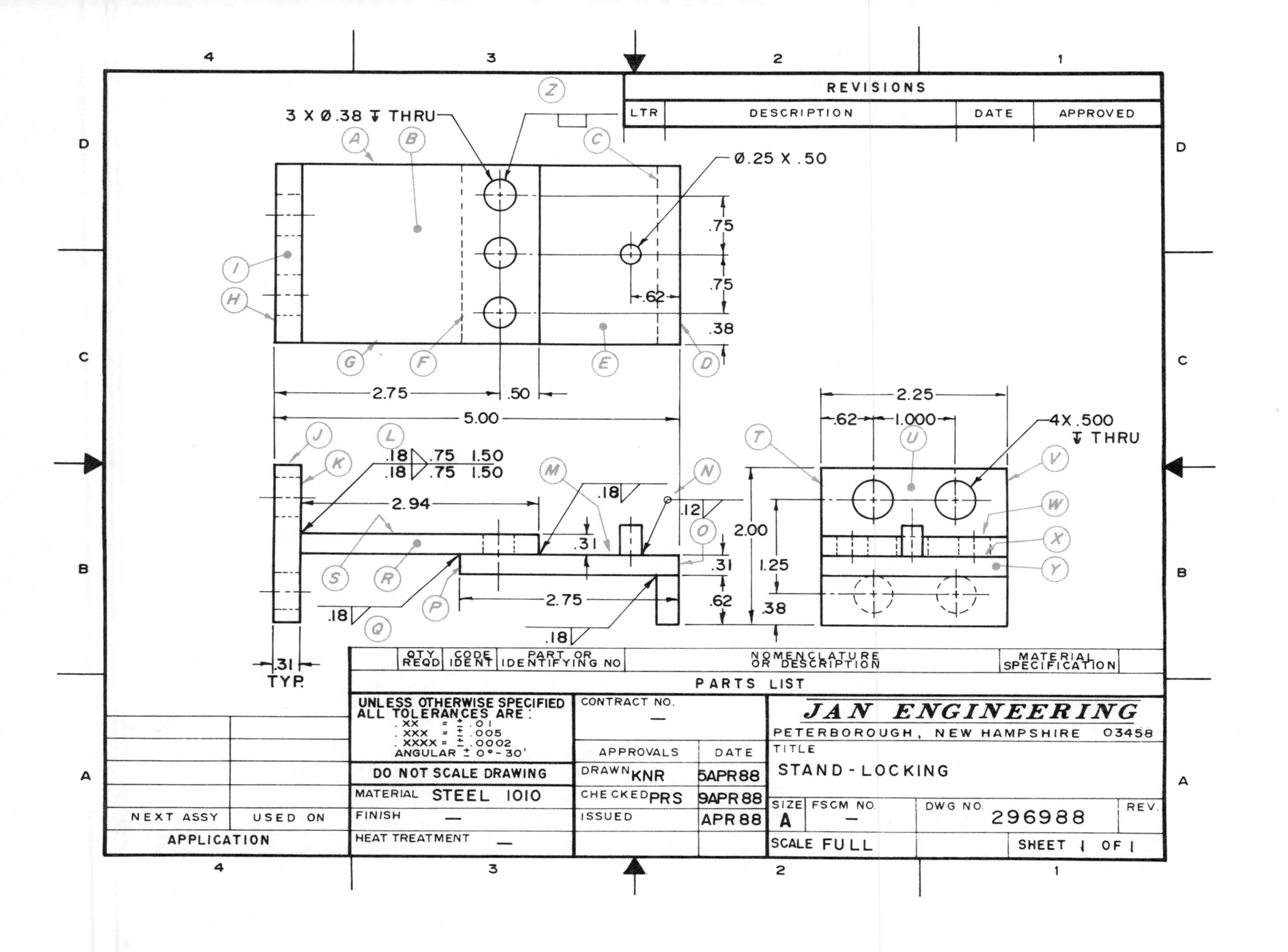
REVISIONS
LTR
DESCRIPTION
DATE
APPROVED
3 X Ø.38 ↧ THRU
Ø.25 X .50
.75
.75
.62
.38
2.75
.50
5.00
.18 .75 1.50
.18 .75 1.50
2.94
.18
.12
.31
.31
.62
2.75
.18
.18
.31
TYP.
2.00
1.25
.38
2.25
.62
1.000
4X .500
↧ THRU
QTY REQD
CODE IDENT
PART OR IDENTIFYING NO
NOMENCLATURE OR DESCRIPTION
MATERIAL SPECIFICATION
PARTS LIST
UNLESS OTHERWISE SPECIFIED ALL TOLERANCES ARE:
.XX = ± .01
.XXX = ± .005
.XXXX = ± .0002
ANGULAR ± 0°-30'
DO NOT SCALE DRAWING
MATERIAL STEEL 1010
FINISH —
HEAT TREATMENT —
CONTRACT NO. —
APPROVALS
DATE
DRAWN KNR 5APR88
CHECKED PRS 9APR88
ISSUED APR 88
JAN ENGINEERING
PETERBOROUGH, NEW HAMPSHIRE 03458
TITLE
STAND-LOCKING
SIZE A
FSCM NO. —
DWG NO. 296988
REV.
SCALE FULL
SHEET 1 OF 1
NEXT ASSY
USED ON
APPLICATION

Geometric Tolerancing

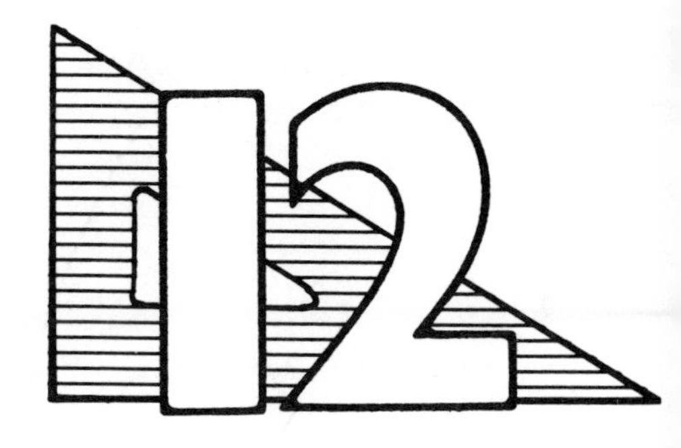

Objective: The reader will know about geometric tolerancing and the various symbols used in industry to call off these tolerances.

Toleranced dimensions as discussed in the preceeding chapters apply only to the *size* of a feature. Geometric tolerancing applies to tolerances of *form, profile, orientation and runout,* and *location.* This is a rather complicated subject and cannot be thoroughly covered in a single chapter; therefore, only highlights of geometric tolerancing will be covered in this book.

All dimensions on a drawing have a tolerance. These tolerances are either noted directly with the dimension, shown as limits, or given in the title block. Remember, as you study geometric tolerancing, dimensions are either *size* dimensions or *location* dimensions.

TOLERANCING

As a review, a tolerance is the total amount of size or location allowed that a dimension may vary (see FIG. 12-1). In the illustration, the *maximum* size (upper limit) could be .510 diameter and the *minimum* size (lower limit) could be .490 diameter.

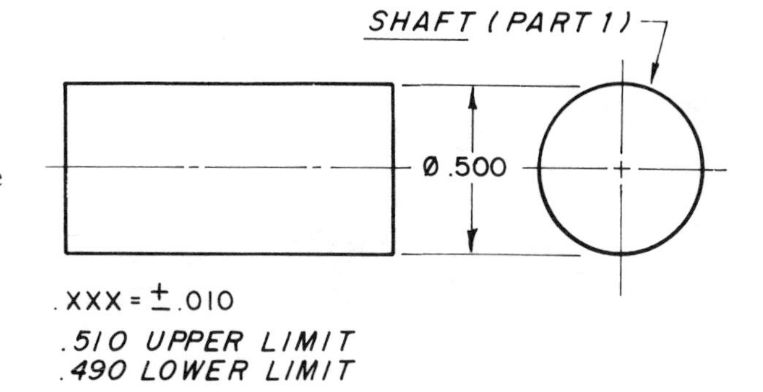

Fig. 12-1. Tolerancing or variable allowed a dimension may vary.

Design Size

The *design size* is the size from which the limits are calculated. In FIG. 12-1, .500 diameter is the design size.

Maximum Material Condition (MMC)

Maximum material condition is the size of the part when it contains the *most material.* This is referred to as MMC. Refer back to FIG. 12-1. When the shaft is as large as its tolerance will allow, .510 diameter, it is said to be at MMC, or maximum material condition.

The MMC of a hole is just the opposite of that of a shaft; when a *hole* is at MMC it is the *smallest* hole (see FIG. 12-2). In this example, the MMC of the hole is the lower limit, that is, .515 diameter.

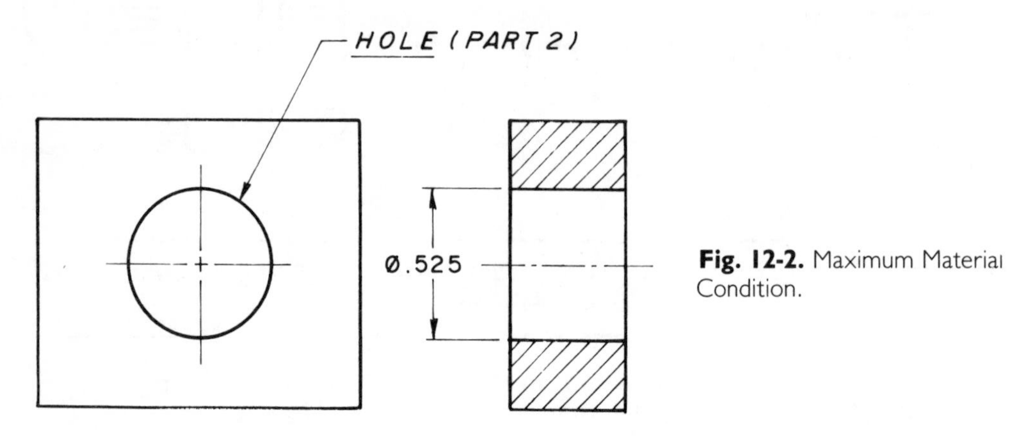

Fig. 12-2. Maximum Material Condition.

Allowance

Allowance is the *tightest fit* allowed for mating parts. It, too, is determined using MMC (see FIG. 12-3). The MMC of the hole is .515 diameter, and the MMC of the shaft is .510. .515 minus .510 equals .005 *allowance,* or the tighest fit.

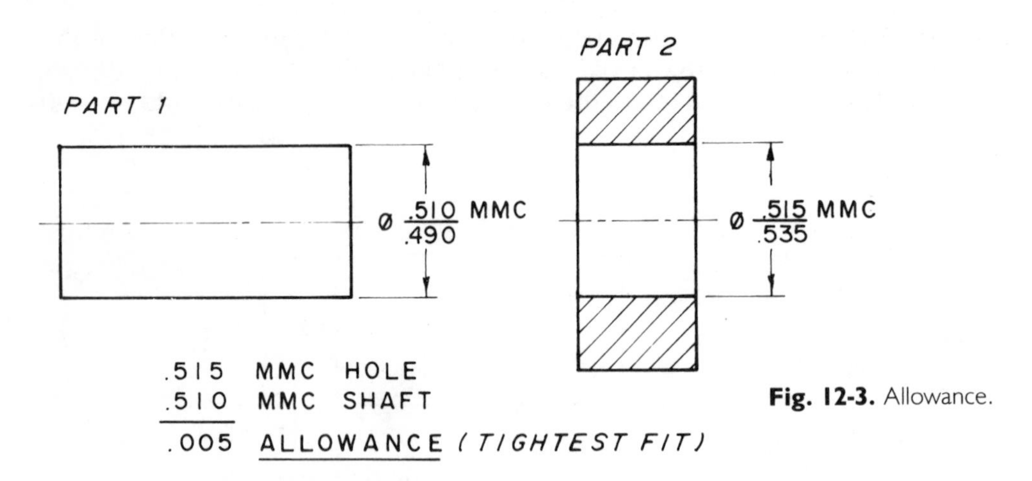

Fig. 12-3. Allowance.

Least Material Condition (LMC)

The *least material conditions* (LMC) is the opposite of maximum material condition (MMC). Refer back to FIGS. 12-1 and 12-2. The LMC of the *shaft* is when it is the *smallest* size, or .490 diameter. The LMC of the *hole* is when it is the *largest* size or .535 diameter.

Clearance

Clearance is the term given when mating parts are at the *loosest* fit. It is the exact opposite of allowance. In calculating clearance, LMC is used in place of MMC. In FIG. 12-4, the clearance between part 1 and 2 is the LMC of the hole

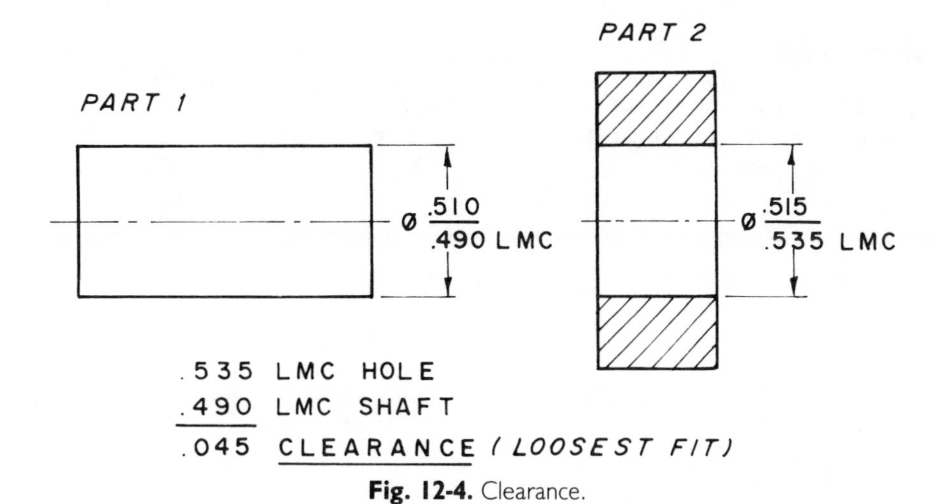

Fig. 12-4. Clearance.

(.535 diameter) minus the LMC of the shaft (.490 diameter), which equals .045 clearance, or the loosest fit.

KINDS OF FITS

There are three standard kinds of fits: clearance fit, interference fit, and transition fit. Maximum material condition (or MMC) is used to calculate each kind of fit.

Clearance Fit

A *clearance fit* is when the mating parts clear each other. In FIG. 12-1, the MMC of the shaft is .510 diameter (the *largest* it can be) In FIG. 12-2, the MMC of the hole was .515 diameter (the *smallest* it can be). In a clearance fit, the shaft is smaller than the hole; therefore, the parts will always fit freely together.

Interference Fit

An *interference fit* is when the two mating parts must be *forced* together because the shaft is larger than the hole (see FIG. 12-5). In this example, the *allowance* is .007, the tightest fit. The *clearance* is .004, the loosest fit.

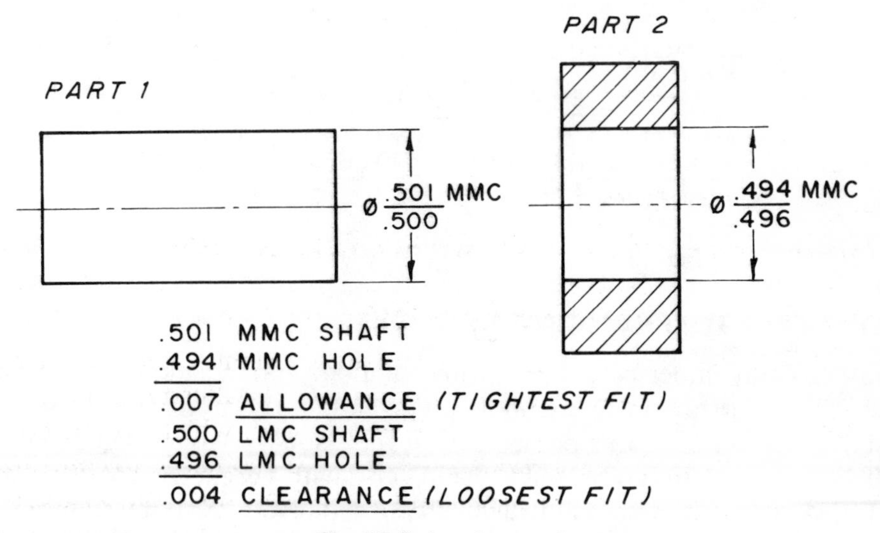

Fig. 12-5. Interference fit.

Transition Fit

Although not used much today, a *transition fit* is a fit where mating parts could be either a clearance fit or an interference fit. In transition fits, allowances, tightest fit and clearances, and loosest fit are calculated exactly the same way as any fit would be calculated. Refer to FIG. 12-6. In this example, the shaft *could* be larger than the hole. In figuring *allowance,* the MMC is used; in figuring the *clearance,* the LMC is used.

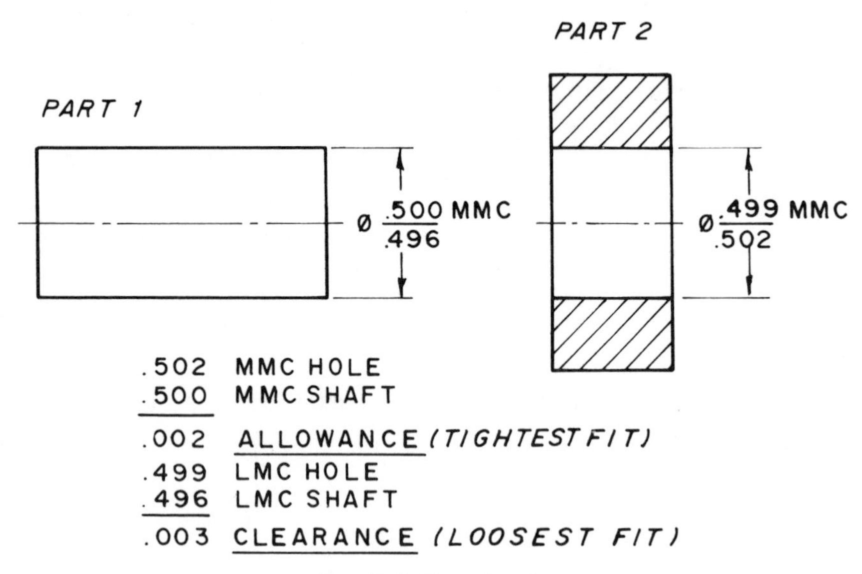

Fig. 12-6. Transition fit.

Formula

In order to calculate fits of mating parts the formula shown in FIG. 12-7 can be used. This formula uses the *basic hole size* method of calculating sizes—that is, starting from the *hole* for all calculations. This formula can be used for other kinds of mating parts such as keyways and keys.

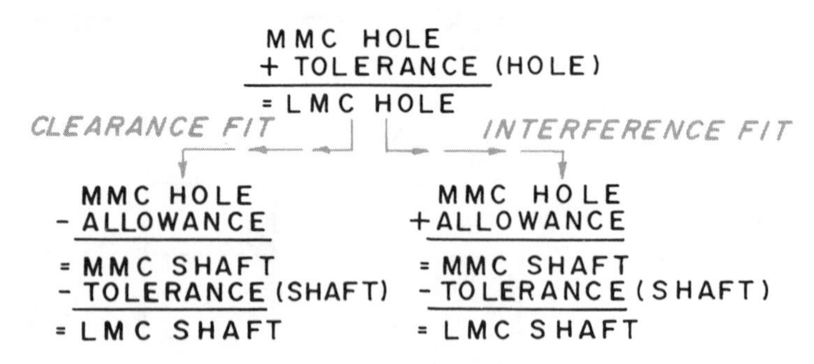

Fig. 12-7. Formula used to figure fits of mating parts.

GEOMETRIC TOLERANCING SYMBOLS

In order to fully understand the drawings of today you must know and have an understanding of the many symbols used. FIGURE 12-8 illustrates most of the symbols covered up to this point—except the last two, which will be covered in this chapter. New symbols associated only with geometric tolerancing are illustrated in FIG. 12-9—each will be explained. Note that the symbols are divided into five types: *form, profile, orientation, location,* and *runout.* These

1	R	RADIUS
2	Ø	DIAMETER
3	□	SQUARE
4	↧	DEPTH
5	∨	COUNTERSINK
6	⌴	COUNTERBORE
7	SF	SPOT FACE *(SOMETIMES ⌴)*
8	SR	SPHERICAL RADIUS
9	SØ	SPHERICAL DIAMETER
10	( )	REFERENCE DIMENSION
11	Ⓟ	PROJECTED TOLERANCE ZONE
12	-A-	DATUM FEATURE

Fig. 12-8. Geometric tolerancing.

SYMBOL		CHARACTERISTRIC	GEOMETRIC TOLERANCE
1	—	STRAIGHTNESS	FORM
2	▱	FLATNESS	
3	○	CIRCULARITY	
4	⌭	CYLINDRICITY	
5	⌒	PROFILE OF A LINE	PROFILE
6	⌓	PROFILE OF A SURFACE	
7	∠	ANGULARITY	ORIENTATION
8	⊥	PERPENDICULARITY	
9	//	PARALLELISM	
10	⌖	TRUE POSITION	LOCATION
11	◎	CONCENTRICITY	
12	⌯ *	SYMMETRY	
13	↗	CIRCULAR RUNOUT	RUNOUT
14	⌰	TOTAL RUNOUT	

* *THE SYMBOL FOR SYMMETRY IS SOMETIMES* ⌖

Fig. 12-9. New symbols associated with geometric tolerancing.

five types are further divided down into fourteen subdivisions, each with its own geometric characteristic symbol.

Modifiers

There are also three *modifiers* used with geometric tolerances (see FIG. 12-10):

Ⓜ for *maximum material condition*
Ⓛ for *least material condition*
Ⓢ meaning *regardless of feature size*

Each will be explained in full.

1	Ⓜ	MAXIMUM MATERIAL CONDITION
2	Ⓛ	LEAST MATERIAL CONDITION
3	Ⓢ	REGARDLESS OF FEATURE SIZE

Fig. 12-10. Modifiers.

Datum Feature Symbol

A *datum feature symbol* is a letter located inside a rectangle with a dash line on either side (see FIG. 12-11). Think of a datum as a starting point. It is selected because of its relationship to the required feature to be toleranced. A datum feature symbol is placed on the surface or on an extension line from the surface as illustrated. (*Note.* The datum means the same surface in the two other views.) The datum feature symbol can also be located under a size dimension, as in FIG. 12-12, when it is to be associated with the center line.

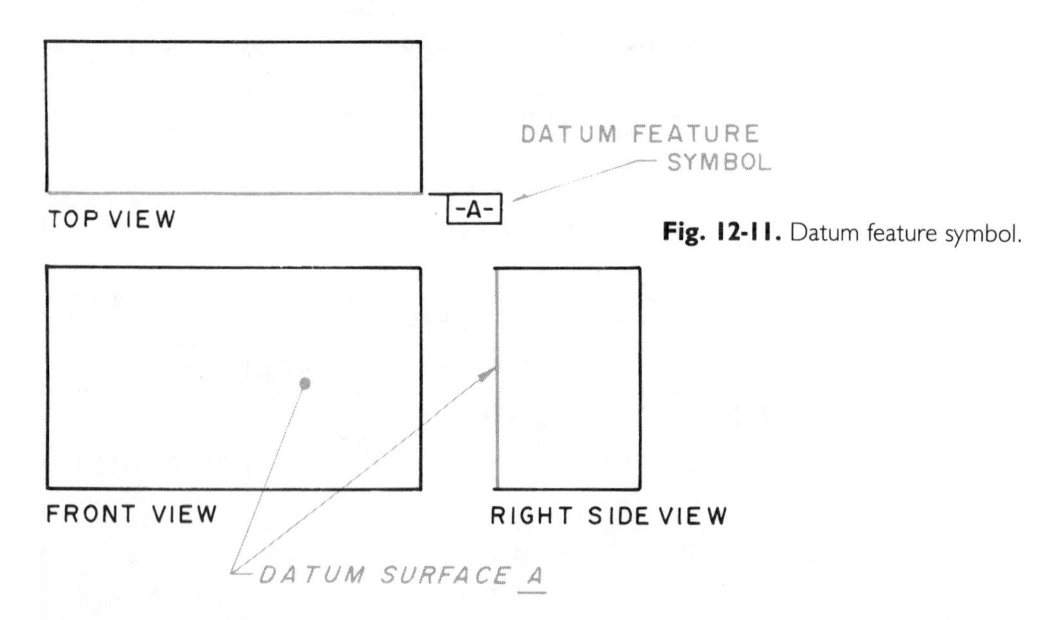

Fig. 12-11. Datum feature symbol.

Basic Dimensions

Any dimension enclosed with a rectangular frame, similar to the above datum feature symbol, is considered a *theoretical exact size*. This dimension *must* be held and has *no* tolerances (see FIG. 12-13). It is used when a perfect size or location *must* be held.

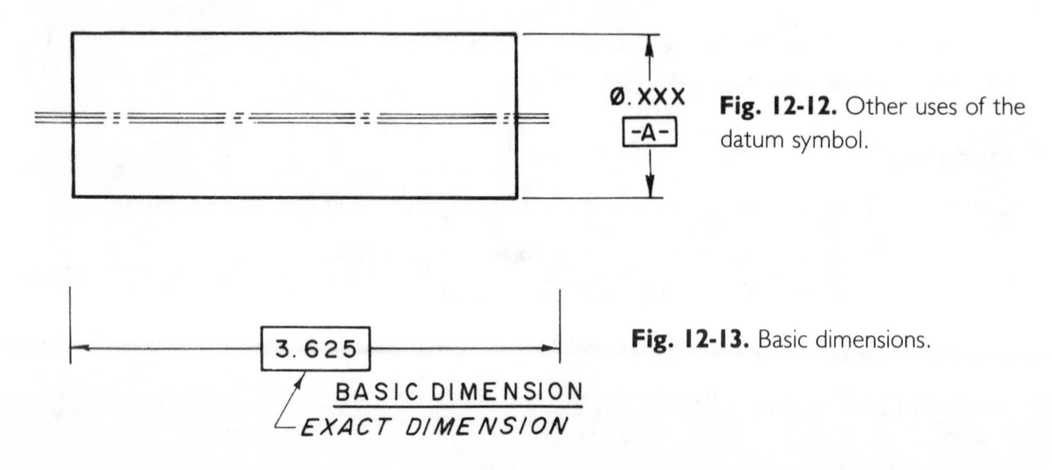

Fig. 12-12. Other uses of the datum symbol.

Fig. 12-13. Basic dimensions.

Feature Control Frame

Feature control frames are a combination of geometric symbols, permissible tolerances, datums, and in some cases, modifiers. FIGURE 12-14 illustrates a simple feature control frame with a geometric symbol adds tolerance. FIGURE 12-15

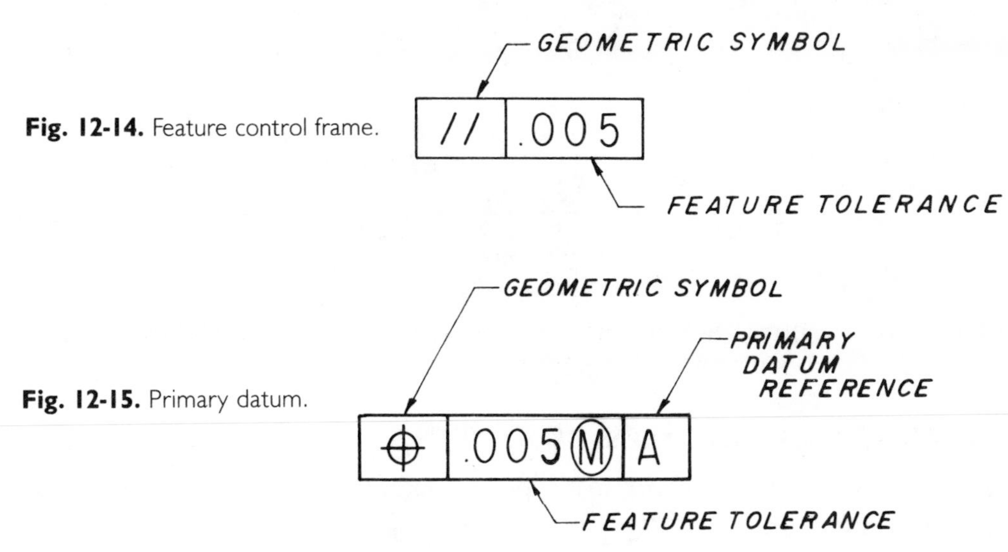

Fig. 12-14. Feature control frame.

Fig. 12-15. Primary datum.

adds the primary datum reference, and FIG. 12-16 adds the secondary datum reference. FIGURE 12-17 adds all three datum references: the first letter, regardless of which letter is the primary datum, the second letter, regardless of which letter is the secondary datum, and the third letter, regardless of which letter is

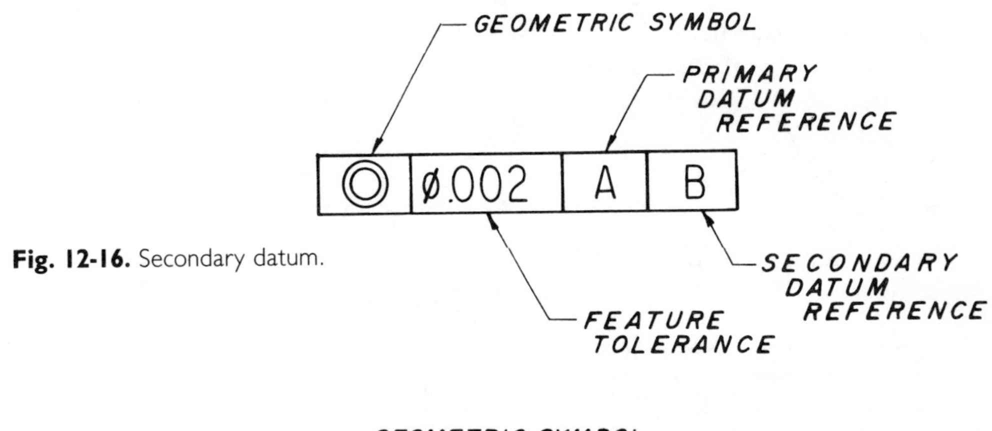

Fig. 12-16. Secondary datum.

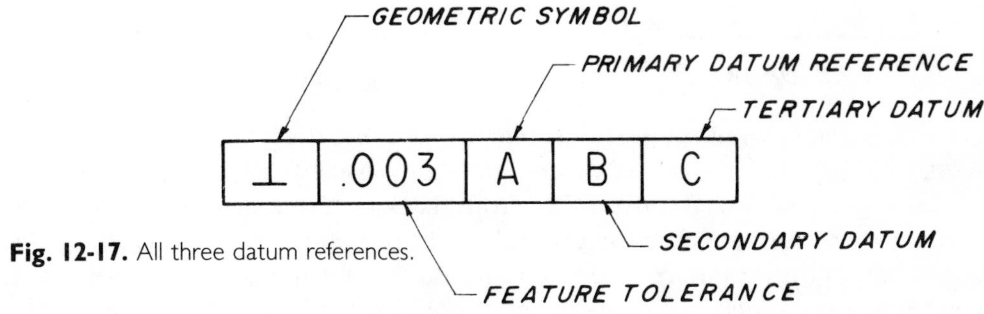

Fig. 12-17. All three datum references.

referred to as the tertiary datum. (*Note.* The letters A, B, and C do not have to be used. Any three letters can be used, and they do not have to be in that order; they could be listed as, C, A, and B.) FIGURE 12-18 illustrates one of the many

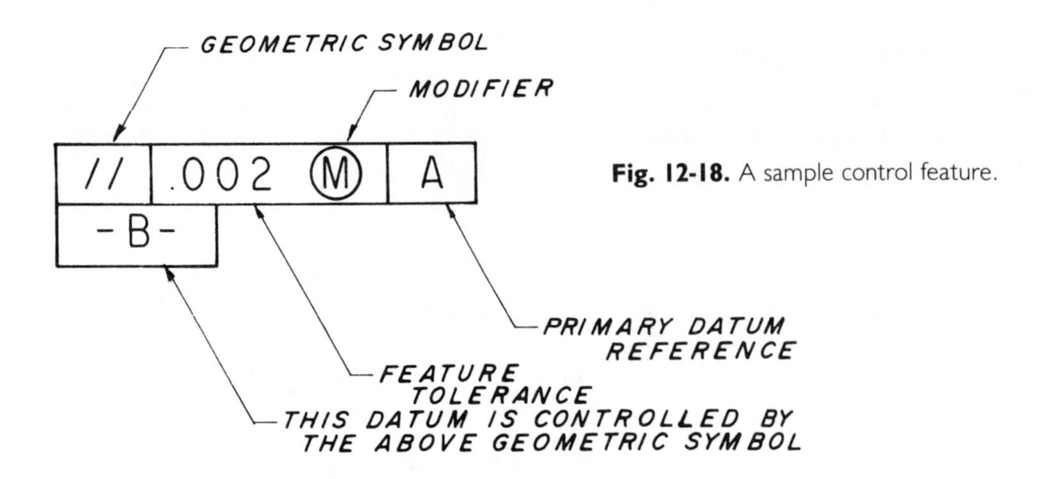

Fig. 12-18. A sample control feature.

other feature control symbols that could be used. In this example, the B is controlled by the above geometric symbol.

GEOMETRIC CHARACTERISTIC SYMBOLS

As mentioned earlier, there are five types of geometric characteristic symbols used: *form, profile, orientation, location,* and *runout.*

Form

There are four feature control symbols used to indicate form: flatness, straightness, circularity, and cylindricity. Datums are *not* specified in the feature control frame.

Flatness. Flatness is a feature control of a surface that requires the entire surface of a part to lie within two hypothetical parallel planes. FIGURE 12-19, illustrates how the flatness symbol is added to the drawing with a .002 tolerance zone added and what it means as drawn. Regardless of the size to which

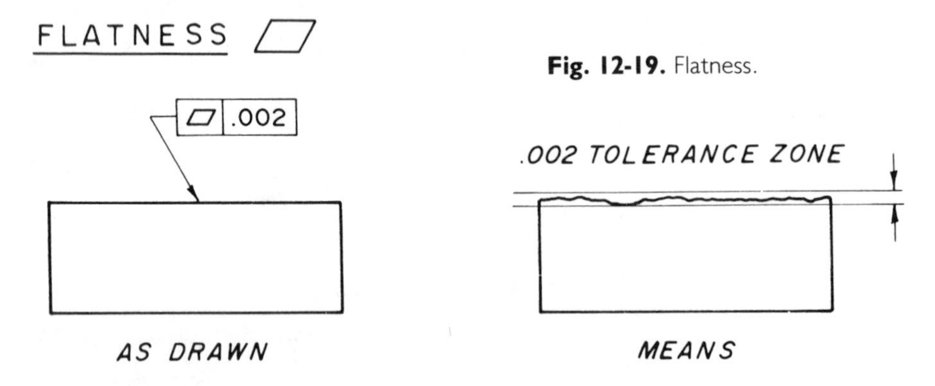

Fig. 12-19. Flatness.

the part is made, within given limits, *all points along its entire surface must lie within the hypothetical parallel lines.* Flatness is usually used for *flat* surfaces.

Straightness. Straightness differs from flatness in that flatness controls the *entire* surface and straightness controls only one element or line along the entire length of the surface (see FIG. 12-20). In the illustration, a .002 tolerance zone has been added but controls only *one element* along the entire length. Straightness is usually used for *round* objects. Adding *modifiers* to the feature control symbol changes the size of the accepted tolerances.

In using conventional tolerancing and *not* taking form into consideration, a simple round shaft, illustrated in FIG. 12-21, is drawn with a .504 upper limit

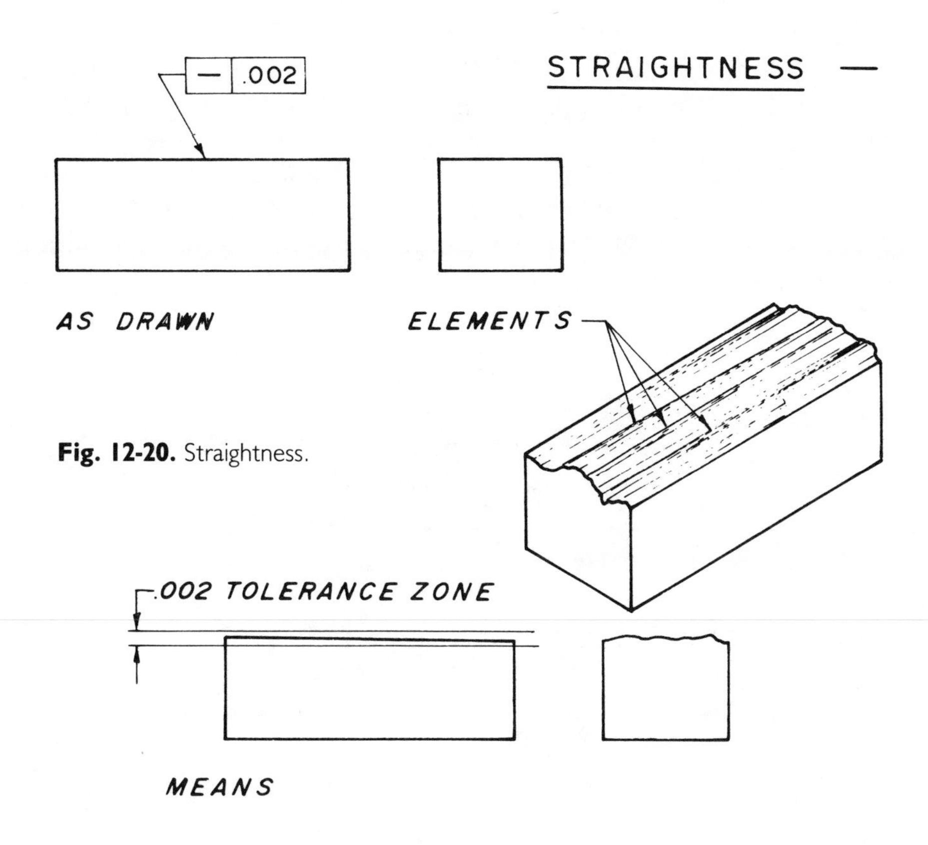

Fig. 12-20. Straightness.

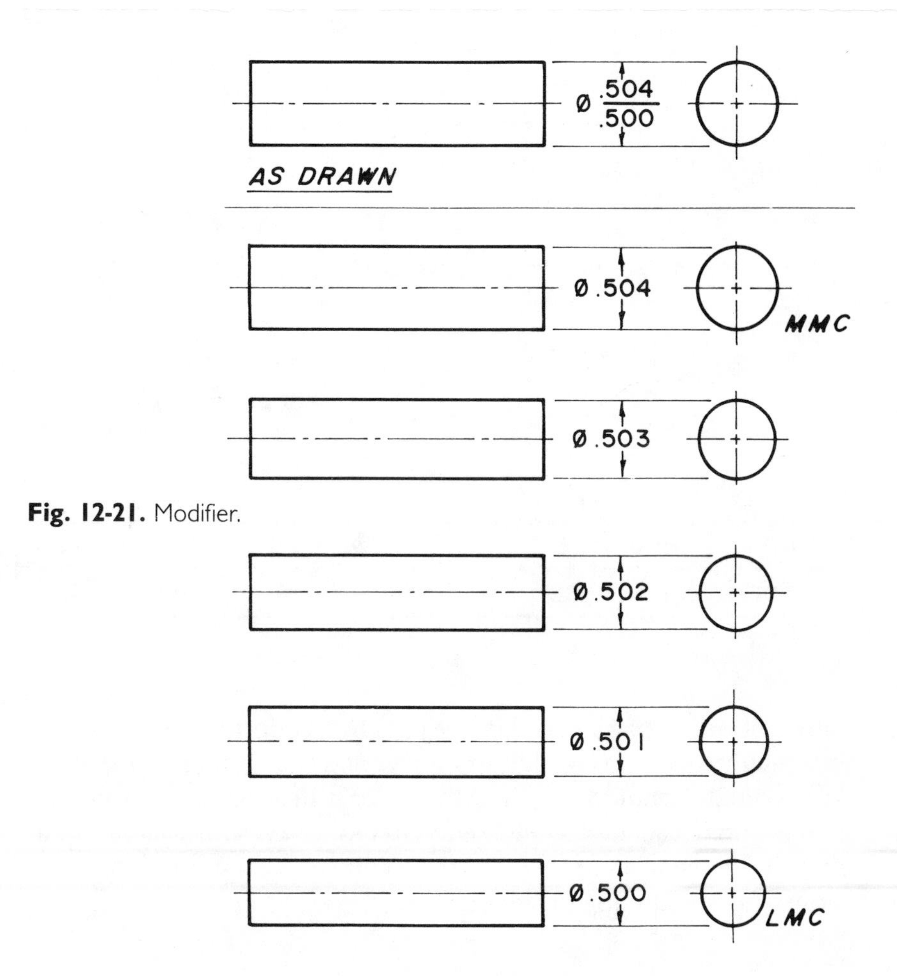

Fig. 12-21. Modifier.

(MMC) and a .500 lower limit (LMC). As shown it can vary in size from the upper limit of .504 diameter to the lower limit of .500 diameter. No regard has been given to its actual final shape. Now refer to FIG. 12-22. If the straight geometric symbol is added to the drawing with a .002 tolerance zone and the M (maximum material condition) modifier added, its shape can vary, as indicated in the block below. At MMC, .504 diameter the part *must* be straight, within .002. At LMC, .500 diameter, the part can vary from straightness up to .006. By adding

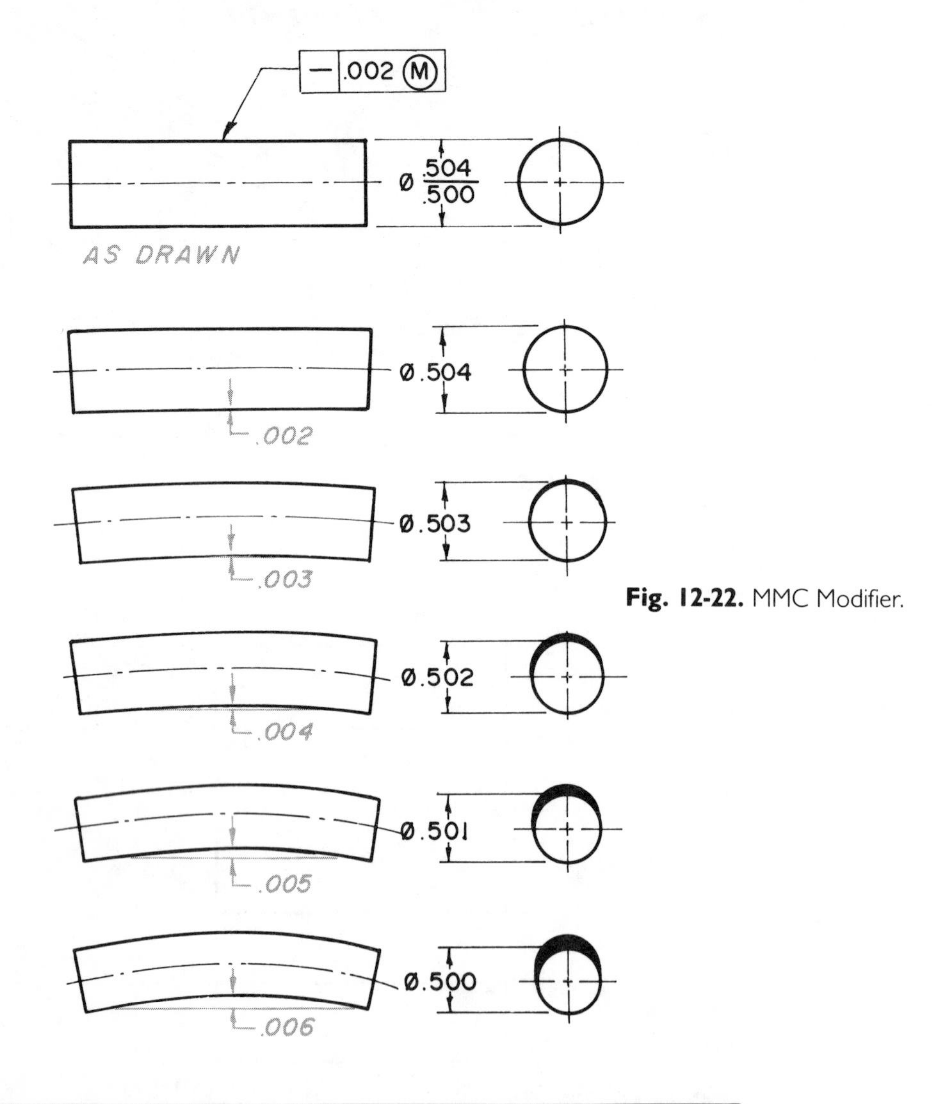

Fig. 12-22. MMC Modifier.

MANUFACTURED SIZE	.504	.503	.502	.501	.500
OUT-OF-STRAIGHTNESS	.002	.003	.004	.005	.006

the M modifier, parts that did *not* pass inspection before *will* pass inspection, thus saving many dollars. If a straight geometric symbol is added to the drawing with a .002 tolerance zone and the L (least material condition) modifier added its shape can also vary as illustrated in FIG. 12-23. This practice is not used very often in industry and is only noted to illustrate how the L modifier effects the geometric tolerancing of the part.

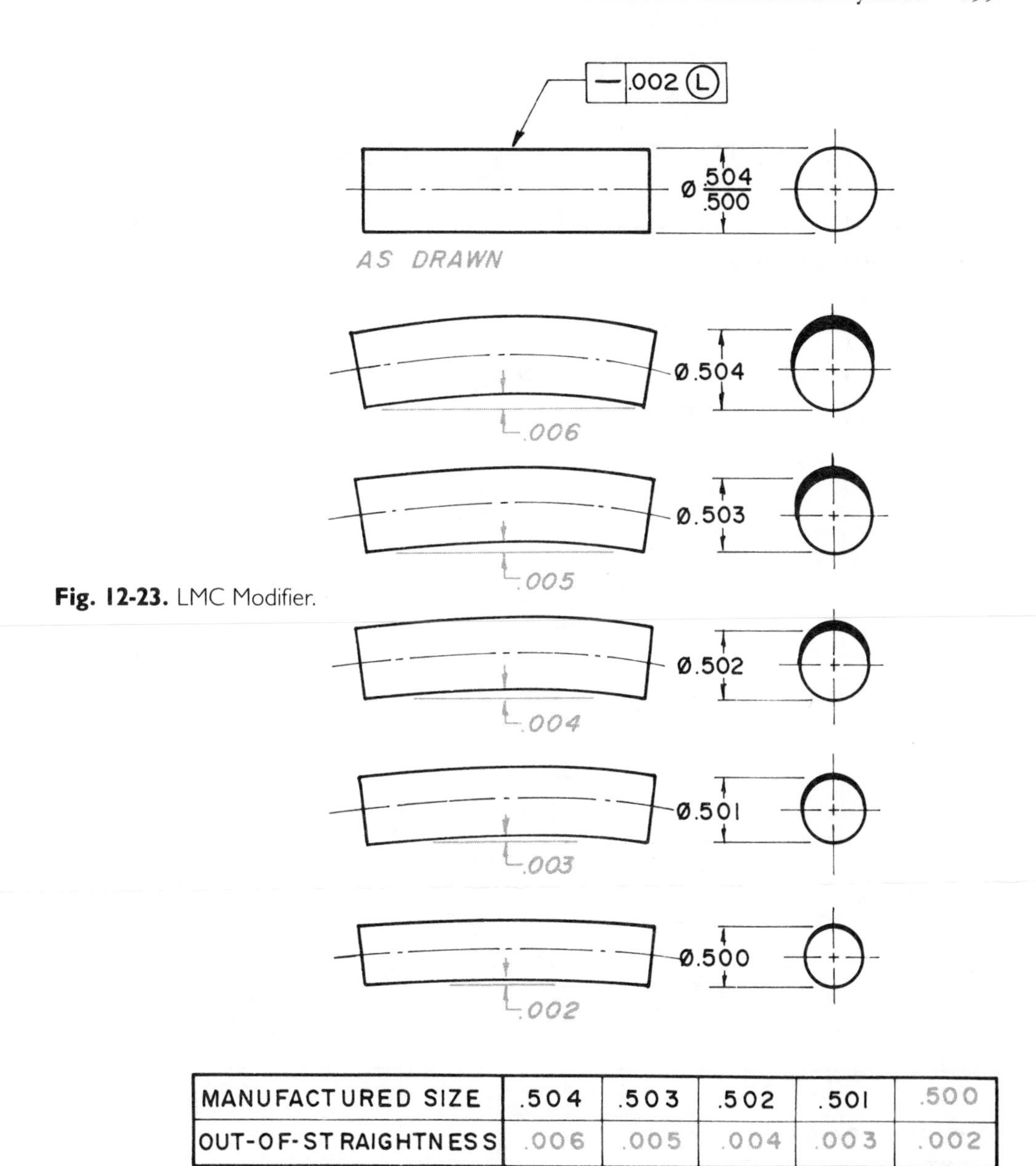

MANUFACTURED SIZE	.504	.503	.502	.501	.500
OUT-OF-STRAIGHTNESS	.006	.005	.004	.003	.002

Fig. 12-23. LMC Modifier.

FIGURE 12-24 illustrates what effect the modifier S has upon the geometric shape of a part. Here a straight geometric symbol has been added to the drawing, with a .002 tolerance zone and the S (regardless of feature size) modifier. In this example, the .002 out of straightness must be held regardless of what dimensional size the part is made. (*Note.* If no modifier is added, it is assumed form *must* be held regardless of feature size.)

Circularity. Circularity is a feature control used to control how *round* a part is to be made. It is used for cylinders, cones, or spheres, where all surfaces meet through a common centerline. It specifies that *all* points on the surface must be equal distance from the centerline. FIGURE 12-25 illustrates how circularity is called off on a drawing, along with the meaning of the call-off. Think of various sections being taken through the part and checked for roundness of each section.

Cylindricity. Cylindricity differs from circularity in that circularity controls roundness at any given section through the part, while *cylindricity* controls roundness along the *entire surface* of the part. The two parallel lines on the symbol itself represent the length of the part (see FIG. 12-26).

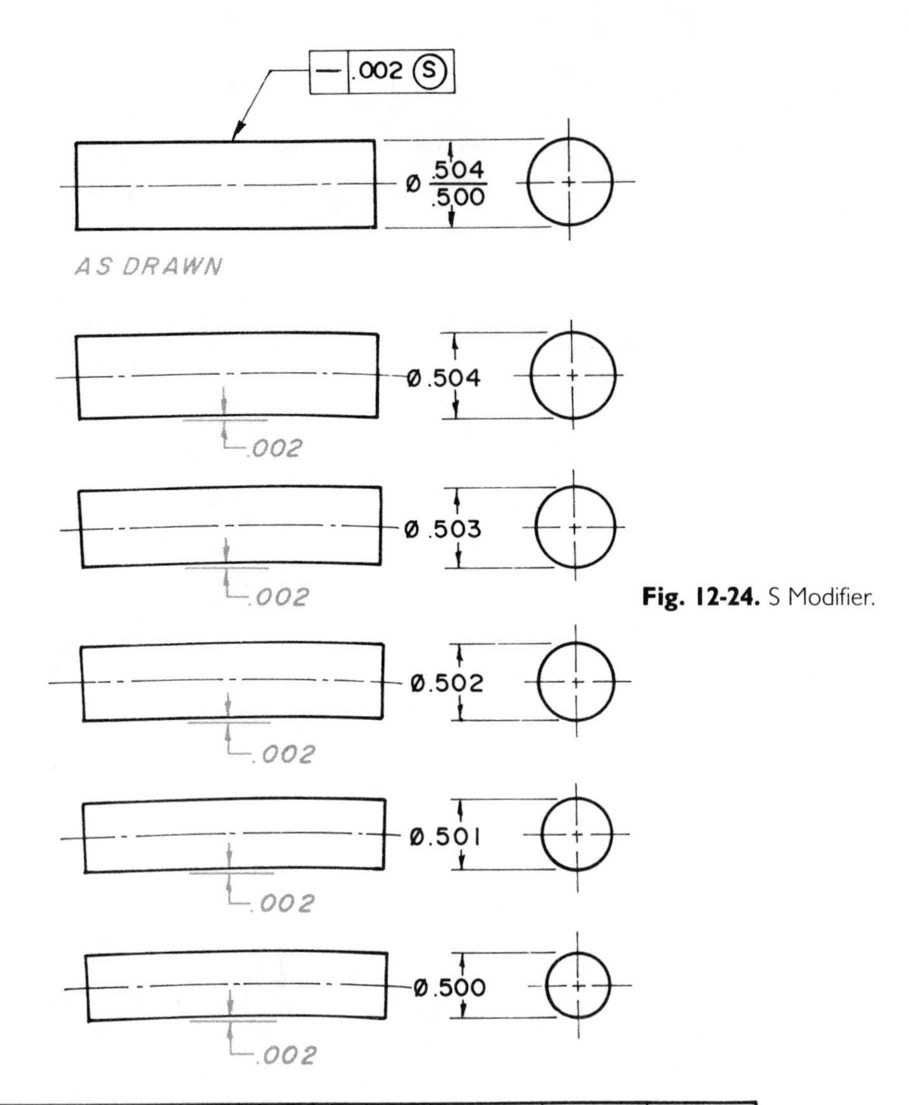

Fig. 12-24. S Modifier.

MANUFACTURED SIZE	.504	.503	.502	.501	.500
OUT-OF-STRAIGHTNESS	.002	.002	.002	.002	.002

Profile

There are two kinds of profile feature control symbols: *profile of a line* and *profile of a surface.* A profile feature control frame specifies the amount of variation allowed for either a line or the entire surface.

Profile of a line controls a specified zone within which the surface, *on that line,* must lie (see FIG. 12-27). This line can be any place on the part.

Profile of a surface is like profile of a line except that the specified tolerance zone encompasses the *entire* surface (see FIG. 12-28).

Orientation

Three geometric characteristic symbols control *orientation* of an object: parallelism, perpendicularity, and angularity.

Parallelism. Parallelism is a control that specifies that all points on a surface, center plane, axis, or line *must* be equal distance (parallel) from a given datum plane or axis within a given tolerance (see FIG. 12-29).

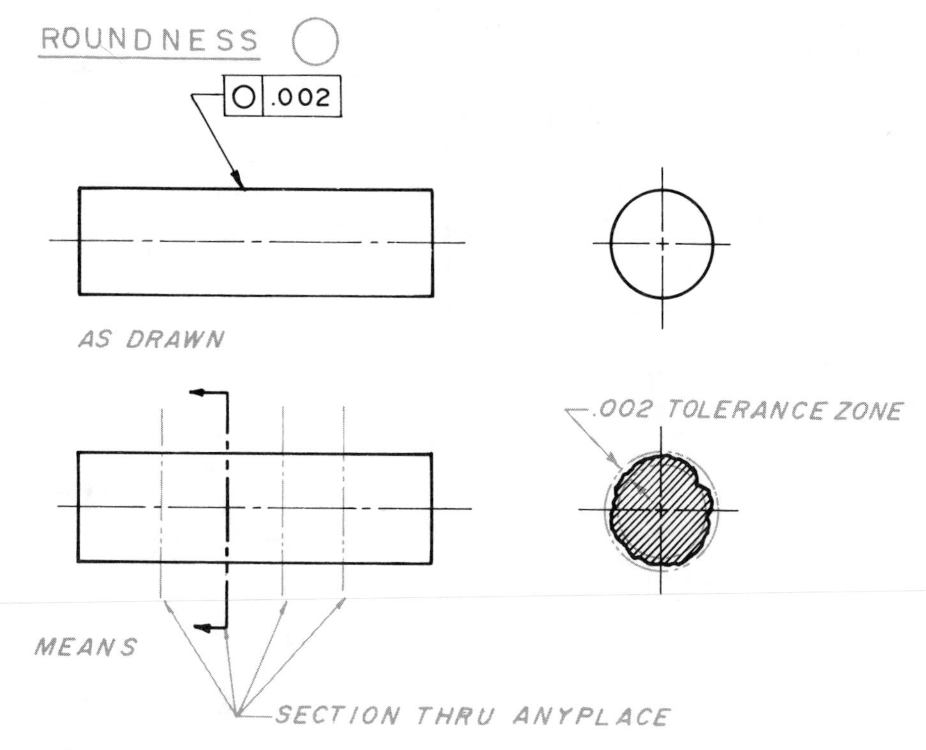

Fig. 12-25. Circularity.

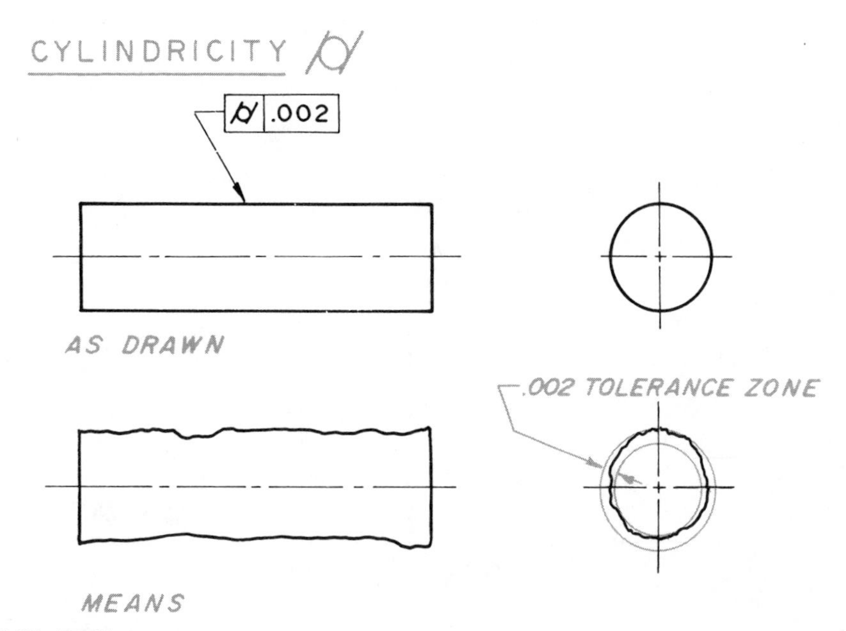

Fig. 12-26. Profile.

Perpendicularity. Perpendicularity is a control that specifies that all points on a surface, axis, or line must be at *right angles* to a datum plane or axis (see FIG. 12-30).

Angularity. Angularity is a control that specifies that all points on a surface or axis must be at the specified angle (other than 90 degrees) from a datum plane or axis (see FIG. 12-31).

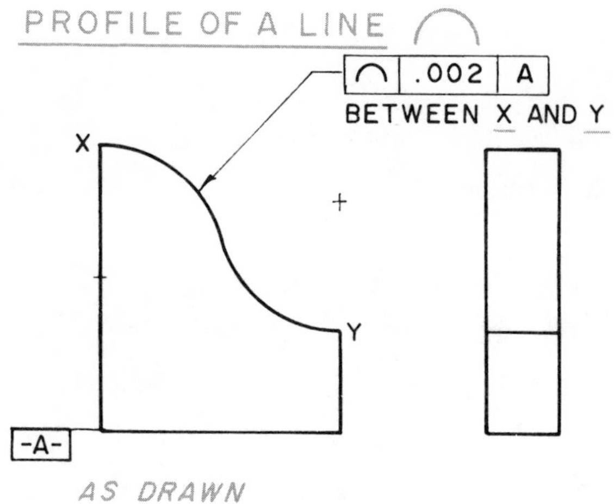

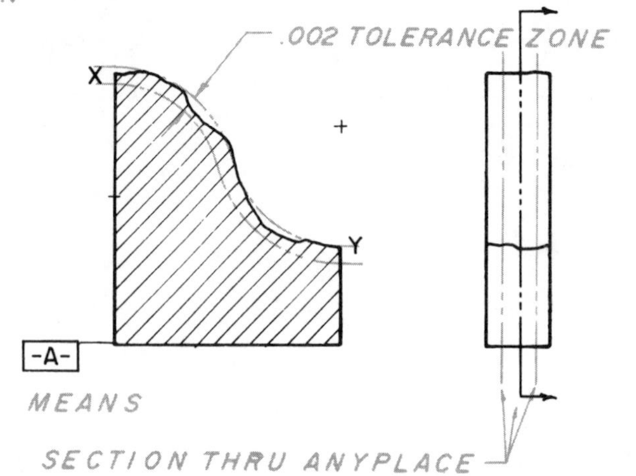

Fig. 12-27. Profile of a line.

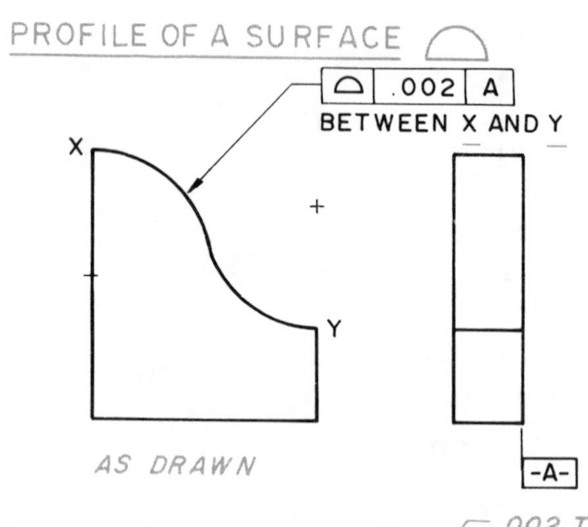

Fig. 12-28. Profile of a surface.

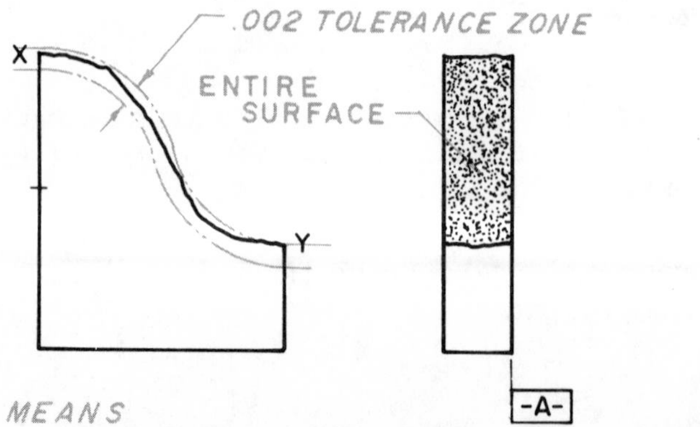

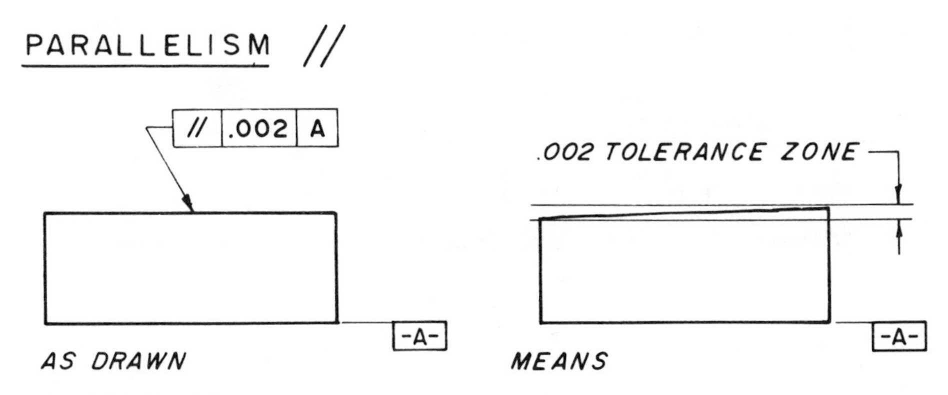

Fig. 12-29. Parallelism.

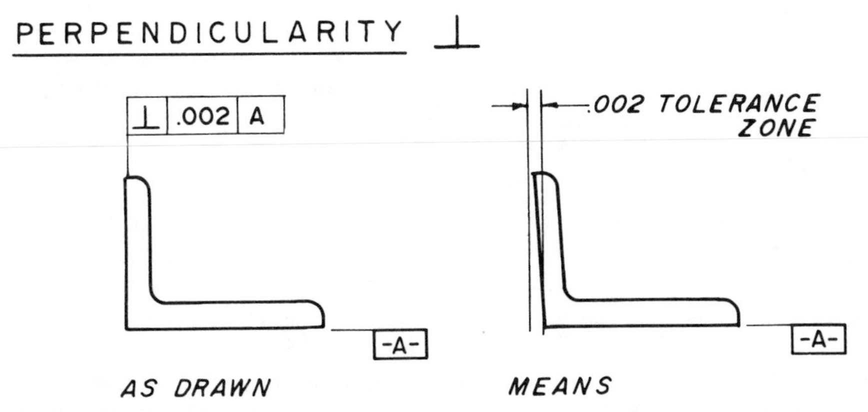

Fig. 12-30. Perpendicularity.

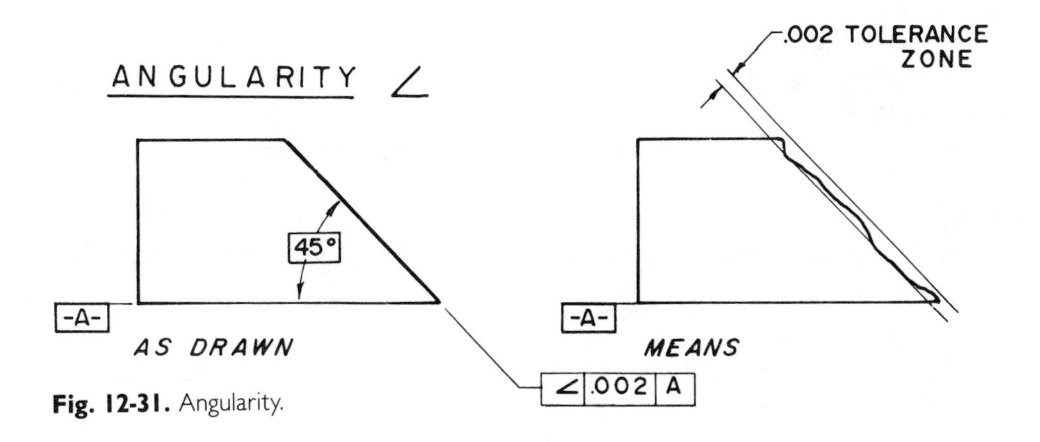

Fig. 12-31. Angularity.

Location

Three feature control symbols control *location* of features on an object: concentricity, true position, and symmetry.

Concentricity. Concentricity is when more than one round, cylinder, cone, or sphere have the same centerline or axis (see FIG. 12-32). Concentricity is a feature control that controls the relationship of two or more cylindrical features that must be in perfect alignment along one axis (see FIG. 12-33). This controls the specified amount of deviation allowed from the centerline or axis.

True position. True position is the *theoretical* exact size or location of a feature on the object. In convention tolerancing, the tolerance zone is a square

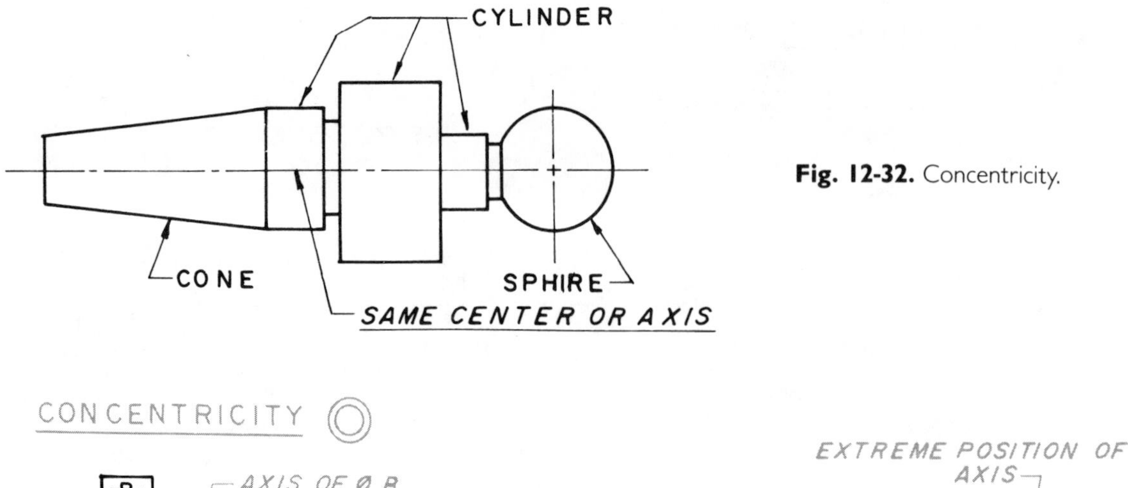

Fig. 12-32. Concentricity.

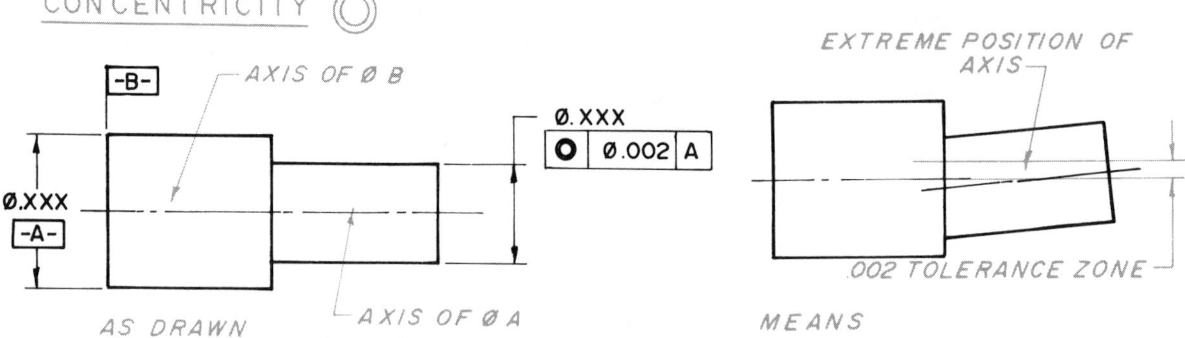

Fig. 12-33. This controls a specified amount of deviation allowed from the centerline or axis.

or rectangle (see FIG. 12-34). In the illustration, the given location dimensions for the hole make up a square tolerance zone with sides of the square equal to .030. This means the center of the hole can lie *any place* within this square.

In *true positioning,* the tolerance zone is a *circle* (see FIG. 12-35). A drawing using true positioning uses *basic dimensions* (theoretical exact dimensions within a rectangular box) to locate the center of the hole. These dimensions *must* be held without any tolerances whatsoever. The true positions symbol is used to indicate there is to be a *round* tolerance zone. The diameter of this zone is then given to the right of the symbol and referenced to surfaces A and B. In FIG. 12-35, the *round* tolerance zone is .042. All centers for the hole must lie within this round tolerance zone. Notice this is a much larger zone than the square zone; many more parts will be within tolerance and pass inspection.

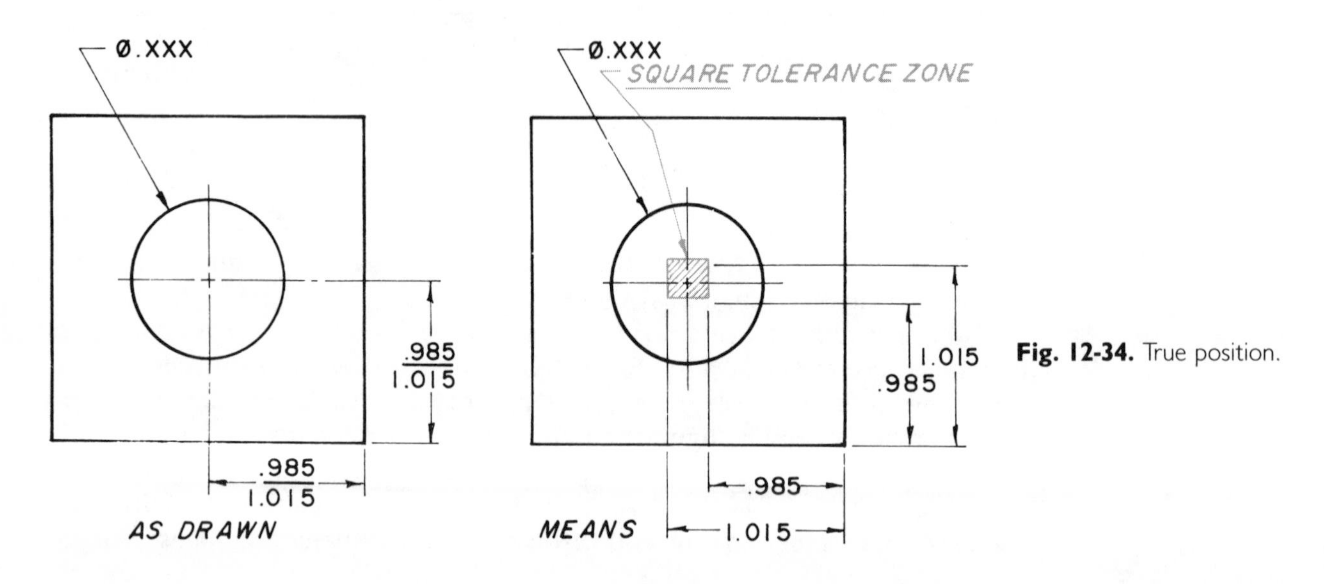

Fig. 12-34. True position.

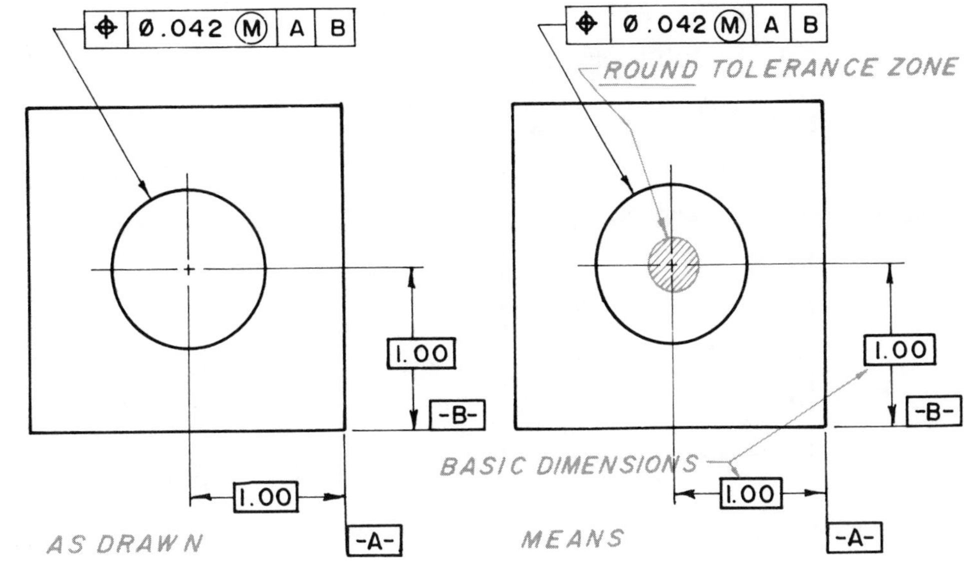

Fig. 12-35. The tolerance zone is a circle.

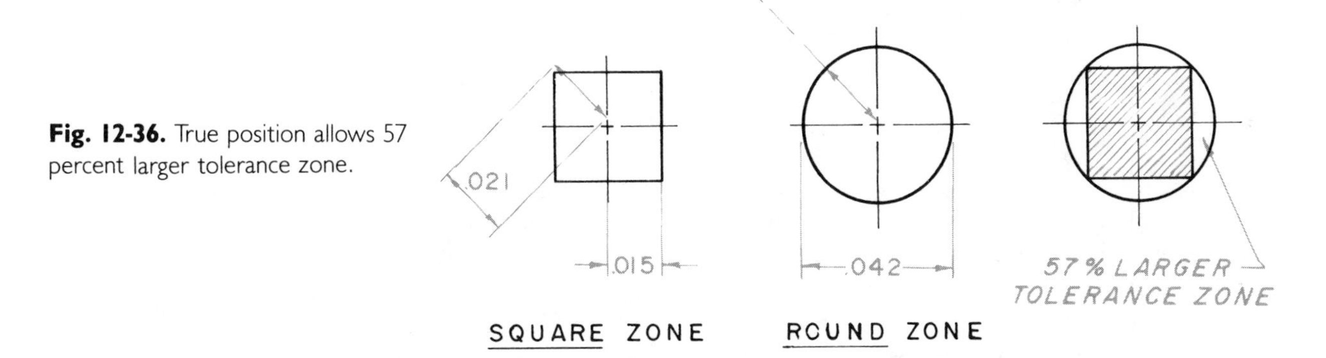

Fig. 12-36. True position allows 57 percent larger tolerance zone.

FIGURE 12-36 illustrates the same tolerance zone—note in the *square* tolerance zone, the sides of the square are .015. Diagonally, this allows *some* centers to actually be .021 from the given center. Using this .021 dimension and applying it for the *radius* of the true position, the *round* tolerance zone a diameter of .042 is derived (see center illustration). This .042 *round* tolerance zone actually allows the same given limits as the .030 square tolerance zone, but allows a 57 percent larger tolerance zone, thus saving manufacturing costs.

Further savings can be achieved by adding a *modifier* to the true position symbol (see FIG. 12-37). This means the .042-diameter tolerance zone must be

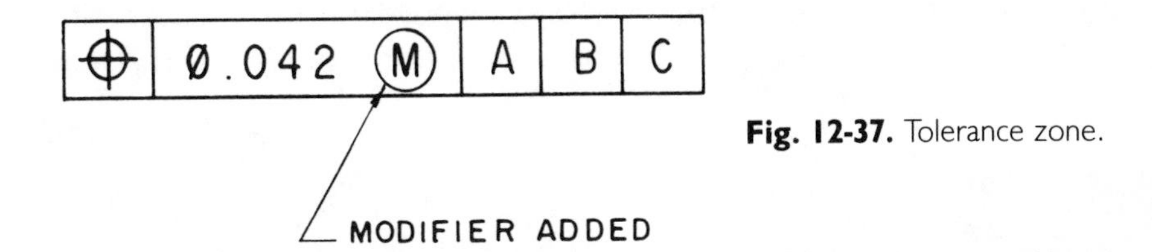

Fig. 12-37. Tolerance zone.

held when the hole is at MMC. When the hole is at LMC, the tolerance zone can be larger, thereby saving even more. Now study FIG. 12-38. The top illustration is as the drawing would be dimensioned—directly below is an exaggerated illustration of the hole at its largest, or LMC, a diameter of .254. If the hole's center is at the furthest allowable distance from the required center, the *top* of the hole is actually .021 plus .127, or .148 distance from the required center. This is

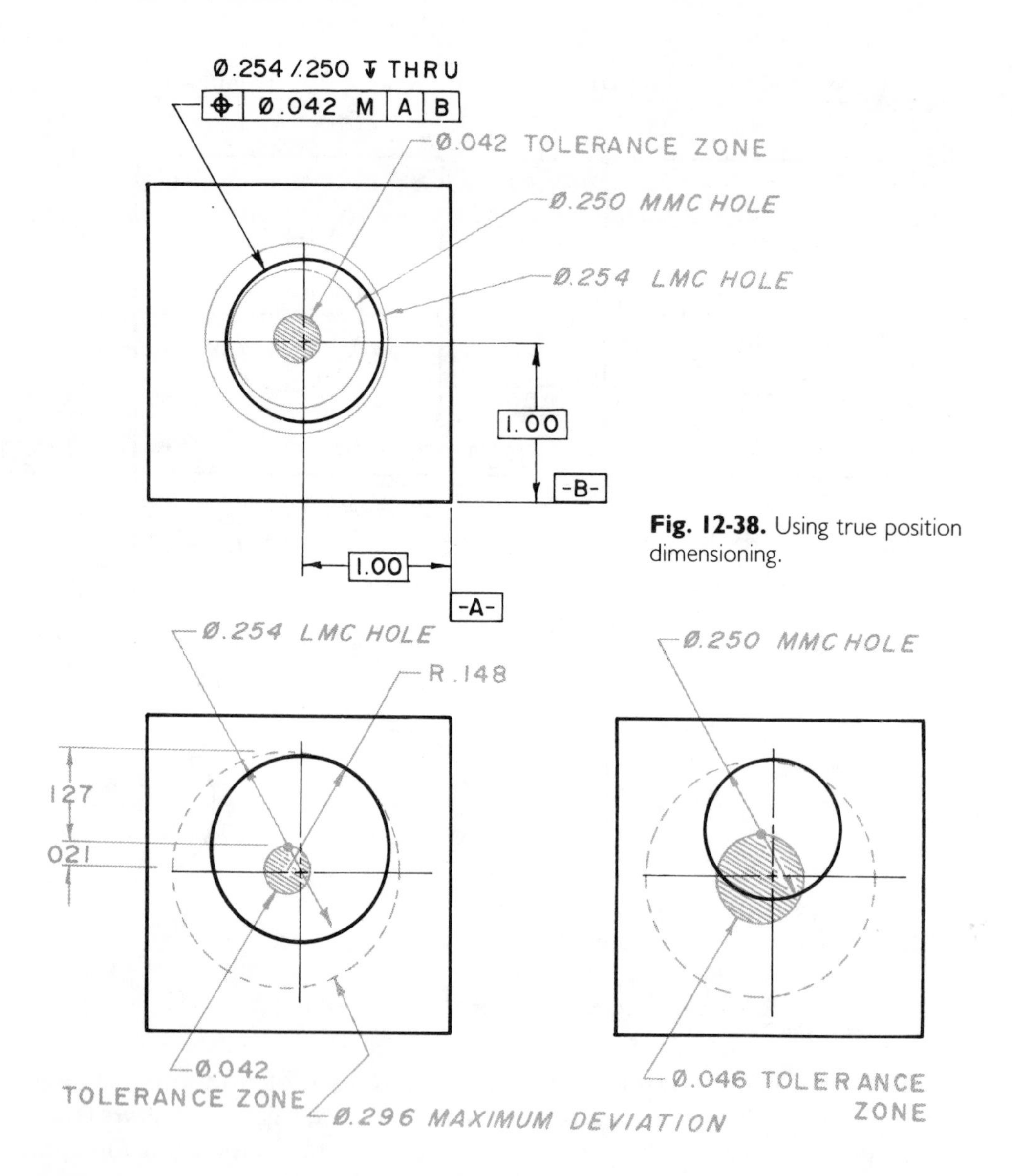

Fig. 12-38. Using true position dimensioning.

actually the ***maximum*** distance allowed. If a circle is drawn (see dashed circle) it allows a .296 ***maximum*** circle of deviation (see bottom-left drawing). If this .296 diameter is applied to the ***smallest*** hole, or a hole at MMC (.250 diameter), the ***tolerance zone*** could be enlarged to .046 diameter and all holes at MMC would still fit within the dashed circle of .296 diameter.

Symmetry. Symmetry is when a feature of a part, must be centrally located ***exactly*** between the center of the part or a given datum (see FIG. 12-39). The symbol for true position is sometimes used instead of the symmetry symbol.

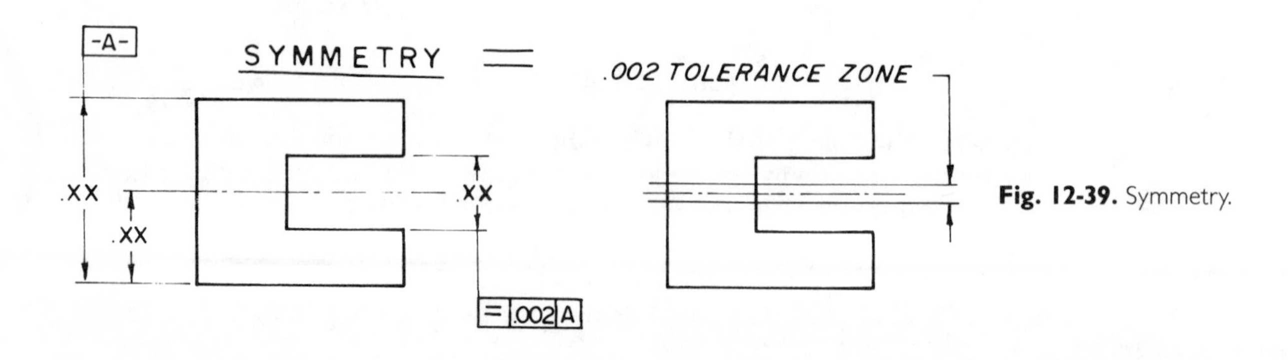

Fig. 12-39. Symmetry.

Runout

There are two types of runout: *circular runout* and *total runout.* Runout controls the amount of deviation from *perfect form* on surfaces that *rotate* about an axis. The axis always is the datum reference. This variation is measured by an instrument with a dial indicator. In testing the part, a finger on the instrument rides on the surface to be measured, and the dial indicates the variation. The term "F.I.R." refers to the reading of this instrument and means *full indicator reading.* Sometimes it is referred to as "FIM," or *full indicator movement.*

Circular runout. Circular runout is the *maximum* variation *at any fixed point around the part* during *one* full revolution of the part about the datum axis (see FIG. 12-40). This measurement can be taken any place along the specified area.

Total runout. Total runout is exactly the same as circular runout except it includes the *entire* surface and is measured in the same way (see FIG. 12-41).

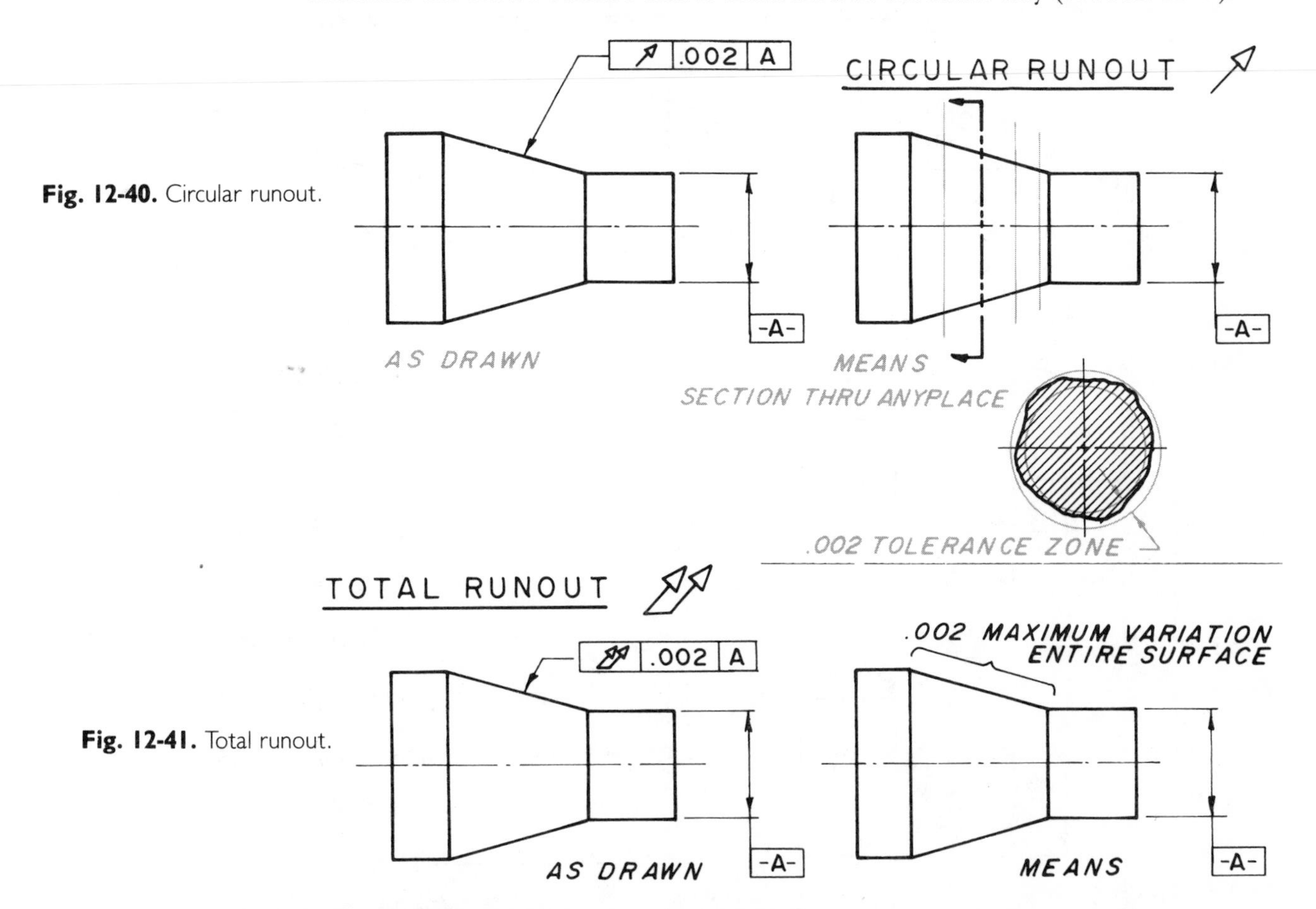

Fig. 12-40. Circular runout.

Fig. 12-41. Total runout.

WHEN GEOMETRIC DIMENSIONING IS REQUIRED

Geometric dimensioning is required when:

- the parts are critical to their function.
- the parts are critical to their interchangeability.
- you wish to avoid scrapping perfectly good parts.
- functional gaging is to be used.
- automated equipment is being used.
- you wish to increase productivity.
- you wish to save manufacturing costs.

WORKSHEET 12-1

Instructions: Fill in all blank spaces, using formula shown in FIG. 12-7. Keep all math on a separate piece of paper. (Answers in Appendix A.)

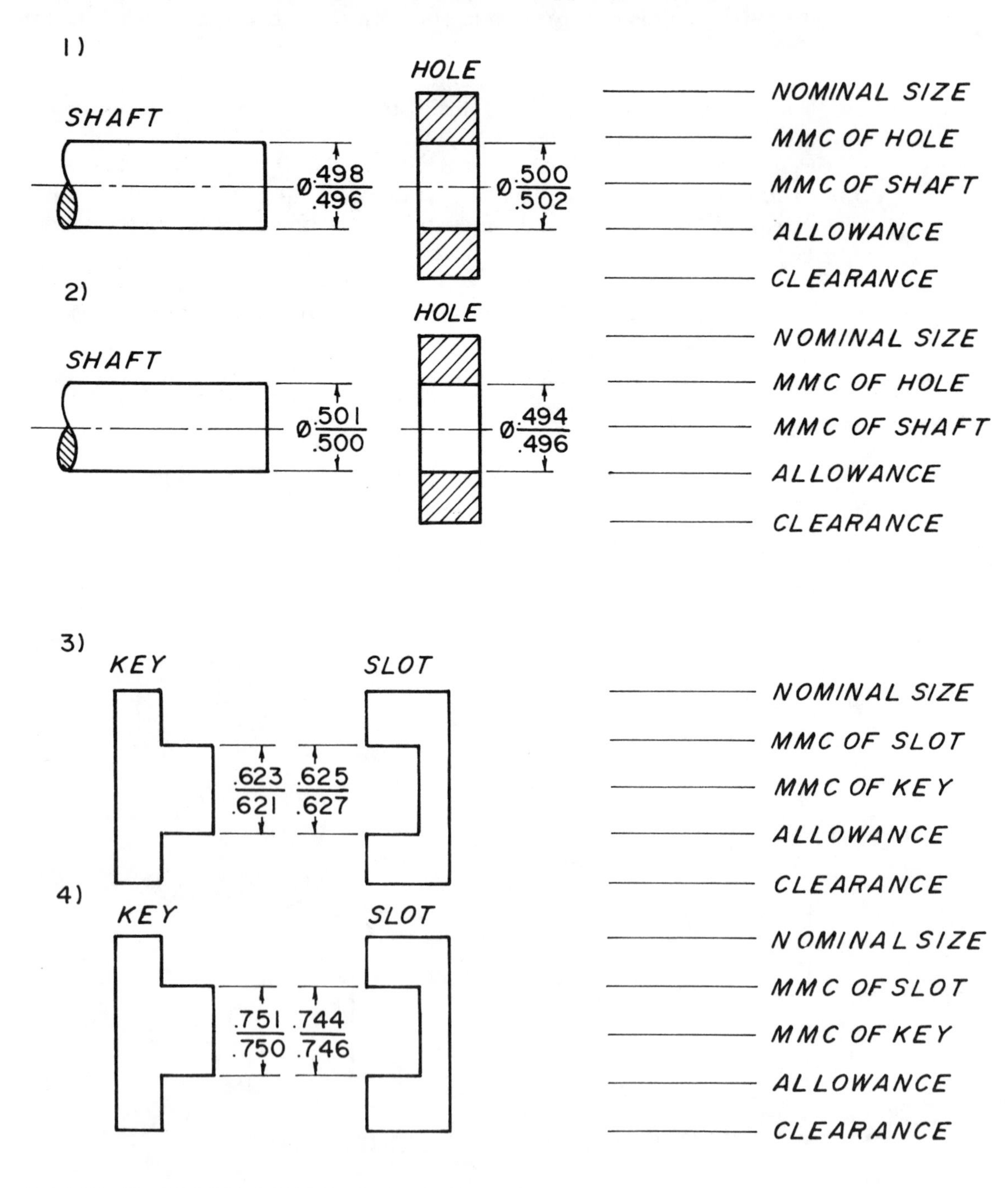

WORKSHEET 12-2

Instructions:

1. Sketch each geometric symbol in the spaces provided and indicate in the spaces to the right of each symbol, which *type* of tolerance it controls. (FORM/ PROFILE/ ORIENTATION/ LOCATION/ RUNOUT)
2. What does each symbol mean? Place your answer in the spaces provided.
3. List in the spaces provided what each symbol, number and letter represents in the feature control block. (Answers in Appendix A.)

1)

	SYMBOL	GEOMETRIC TOLERANCE
ANGULARITY	______	______
TRUE POSITION	______	______
FLATNESS	______	______
PROFILE OF A SURFACE	______	______
PERPENDICULARITY	______	______
CIRCULAR RUNOUT	______	______
STRAIGHTNESS	______	______
TOTAL RUNOUT	______	______
PROFILE OF A LINE	______	______
CYLINDRICITY	______	______
CIRCULARITY	______	______

2)

SYMBOL		SYMBOL	
Ⓜ	______	◎	______
-A-	______	⌭	______
( )	______	Ø	______
R	______	Ⓢ	______
↧	______	.25 (boxed)	______
∨	______	⊔	______

3)

⊥	.005 Ⓜ	A	B	C Ⓜ

______
______
______
______
______
______
______

WORKSHEET 12-3

Instructions: Using the drawing A751291 on p. 211, answer the following questions in the spaces provided. (Answers in Appendix A.)

1. Surface M in the right-side view is what surface in the front view?

2. The 2.00 diameter *must* be perpendicular to datum -A- within what tolerance?

3. What is the diameter at N in the right-side view?

4. How far is it from surface A in the front view to surface R in the right-side view?

5. What is the surface finish at Q in the right-side view?

6. How many total finished surfaces are there?

7. Explain in full what the feature control frame at J means.

8. The 1.50 diameter *must* be concentric to which diameter circle within .001?

9. How many *different* 16 microinch surfaces are there?

10. What is the diameter at P in the right-side view?

11. How far apart are the two 32 microinch surfaces?

12. What is the rectangle at G, with the -B- enclosed, mean?

13. What are the upper and lower limits of the diameter at O in the right-side view?

14. How far is surface D from surface F?

15. What is dimension S?

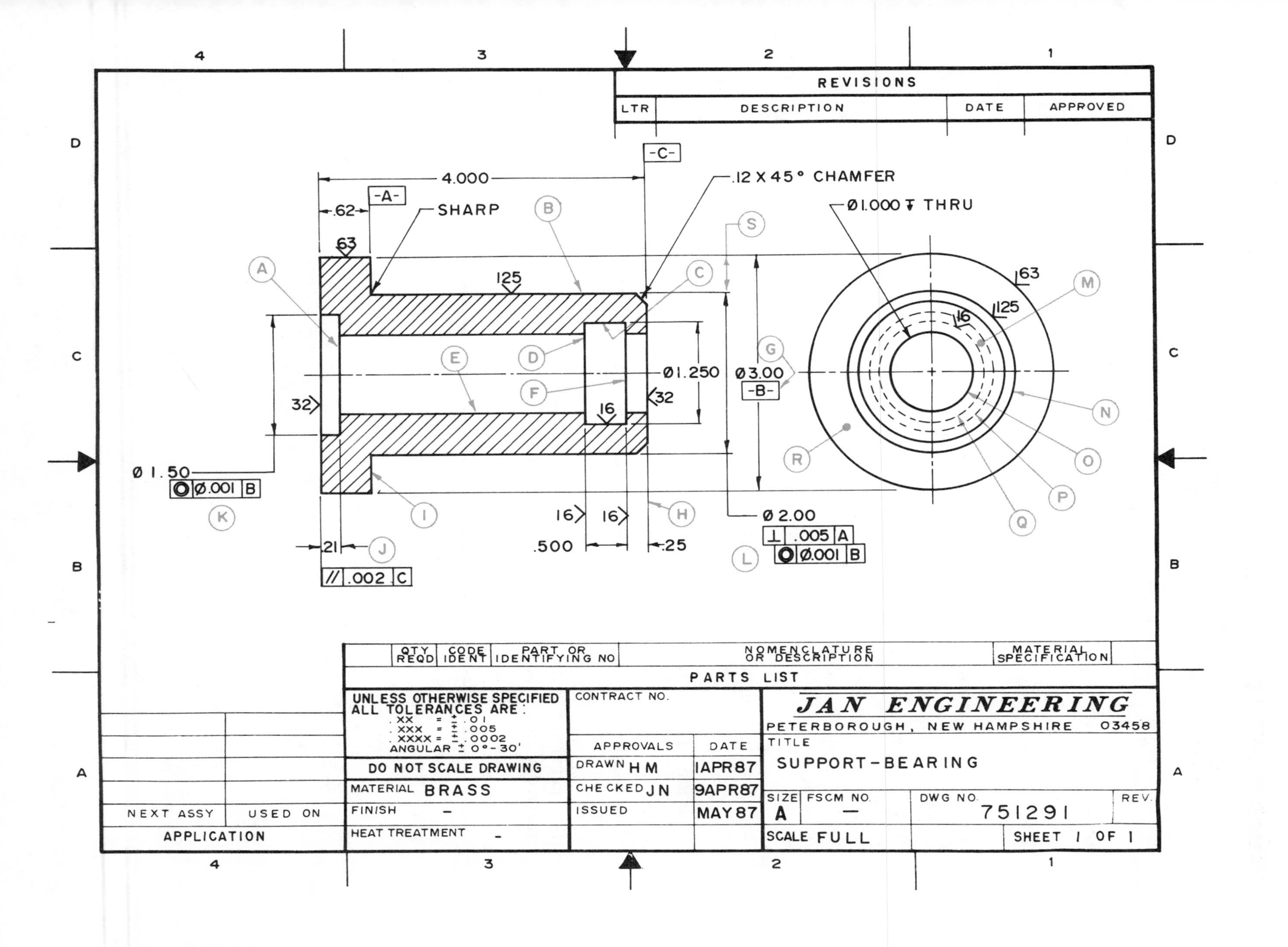
REVISIONS
LTR
DESCRIPTION
DATE
APPROVED
-C-
.12 X 45° CHAMFER
Ø1.000 ↧ THRU
4.000
SHARP
-A-
.62
63
125
Ø3.00
-B-
Ø2.00
⊥ .005 A
◎ Ø.001 B
Ø1.250
32
16
.500
.25
.21
// .002 C
Ø 1.50
◎ Ø.001 B
PARTS LIST
QTY REQD
CODE IDENT
PART OR IDENTIFYING NO
NOMENCLATURE OR DESCRIPTION
MATERIAL SPECIFICATION
JAN ENGINEERING
PETERBOROUGH, NEW HAMPSHIRE 03458
TITLE
SUPPORT-BEARING
SIZE A
FSCM NO. —
DWG NO. 751291
REV.
SCALE FULL
SHEET 1 OF 1
UNLESS OTHERWISE SPECIFIED ALL TOLERANCES ARE:
.XX = ±.01
.XXX = ±.005
.XXXX = ±.0002
ANGULAR ± 0°-30'
DO NOT SCALE DRAWING
MATERIAL BRASS
FINISH -
HEAT TREATMENT -
CONTRACT NO.
APPROVALS
DATE
DRAWN HM 1APR87
CHECKED JN 9APR87
ISSUED MAY 87
NEXT ASSY
USED ON
APPLICATION

WORKSHEET 12-4

Instruction: Using drawing A15575 on p. 213, answer the following questions in the spaces provided. (Answers in Appendix A.)

1. What kind of steel is used in this part?

2. What is the MMC of the four .750 diameter holes?

3. What do the dimensions within the rectangles indicated?

4. What do the *letters* within the rectangles indicate?

5. How far apart are the two 63 microinch finished surfaces?

6. Explain what the information and symbols within the feature control frame at A mean.

7. What is the carbon content of this part?

8. What is the LMC of the largest hole?

9. The -A- within the rectangle under the 6.50 dimension indicates what?

10. How much *could* the .010 diameter true position tolerance zone be enlarged when the hole is at MMC and still be within tolerance?

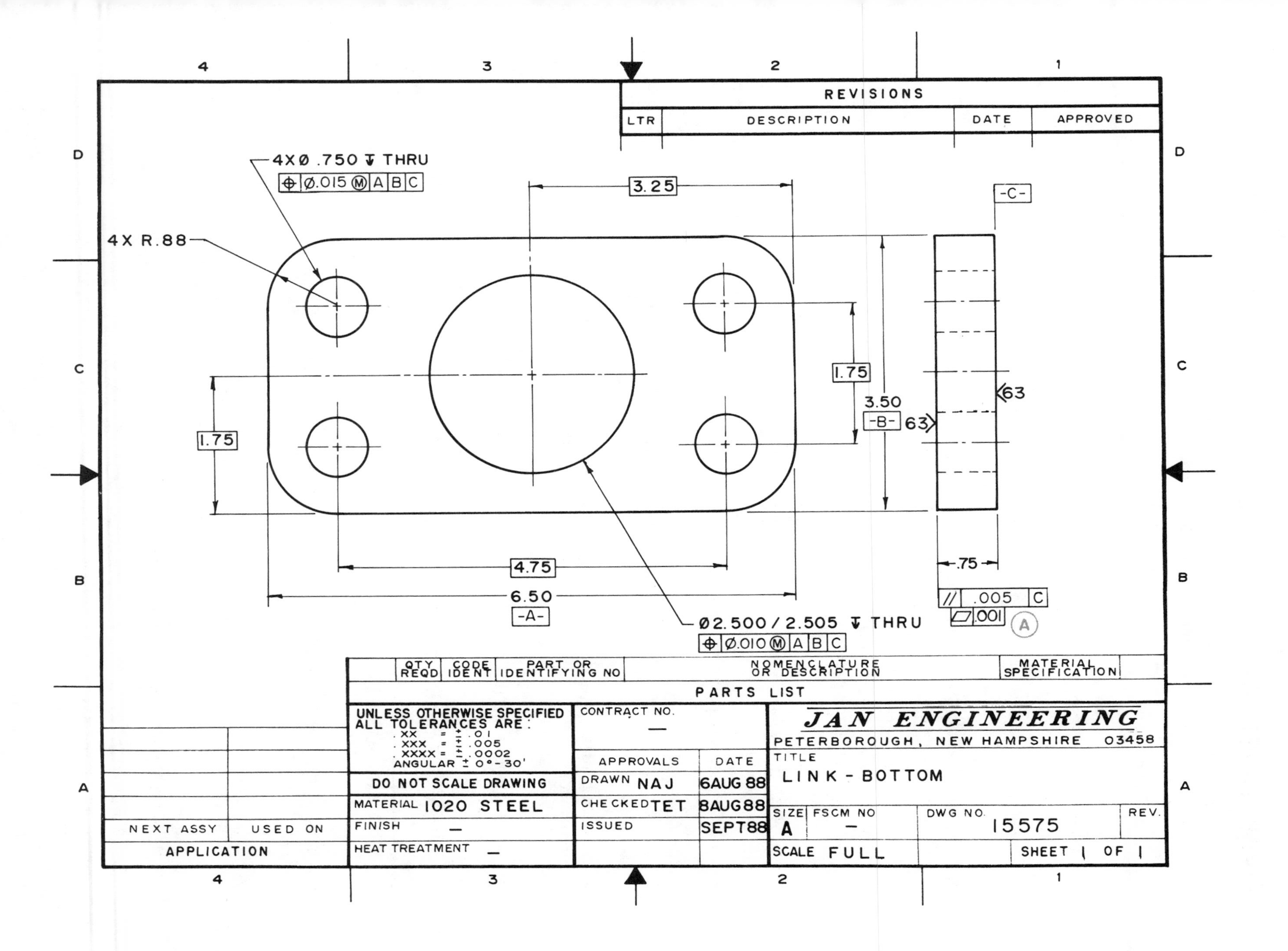
REVISIONS
LTR
DESCRIPTION
DATE
APPROVED
4XØ .750 ↧ THRU
⌖ Ø.015 Ⓜ A B C
4 X R .88
3.25
-C-
1.75
3.50
-B-
63
63
1.75
4.75
6.50
-A-
.75
// .005 C
▱ .001
A
Ø2.500 / 2.505 ↧ THRU
⌖ Ø.010 Ⓜ A B C
QTY REQD
CODE IDENT
PART OR IDENTIFYING NO
NOMENCLATURE OR DESCRIPTION
MATERIAL SPECIFICATION
PARTS LIST
UNLESS OTHERWISE SPECIFIED ALL TOLERANCES ARE:
.XX = ± .01
.XXX = ± .005
.XXXX = ± .0002
ANGULAR ± 0°-30'
DO NOT SCALE DRAWING
MATERIAL 1020 STEEL
FINISH —
HEAT TREATMENT —
CONTRACT NO. —
APPROVALS
DATE
DRAWN NAJ 6AUG 88
CHECKED TET 8AUG88
ISSUED SEPT88
JAN ENGINEERING
PETERBOROUGH, NEW HAMPSHIRE 03458
TITLE
LINK - BOTTOM
SIZE A
FSCM NO —
DWG NO. 15575
REV.
SCALE FULL
SHEET 1 OF 1
NEXT ASSY
USED ON
APPLICATION

WORKSHEET 12-5

Instruction: Using drawing A3371141 on p. 215, answer the following questions in the spaces provided. (Answers in Appendix A.)

1. Is the thread at A coarse or fine?

2. What is dimension B?

3. What is the tolerance on the .6244/.6240 diameter?

4. The key used for this part is a square key—what is the NOMINAL SIZE of the key and approximately, how long should it be?

5. Refer to feature control at D. How far from "straightness" is allowed when the diameter of the part is .6241?

6. How many finished surfaces are there on this part?

7. What finish is required on datum -A-?

8. The surface at E *must* be held parallel to datum -A- within what tolerance?

9. Surface E MUST be held perpendicular to datum -B- diameter within what tolerance?

10. What diameter has the *smoothest* finish—how can you tell?

11. What kind of steel is called-off for this part?

12. What is the MAXIMUM size diameter C can be and still be within limits?

13. What is the MINIMUM distance the two 63 microinch surfaces can be apart?

14. What does the term "TYP" mean?

15. What are the limit dimensions for datum -B-?

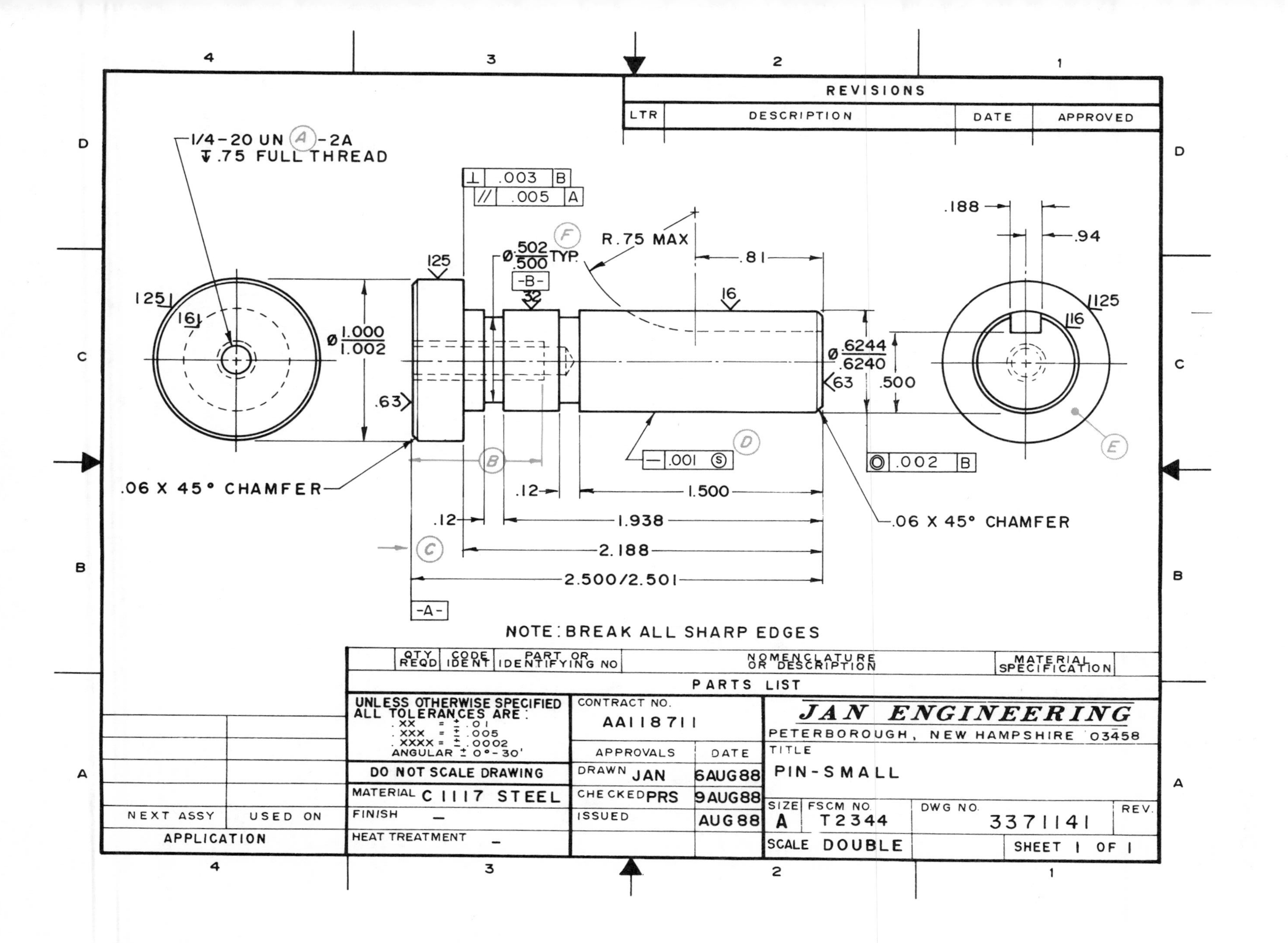
REVISIONS
LTR
DESCRIPTION
DATE
APPROVED
1/4-20 UN-2A
↧.75 FULL THREAD
⊥ .003 B
// .005 A
R.75 MAX
Ø .502/.500 TYP.
-B-
.81
.188
.94
Ø 1.000/1.002
Ø .6244/.6240
.500
.06 X 45° CHAMFER
.001 Ⓢ
.002 B
.06 X 45° CHAMFER
.12
1.500
1.938
2.188
2.500/2.501
-A-
NOTE: BREAK ALL SHARP EDGES
QTY REQD
CODE IDENT
PART OR IDENTIFYING NO
NOMENCLATURE OR DESCRIPTION
MATERIAL SPECIFICATION
PARTS LIST
UNLESS OTHERWISE SPECIFIED ALL TOLERANCES ARE:
.XX = ± .01
.XXX = ± .005
.XXXX = ± .0002
ANGULAR ± 0°-30'
DO NOT SCALE DRAWING
MATERIAL C 1117 STEEL
FINISH -
HEAT TREATMENT -
CONTRACT NO.
AA118711
APPROVALS
DATE
DRAWN JAN 6AUG88
CHECKED PRS 9AUG88
ISSUED AUG88
JAN ENGINEERING
PETERBOROUGH, NEW HAMPSHIRE 03458
TITLE
PIN-SMALL
SIZE A
FSCM NO T2344
DWG NO 3371141
REV.
SCALE DOUBLE
SHEET 1 OF 1
NEXT ASSY
USED ON
APPLICATION

Kinds of Drawings

Objective: The reader will identify the various kinds of drawings used in industry today, be able to use a master parts list, and know the engineering change procedure.

There are many different types of drawings used in industry today. Four of the most widely used types in manufacturing are design, layouts, assembly drawings, subassembly drawings, and detail drawings. It is important for you to be able to identify each type of drawing and understand the differences between each of them.

DESIGN LAYOUT DRAWINGS

The *design layout drawing* is usually done by a designer or engineer. This type of drawing is usually drawn to exact size or as large as space will allow. Usually a design layout is *not* dimensioned except for a few general overall dimensions. This drawing is used to design the product, to check the design function, and to price out the cost of the new product. As a rule, no one outside of the engineering department uses the design layout drawing. It is from this drawing that the drafters develop all the actual working drawings used to manufacture the part.

ASSEMBLY DRAWINGS

Any finished product that has more than one part *must* have an assembly drawing. The assembly drawing illustrates three things: 1. what the finished product will look like; 2. how the finished product is assembled; and 3. what parts are used to make the finished product. For a better understanding, refer ahead to Worksheet 13-1 on p. 224.

Because the assembly drawing is used to show *how* the parts are assembled and *where* the parts are located, many times a full section view is used. There is only *one* final assembly drawing for any given product. As a general rule, an assembly drawing is not dimensioned, as each of the parts are fully dimensioned themselves. If any dimensions are added they usually are very general overall dimensions. Hidden lines are almost always omitted from an assembly drawing unless they are absolutely necessary to illustrate some important hidden feature that might be otherwise missed. (See FIG. 13-1)

An assembly drawing must call off each and every part that is used to make up the assembly. Each part is listed by its part number, code identification (if required), title, and the total number of each part required to assemble one complete assembly. Because individual parts could be made of different kinds of materials, "as noted" is written in the title block.

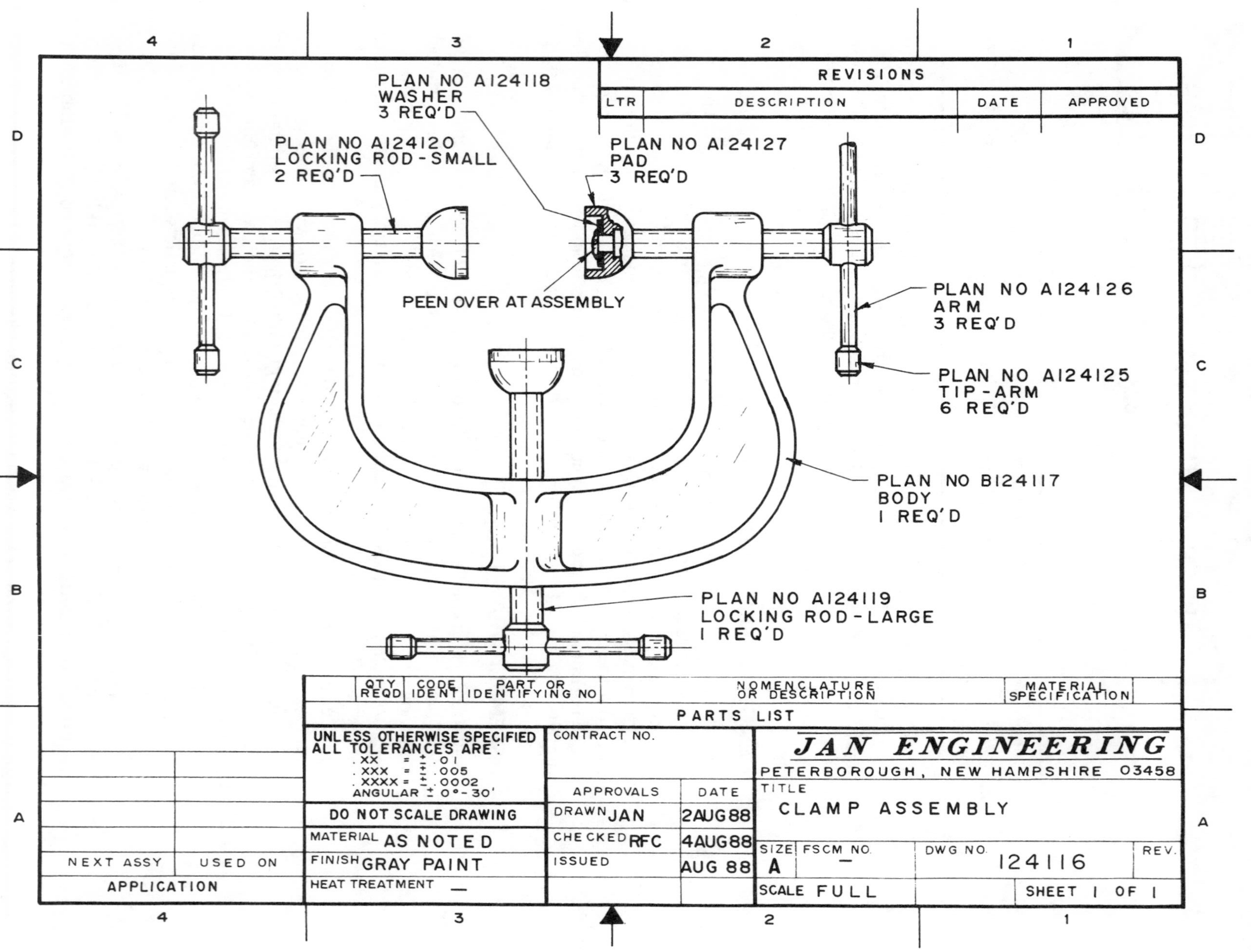

Fig. 13-1. Assembly drawing.

SUBASSEMBLY DRAWINGS

A subassembly drawing is very similar to an assembly drawing and follows all the above drawing practices or rules. Subassemblies are made up of two or more parts that are *permanently* fastened together.

Some subassemblies require special processes done to them after the parts are assembled. In FIG. 13-2, a bushing (part 1) is first pressed into part 2, and then a .750/.755 diameter hole is drilled through both. It would have been almost impossible to drill the holes into each individual part and have the holes line up exactly. Even if it could be done, it would be very expensive.

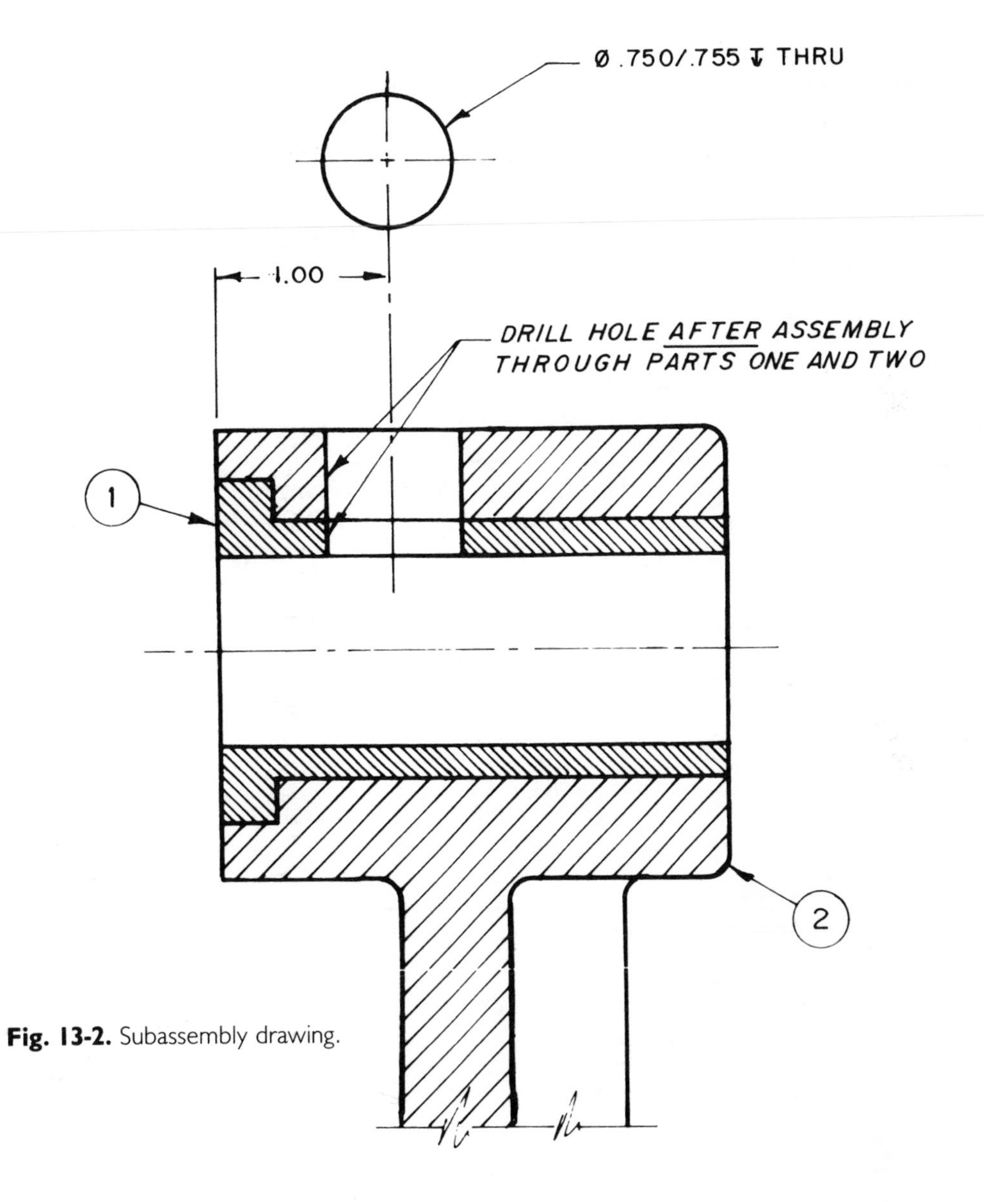

Fig. 13-2. Subassembly drawing.

Any one assembly drawing could have many subassemblies. For example, think of an automobile: the engine is one subassembly, the transmission another, the water pump another, and so on. Each of these subassemblies, along with many individual parts, make up the assembly of the finished automobile.

A subassembly is usually stocked and purchased as a unit. (Refer to Worksheet 13-5 for an example of a subassembly.)

DETAIL DRAWINGS

Each and every part must have its own fully dimensioned detail drawing—each with its own drawing number, code identification (if required), and title or name. It must include *all* information required to fully manufacture the part without any question whatsoever—anywhere in the world.

In general practice in the manufacturing world, only one part is drawn on a sheet of paper, regardless of how small the part is. All worksheet drawings used throughout this book have been of detail drawings.

PURCHASED PARTS

A manufacturing company cannot economically manufacture each and every part used in its product. Standard items such as screws, nuts, bolts, washers, ball bearings, and the like are less expensive from companies that specialize in making these products. Purchased parts do *not* have a company drawing number; therefore, in place of calling off a drawing number, the abbreviation "PURCH" is added. All important technical information must be given when calling for a purchased part.

Modified Purchased Parts

Any purchased part that needs to be changed, however slightly, must be made into a detail drawing and the modified area fully dimensioned (see FIG. 13-3). Under the "material" category in the title block, the standard purchased part is listed. In FIG. 13-3 a standard hex head screw has a modified end with an undercut and a .25 diameter hole added. This modified part is now a standard detail drawing with its own number, code identification (if required), and title.

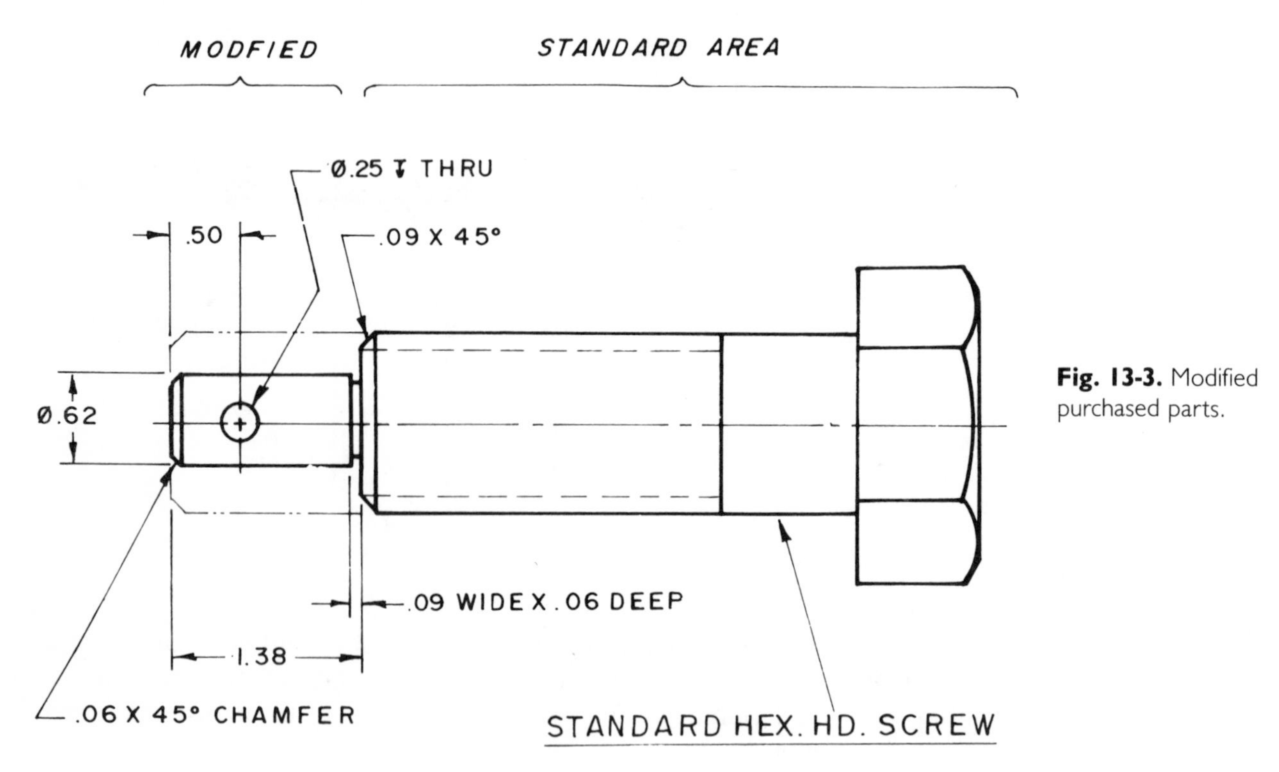

Fig. 13-3. Modified purchased parts.

REVISION OF DRAWINGS

Anyone with a suggestion about how to improve an existing part, how to correct an error on an existing part, or how to manufacture it using a less expensive method, must submit his or her idea to an *engineering change committee.* This idea is submitted to the committee through an *engineering change request* (E.C.R.). This committee is made up of representatives from the various departments throughout the company (see FIG. 13-4). As soon as a drawing has been released to the factory, *no* changes can be made to it without approval of this committee. Even the drafter who originally drew the drawing *cannot* make a change upon this drawing—even if there is an error on it. The E.C.R. must include what change is to be made, why it is to be made, whether it affects

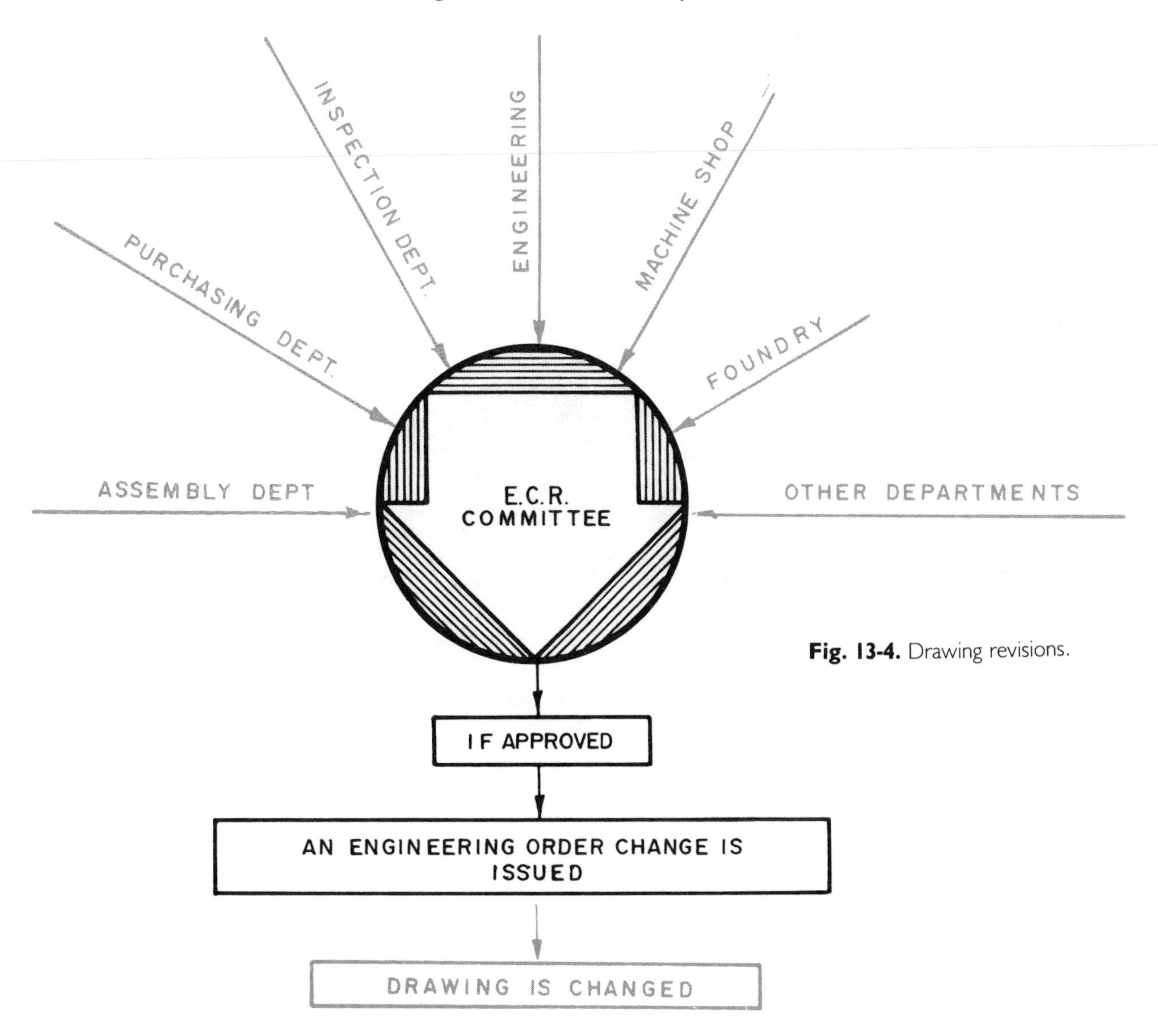

Fig. 13-4. Drawing revisions.

other parts, who suggested the change, the date of request, whether the change is interchangeable with the existing part, and any other important information. Each E.C.R. is numbered and recorded. If the committee approves the change, an *engineering change order* (E.C.O.) is released to the engineering department and a change is made per the order. The E.C.O. includes the information as noted on the E.C.R., the name of the person requesting the change, and the date of the change. Each E.C.O. has its own number, and it, too, is recorded. If the change is extensive, the drawing must be redrawn. The original drawing is *never* destroyed; it is stamped "OBSOLETE" and filed in the obsolete file.

Each company has its own method of indicating where the change has been made on the drawing. Most, however, indicate the change by a small balloon near the dimension or feature that has been changed. When a change has been made, it must be recorded in the revision block and the letter added to the right of the drawing number. In many companies, the *last* revision letter becomes part of the drawing number.

PARTS LIST

A parts list is a bill of materials that itemizes the parts needed to make up one complete unit. Parts lists vary greatly from company to company and are referred to by many different names. For this book, the parts list will be referred to as a *master parts list* (M.P.L.). A parts list must include all drawing numbers, all drawing titles, all code identifications (if required), purchased parts and the required number of each to make up one complete unit of the product. The drafter usually makes up the M.P.L. and the parts usually are listed in the *suggested* order the parts are to be assembled. The parts are usually listed from the *bottom* to the *top* (see FIG. 13-5).

Some companies use a system of *indents*. These indents indicate what kind of a drawing it is and which parts are to be purchased. The first indent lists the one assembly drawing, the second indent indicates all subassemblies, the third indent lists all detail drawings, and the fourth indent lists all purchased parts. Refer again to FIG. 13-5. Note that it is read from the bottom to the top, and lists the parts in the *suggested* order of assembly. It is easy to identify all subassemblies, all detail drawings, and all purchased parts. Not all companies use this system and each has its own methods to call out the parts of an assembly.

REVISIONS

LTR	DESCRIPTION	DATE	APPROVED

LINE	QTY REQD	CODE IDENT	PART OR IDENTIFYING NO	NOMENCLATURE OR DESCRIPTION	MATERIAL SPECIFICATION
17	4		PURCH.	NUT-HEX 1/2-13UNC	STEEL
16	4		PURCH.	BOLT 1/2-13UNC-4.00 LG.	STEEL
15	2		A77841	KEY-SPECIAL VICE	STEEL
14	1		A77962	COLLAR	STEEL
13	4		PURCH.	SCREW 1/4-20UNC-1.00LG.	STEEL
12	2		A77961	PLATE-JAW	STEEL
11	2		A77956	BALL-HANDLE	STEEL
10	1		A77953	ROD-HANDLE	STEEL
9	1		A77954	SCREW-VICE	STEEL
8	1		C77955	JAW SLIDING	STEEL
7	1		PURCH.	NUT-HEX 1/2-13 UNC	STEEL
6	1		PURCH.	BOLT 1/2-13UNC-2.00LG.	STEEL
5	1		A77946	SPACER-BASE	STEEL
4	1		A77951	BASE-UPPER	C.I.
3	1		B77952	BASE-LOWER	C.I.
2	1		C77947	BASE-VICE ASSEMBLY	AS NOTED
1	1		D77942	VICE ASSEMBLY-MACHINE	AS NOTED

PARTS LIST

INDENT (4) PURCHASED PARTS
INDENT (3) DETAIL DRAWINGS
INDENT (2) SUB-ASSEMBLY DRAWINGS
INDENT (1) ASSEMBLY DRAWINGS

UNLESS OTHERWISE SPECIFIED ALL TOLERANCES ARE:
.XX = ± .01
.XXX = ± .005
.XXXX = ± .0002
ANGULAR ± 0°-30'

DO NOT SCALE DRAWING

MATERIAL	—
FINISH	—
HEAT TREATMENT	—

CONTRACT NO. —

APPROVALS		DATE
DRAWN	JAN	6AUG88
CHECKED	RFC	7AUG88
ISSUED		10AUG88

JAN ENGINEERING
PETERBOROUGH, NEW HAMPSHIRE 03458

TITLE: VICE ASSEMBLY-MACHINE M.P.L.

SIZE	FSCM NO.	DWG NO.	REV.
A	—	395711	

SCALE — SHEET 1 OF 1

NEXT ASSY	USED ON
APPLICATION	

Fig. 13-5. Parts list.

WORKSHEET 13-1

Instructions: Using drawing A13169 on p. 225, answer the following questions in the spaces provided.

1. What kind of drawing is this?

2. What are the dimensions called in the columns at the right side of the drawing?

3. How long is part number A13169-5?

4. What is the carbon content of the part?

5. This drawing was drawn to what scale?

6. List what features *all* parts on this drawing have in common.

7. What is the diameter of part number A13169-2?

8. If part A13169-3 is made .253 diameter, will it pass inspection?

9. If part A13169-6 is made .389 diameter, how *far* from straightness can it be and still be within tolerance?

10. What is the required surface finish of this part?

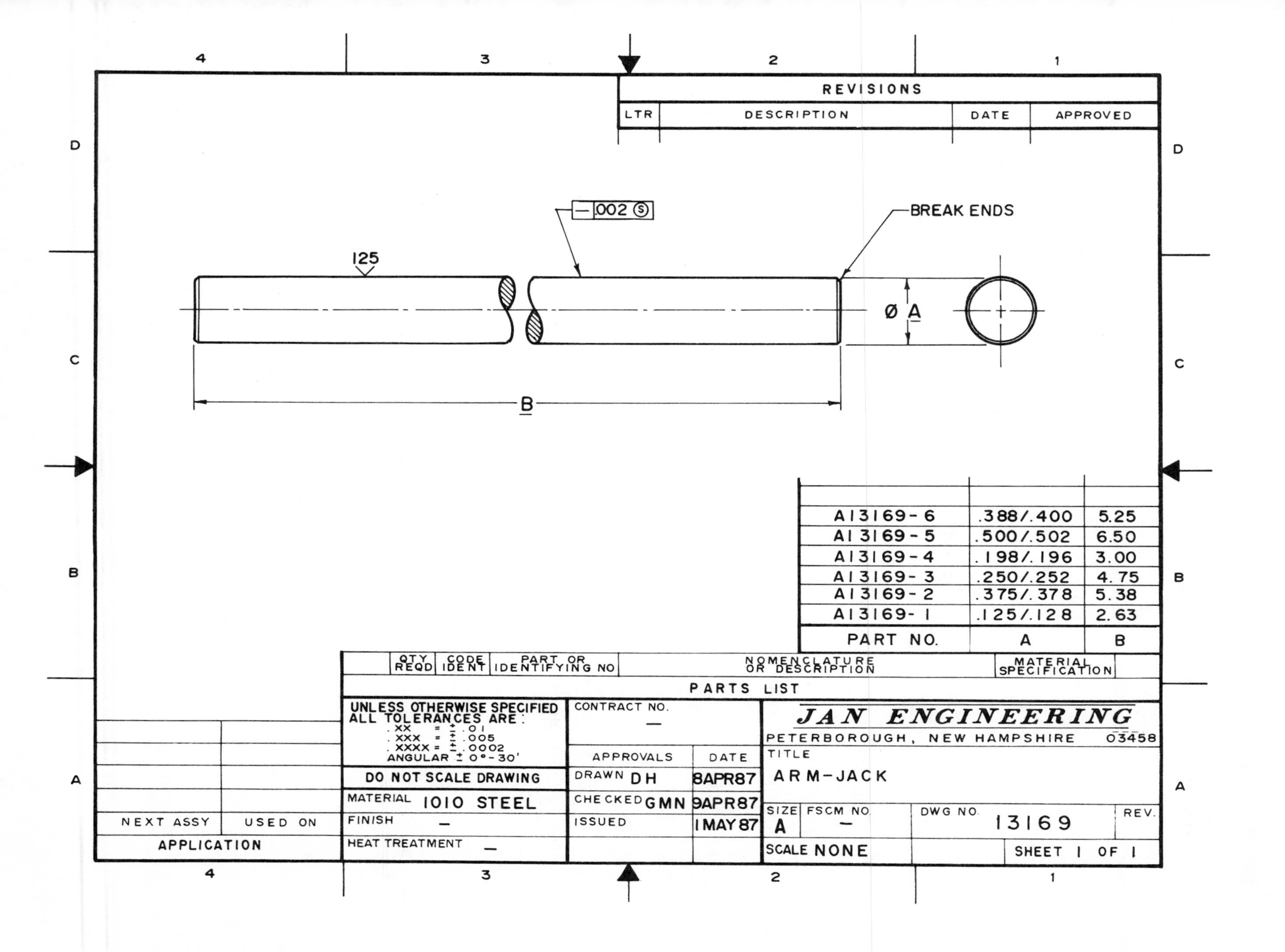
REVISIONS
LTR
DESCRIPTION
DATE
APPROVED
.002 Ⓢ
BREAK ENDS
125
Ø A
B
A13169-6 .388/.400 5.25
A13169-5 .500/.502 6.50
A13169-4 .198/.196 3.00
A13169-3 .250/.252 4.75
A13169-2 .375/.378 5.38
A13169-1 .125/.128 2.63
PART NO. A B
QTY REQD
CODE IDENT
PART OR IDENTIFYING NO
NOMENCLATURE OR DESCRIPTION
MATERIAL SPECIFICATION
PARTS LIST
UNLESS OTHERWISE SPECIFIED ALL TOLERANCES ARE:
.XX = ± .01
.XXX = ± .005
.XXXX = ± .0002
ANGULAR ± 0°-30'
DO NOT SCALE DRAWING
MATERIAL 1010 STEEL
FINISH —
HEAT TREATMENT —
CONTRACT NO. —
APPROVALS
DATE
DRAWN DH 8APR87
CHECKED GMN 9APR87
ISSUED 1MAY87
JAN ENGINEERING
PETERBOROUGH, NEW HAMPSHIRE 03458
TITLE
ARM-JACK
SIZE A
FSCM NO. —
DWG NO. 13169
REV.
SCALE NONE
SHEET 1 OF 1
NEXT ASSY
USED ON
APPLICATION

WORKSHEET 13-2

Instructions: Using drawing A16831 on p. 227, answer the following questions in the spaces provided.

1. What kind of a drawing is this?

2. What kind of a section view is used in this drawing?

3. How far apart are the two finished surfaces?

4. The dimension at the round change-balloon letter A, was what size when the drawing was first issued?

5. The .750 diameter must be concentric to the 1.250 diameter within what tolerance?

6. What is the part made of?

7. The 1.375 diameter *must* be perpendicular to datum -A- within what tolerance?

8. How many finished surfaces are required on this part?

9. The right end of the front view must be parallel to the left end within what tolerance?

10. Is the 3/8-24 thread coarse or fine?

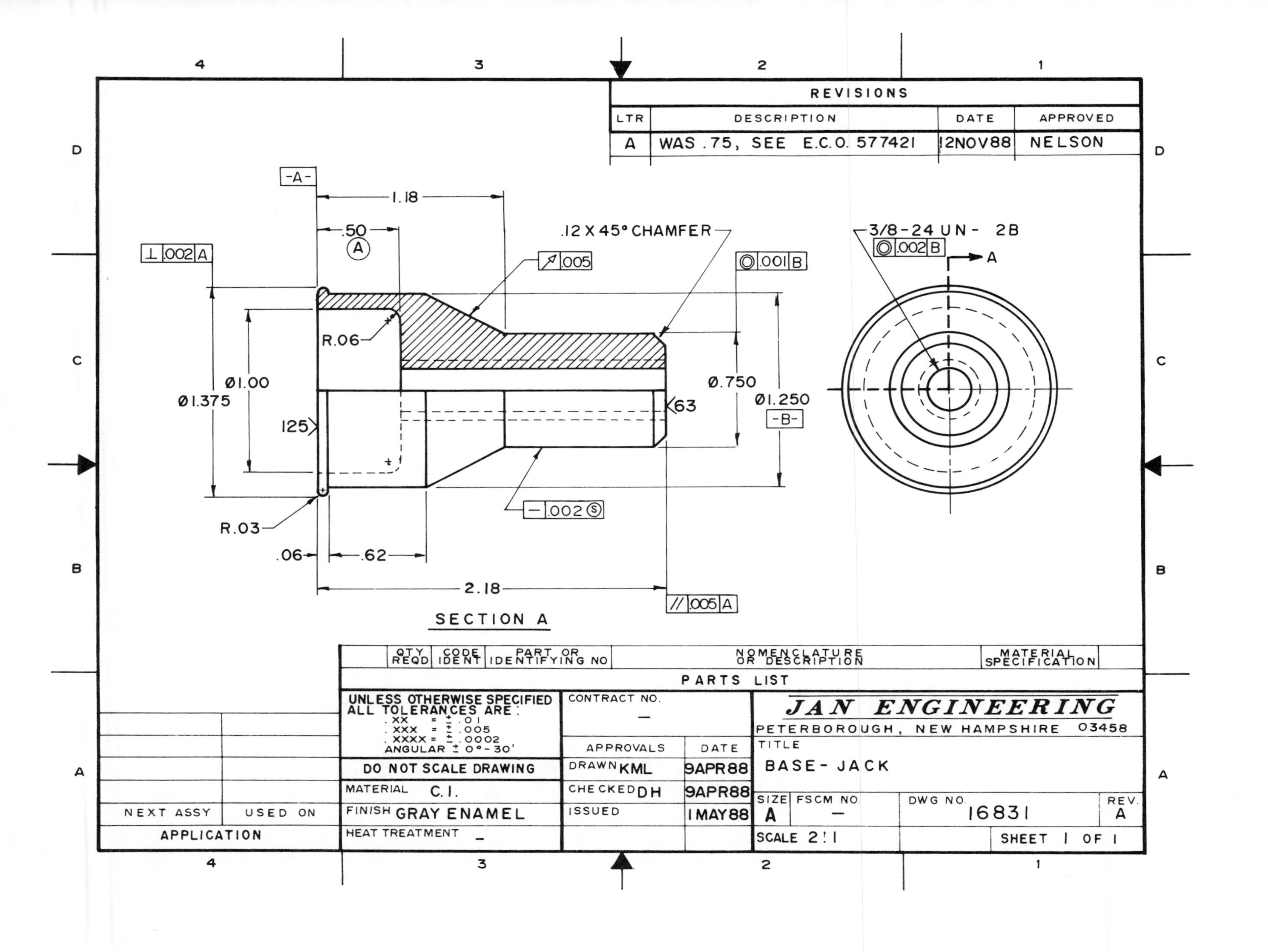
REVISIONS
LTR
DESCRIPTION
DATE
APPROVED
A
WAS .75, SEE E.C.O. 577421
12NOV88
NELSON
-A-
1.18
.50
A
⊥ .002 A
.12 X 45° CHAMFER
.005
.001 B
3/8-24 UN- 2B
.002 B
A
R.06
Ø1.00
Ø1.375
125
Ø.750
63
Ø1.250
-B-
— .002 S
R.03
.06
.62
2.18
// .005 A
SECTION A
QTY REQD
CODE IDENT
PART OR IDENTIFYING NO
NOMENCLATURE OR DESCRIPTION
MATERIAL SPECIFICATION
PARTS LIST
UNLESS OTHERWISE SPECIFIED ALL TOLERANCES ARE:
.XX = ± .01
.XXX = ± .005
.XXXX = ± .0002
ANGULAR ± 0° - 30'
DO NOT SCALE DRAWING
MATERIAL C.I.
FINISH GRAY ENAMEL
HEAT TREATMENT -
CONTRACT NO.
-
APPROVALS
DATE
DRAWN KML
9APR88
CHECKED DH
9APR88
ISSUED
1MAY88
JAN ENGINEERING
PETERBOROUGH, NEW HAMPSHIRE 03458
TITLE
BASE - JACK
SIZE
A
FSCM NO.
-
DWG NO.
16831
REV.
A
SCALE 2:1
SHEET 1 OF 1
NEXT ASSY
USED ON
APPLICATION
1
2
3
4
A
B
C
D

WORKSHEET 13-3

Instructions: Using drawing A16832 on p. 229, answer the following questions in the spaces provided.

1. What kind of drawing is this drawing?

2. How many finished surfaces are called for on this part?

3. What is the tolerance on the .203/.200 diameter hole?

4. How deep are the full 8-36 UNF 2-B threads?

5. Explain in full the meanings of the thread call-off:

   8 ____________________

   36 ____________________

   UN ____________________

   F ____________________

   2 ____________________

   B ____________________

6. What diameter is datum -A-?

7. How *wide* is the undercut?

8. How far apart are the two 63 microinch finished surfaces?

9. The .250/.245 diameter must be concentric to datum -A- within what tolerance?

10. The two 63-microinch surfaces *must* be parallel to each other within what tolerance?

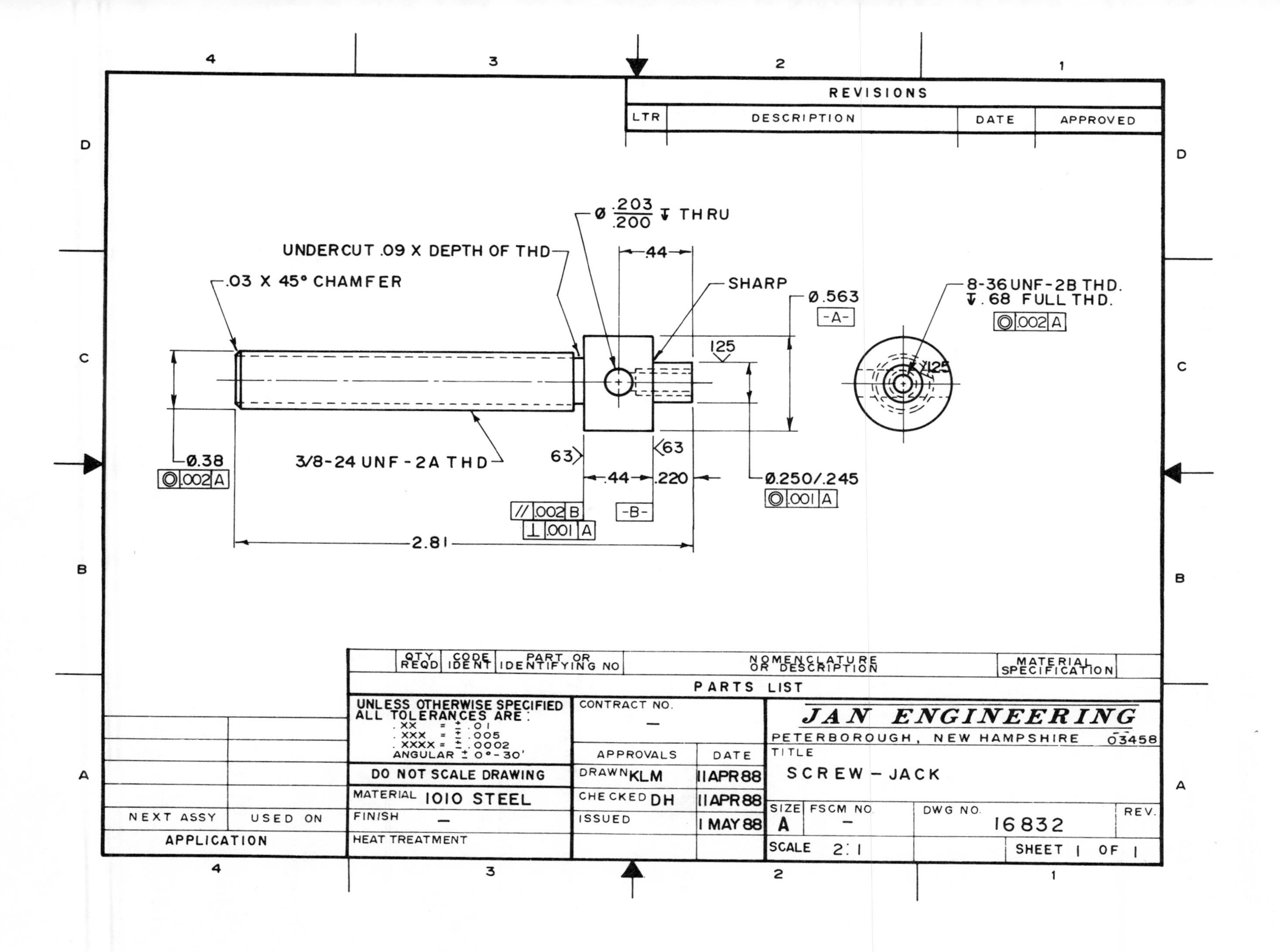
REVISIONS
LTR
DESCRIPTION
DATE
APPROVED
Ø .203/.200 ▼ THRU
UNDERCUT .09 X DEPTH OF THD
.44
.03 X 45° CHAMFER
SHARP
Ø .563
-A-
8-36 UNF-2B THD.
▼ .68 FULL THD.
◎ .002 A
125
125
63
63
Ø.38
◎ .002 A
3/8-24 UNF-2A THD
.44
.220
Ø.250/.245
◎ .001 A
// .002 B
⊥ .001 A
-B-
2.81
QTY REQD
CODE IDENT
PART OR IDENTIFYING NO
NOMENCLATURE OR DESCRIPTION
MATERIAL SPECIFICATION
PARTS LIST
UNLESS OTHERWISE SPECIFIED ALL TOLERANCES ARE:
.XX = ± .01
.XXX = ± .005
.XXXX = ± .0002
ANGULAR ± 0°-30'
DO NOT SCALE DRAWING
MATERIAL 1010 STEEL
FINISH —
HEAT TREATMENT
CONTRACT NO. —
APPROVALS
DATE
DRAWN KLM
11 APR 88
CHECKED DH
11 APR 88
ISSUED
1 MAY 88
JAN ENGINEERING
PETERBOROUGH, NEW HAMPSHIRE 03458
TITLE
SCREW - JACK
SIZE A
FSCM NO. —
DWG NO. 16832
REV.
SCALE 2:1
SHEET 1 OF 1
NEXT ASSY
USED ON
APPLICATION

WORKSHEET 13-4

Instructions: Using drawing A16833 on p. 231, answer the following questions in the spaces provided.

1. What kind of a drawing is this?
2. What is the carbon content of this part?
3. The ends must be parallel to each other within what tolerance?
4. How deep is the .195/.193 hole?
5. How wide is the knurl surface to be?
6. What style knurl is called for and what is its *pitch?*
7. How many surfaces are to be finished?
8. How long was the part when the drawing was first issued?
9. What is the *tolerance* on all three-placed dimensions?
10. After the change, is the knurled pattern centered within the .38 length?

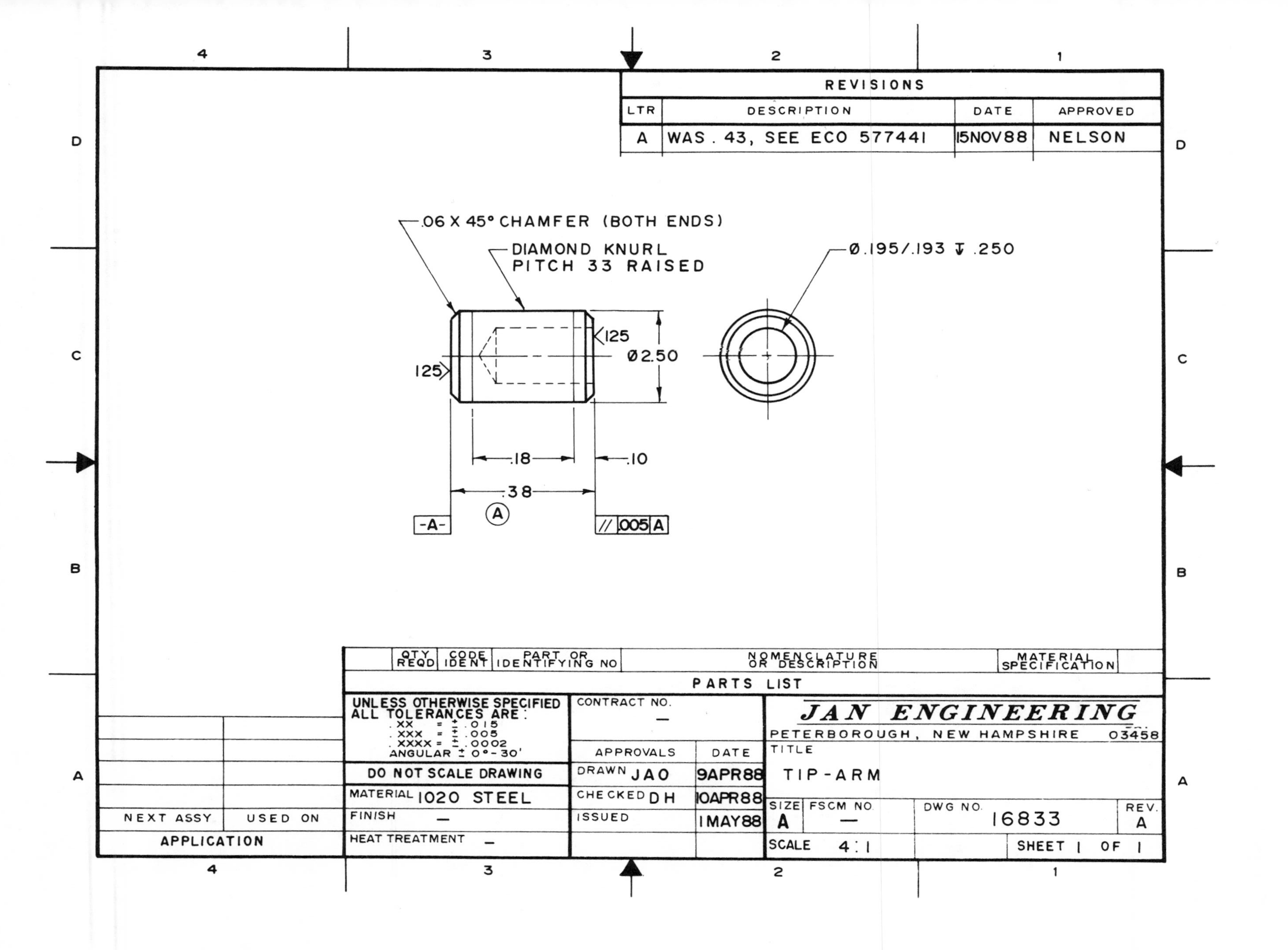
REVISIONS
LTR
DESCRIPTION
DATE
APPROVED
A
WAS .43, SEE ECO 577441
15NOV88
NELSON
.06 X 45° CHAMFER (BOTH ENDS)
DIAMOND KNURL PITCH 33 RAISED
Ø.195/.193 ↧ .250
125
125
Ø2.50
.18
.10
.38
Ⓐ
-A-
// .005 A
QTY REQD
CODE IDENT
PART OR IDENTIFYING NO
NOMENCLATURE OR DESCRIPTION
MATERIAL SPECIFICATION
PARTS LIST
UNLESS OTHERWISE SPECIFIED ALL TOLERANCES ARE:
.XX = ± .015
.XXX = ± .005
.XXXX = ± .0002
ANGULAR ± 0°-30'
DO NOT SCALE DRAWING
MATERIAL 1020 STEEL
FINISH —
HEAT TREATMENT —
CONTRACT NO. —
APPROVALS
DATE
DRAWN JAO
9APR88
CHECKED DH
10APR88
ISSUED
1MAY88
JAN ENGINEERING
PETERBOROUGH, NEW HAMPSHIRE 03458
TITLE
TIP-ARM
SIZE A
FSCM NO. —
DWG NO. 16833
REV. A
SCALE 4:1
SHEET 1 OF 1
NEXT ASSY
USED ON
APPLICATION

WORKSHEET 13-5

Instructions: Using drawing A16834 on p. 233, answer the following questions in the spaces provided.

1. What kind of a drawing is this?
2. How many *different* parts are called for on this drawing?
3. How many *total* parts are called for on this drawing?
4. Fill in the *missing* dimension.
5. What scale was used on this drawing?
6. What do the dash lines on this drawing indicate?
7. What material is called for, for this part?
8. Have any changes been made on this drawing?
9. Explain why hidden lines are *not* used on this drawing.
10. What is the drawing number for the *jack arm?*

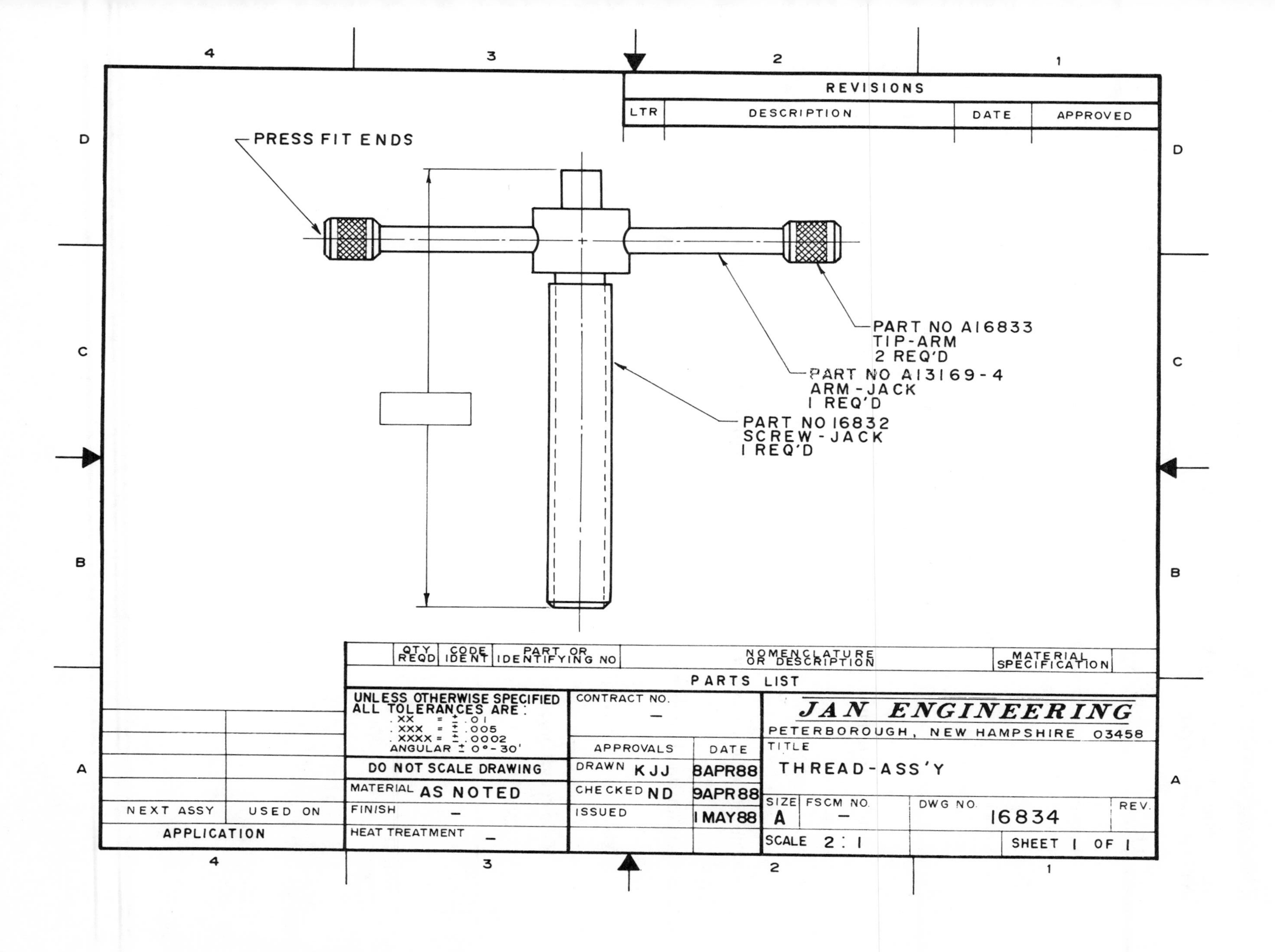

REVISIONS
LTR
DESCRIPTION
DATE
APPROVED
PRESS FIT ENDS
PART NO A16833
TIP-ARM
2 REQ'D
PART NO A13169-4
ARM-JACK
1 REQ'D
PART NO 16832
SCREW-JACK
1 REQ'D
QTY REQD
CODE IDENT
PART OR IDENTIFYING NO
NOMENCLATURE OR DESCRIPTION
MATERIAL SPECIFICATION
PARTS LIST
UNLESS OTHERWISE SPECIFIED ALL TOLERANCES ARE:
.XX = ±.01
.XXX = ±.005
.XXXX = ±.0002
ANGULAR ± 0°-30'
DO NOT SCALE DRAWING
MATERIAL AS NOTED
FINISH —
HEAT TREATMENT —
CONTRACT NO. —
APPROVALS
DATE
DRAWN KJJ 8APR88
CHECKED ND 9APR88
ISSUED 1 MAY88
JAN ENGINEERING
PETERBOROUGH, NEW HAMPSHIRE 03458
TITLE
THREAD-ASS'Y
SIZE A
FSCM NO. —
DWG NO. 16834
REV.
SCALE 2:1
SHEET 1 OF 1
NEXT ASSY
USED ON
APPLICATION

WORKSHEET 13-6

Instructions: Using drawing A16835 on p. 235, answer the following questions in the spaces provided.

1. What kind of a drawing is this?

2. How many finished surfaces are called for and which surface (s) is the smoothest?

3. What kind of a section view is this?

4. How many notches are there and what is nominal size of each?

5. There are two dimensions within parentheses; what does this mean?

6. What is datum -B- ?

7. The .563 diameter *must* be concentric to datum -B- within that tolerance?

8. The top and bottom surfaces must be parallel to what tolerance?

9. What is the tolerance called for on all two-placed decimals?

10. If the center hole was made .253 diameter, would it pass inspection?

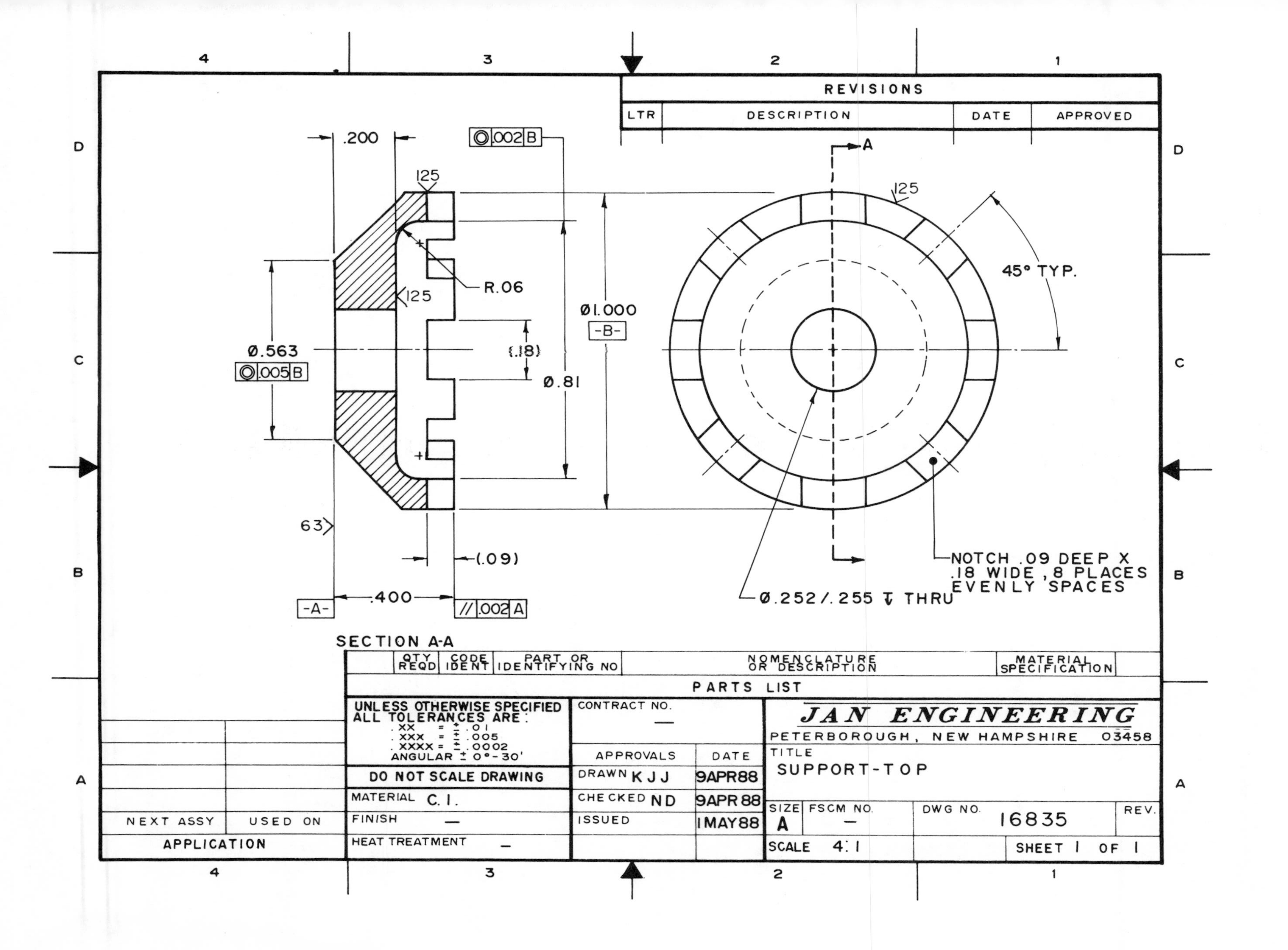
REVISIONS
LTR
DESCRIPTION
DATE
APPROVED
.200
.002 B
125
R.06
125
Ø1.000
-B-
Ø.563
.005 B
(.18)
Ø.81
A
125
45° TYP.
63
(.09)
-A-
.400
// .002 A
NOTCH .09 DEEP X .18 WIDE, 8 PLACES EVENLY SPACES
Ø.252/.255 ↧ THRU
SECTION A-A
QTY REQD
CODE IDENT
PART OR IDENTIFYING NO
NOMENCLATURE OR DESCRIPTION
MATERIAL SPECIFICATION
PARTS LIST
UNLESS OTHERWISE SPECIFIED ALL TOLERANCES ARE:
.XX = ± .01
.XXX = ± .005
.XXXX = ± .0002
ANGULAR ± 0°-30'
DO NOT SCALE DRAWING
MATERIAL C.I.
FINISH —
HEAT TREATMENT —
CONTRACT NO. —
APPROVALS
DATE
DRAWN KJJ 9APR88
CHECKED ND 9APR88
ISSUED 1MAY88
JAN ENGINEERING
PETERBOROUGH, NEW HAMPSHIRE 03458
TITLE
SUPPORT-TOP
SIZE A
FSCM NO. —
DWG NO. 16835
REV.
SCALE 4:1
SHEET 1 OF 1
NEXT ASSY
USED ON
APPLICATION

WORKSHEET 13-7

Instructions: Using M.P.L. A16836 and drawing A16836 on pp. 237 and 239, answer the following questions.

1. What kind of drawing is this?
2. Fill in the missing drawing.
3. How many *different* parts are used on this drawing?
4. How many *total* parts are used on this drawing?
5. What is the designed capacity (movement) of this *jack?*
6. Explain what holds the top support, A16835, in place?
7. Why are parts A16833, A13169-4, and A16832 *not listed?*
8. How thick is the 1/4 washer?
9. How many detail drawings are used on this assembly?
10. How many parts are purchased?

REVISIONS

LTR	DESCRIPTION	DATE	APPROVED

PARTS LIST

LINE	QTY REQD	CODE IDENT	PART OR IDENTIFYING NO	NOMENCLATURE OR DESCRIPTION	MATERIAL SPECIFICATION	
15						
14	1		PURCH.	SCREW - PAN HD. MACH	-	
13				NO. 8-36UNF-2AX.50LG		
12	1		PURCH.	WASHER-PLAIN	-	
11				1/4 SIZE- (.065 TK.)		
10	1		A16831	BASE-JACK	C.I.	
9						
8	1		A13169-4	ARM - JACK	1010 STEEL	
7	1		A16832	SCREW - JACK	1010 STEEL	
6	2		A16833	TIP-ARM	1020 STEEL	
5	1		A16834	THREAD - ASSEMBLY	AS NOTED	
4						
3	1		A16835	SUPPORT - TOP	C. I.	
2						
1	1		A16836	JACK ASSEMBLY	AS NOTED	

UNLESS OTHERWISE SPECIFIED ALL TOLERANCES ARE:
.XX = ± .01
.XXX = ± .005
.XXXX = ± .0002
ANGULAR ± 0° - 30'

DO NOT SCALE DRAWING

MATERIAL —
FINISH —
HEAT TREATMENT —

CONTRACT NO. —

APPROVALS		DATE
DRAWN	LMP	9APR88
CHECKED	DH	9APR88
ISSUED		1MAY88

JAN ENGINEERING
PETERBOROUGH, NEW HAMPSHIRE 03458

TITLE
JACK ASSEMBLY M.P.L.

SIZE	FSCM NO.	DWG NO.	REV.
A	—	16836	

SCALE NONE SHEET 1 OF 1

NEXT ASSY	USED ON
APPLICATION	

WORKSHEET 13-8

Instructions: Using M.P.L. A16836 and the assembly drawing A16836 on pp. 237 and 239, and all required detailed drawings, answer the following questions in the spaces provided.

1. What kind of a *fit* is between parts A13169-4 and A16833?

2. What is the *allowance* between parts A16832 *(jack screw)* and A16835 *(top support)?*

3. How far *above* the 125 microinch finish surface of part A16835 *(top surface)* will the .250/.245 diameter feature of part A16832 *(jack screw)* extend?

4. What are the *allowance* and *clearance* between parts A16832 *(jack screw)* and part A13169-4 *(jack arm)?*

5. The *thickness* of the head of a number 8-36 UNF *pan head machine screw* is .096 with the washer and screw in place. Will the top of the head screw extend *above* or *below* the *top support,* A16835 and by how much?

6. Using the given parallel geometric tolerancing of parts A16831 *(jack base),* A16832 *(jack screw)* and A16835 *(top support)*—at *worst condition,* how far from parallel could the bottom surface of the *jack base* be from the *top support* and still be within tolerance?

7. Refer to part A16832 *(jack screw)* why was the undercut added?

8. If the *jack arm,* part of A13169-4 is pushed to the *right* as far as possible, what is the distance from the center of the *jack* to the right tip of part A16833, *arm tip?*

9. With part A16832 *(jack screw)* in its *lowest* position into part A16831 *(jack base),* how much will the threads of part A16832 extend into the hollow of part A16831?

10. With the *jack arm,* part A13169-4 rotated 12 *full turns* counter-clockwise, how far *up* will the *top support,* A16835, raise?

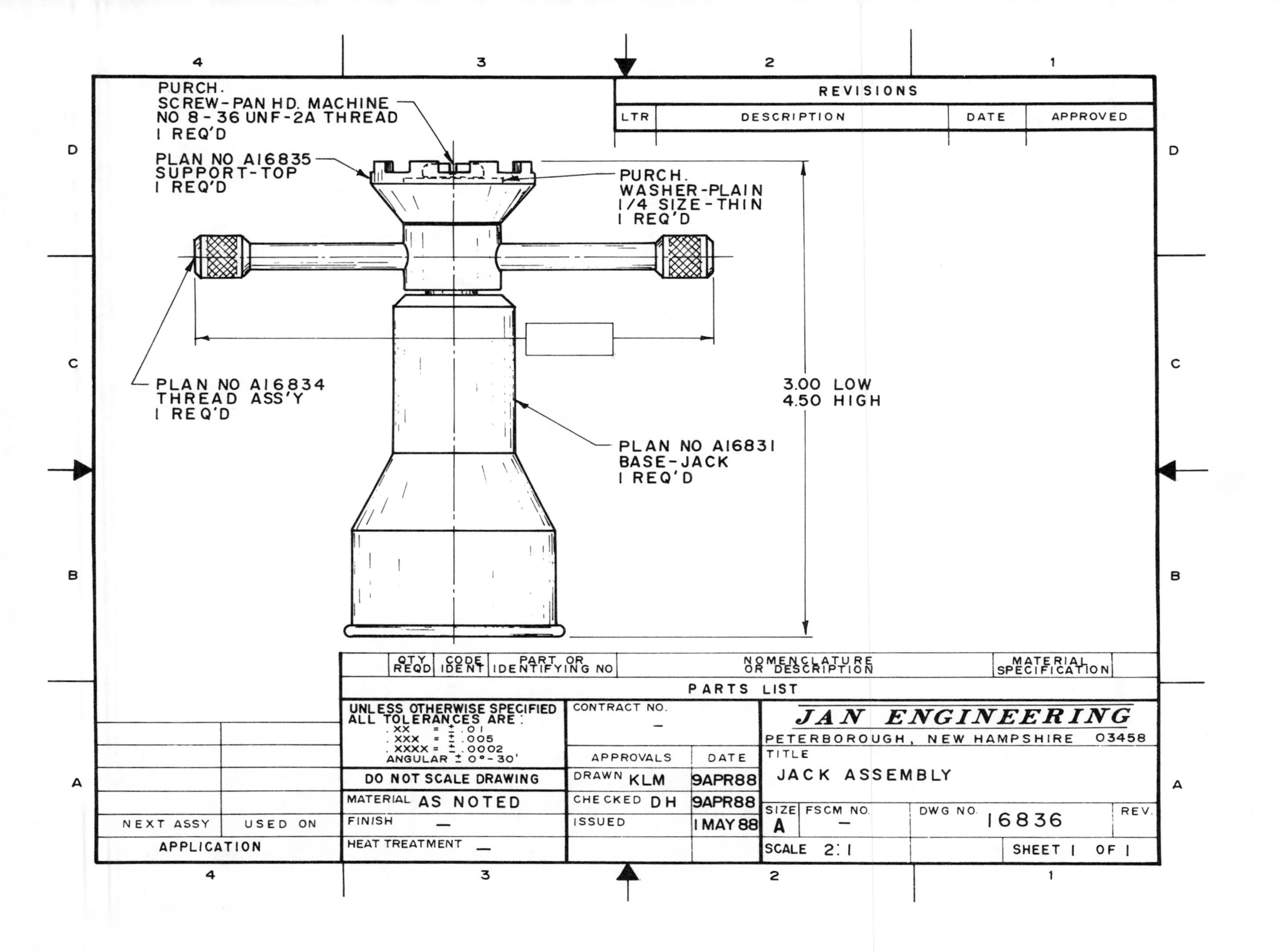
PURCH.
SCREW-PAN HD. MACHINE
NO 8-36 UNF-2A THREAD
1 REQ'D
PLAN NO A16835
SUPPORT-TOP
1 REQ'D
PURCH.
WASHER-PLAIN
1/4 SIZE-THIN
1 REQ'D
PLAN NO A16834
THREAD ASS'Y
1 REQ'D
3.00 LOW
4.50 HIGH
PLAN NO A16831
BASE-JACK
1 REQ'D
REVISIONS
LTR
DESCRIPTION
DATE
APPROVED
QTY REQD
CODE IDENT
PART OR IDENTIFYING NO
NOMENCLATURE OR DESCRIPTION
MATERIAL SPECIFICATION
PARTS LIST
UNLESS OTHERWISE SPECIFIED ALL TOLERANCES ARE:
.XX = ±.01
.XXX = ±.005
.XXXX = ±.0002
ANGULAR ± 0°-30'
DO NOT SCALE DRAWING
MATERIAL AS NOTED
FINISH —
HEAT TREATMENT —
CONTRACT NO. —
APPROVALS
DATE
DRAWN KLM 9APR88
CHECKED DH 9APR88
ISSUED 1 MAY 88
JAN ENGINEERING
PETERBOROUGH, NEW HAMPSHIRE 03458
TITLE
JACK ASSEMBLY
SIZE A
FSCM NO. —
DWG NO. 16836
REV.
SCALE 2:1
SHEET 1 OF 1
NEXT ASSY
USED ON
APPLICATION

Worksheet Answers

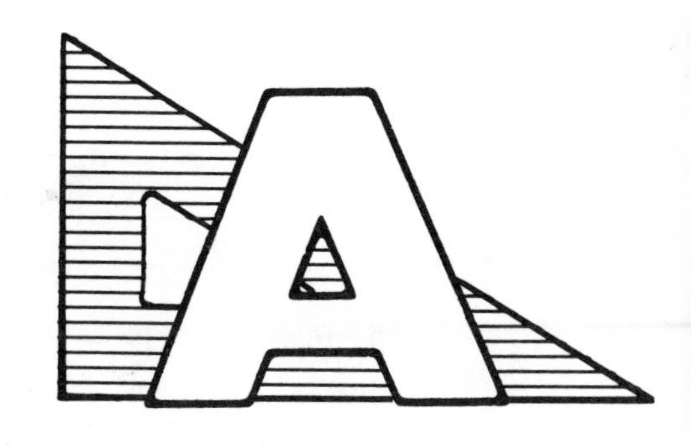

WORKSHEET 1-1

PROB.	DIMENSION	GIVEN TOLERANCE	UPPER LIMIT	LOWER LIMIT	TOLERANCE
1	.75	.XX ± .01	.76	.74	.02
2	.62	.XX ± .02	.64	.60	.04
3	.720	.XXX ± .00	.725	.715	.010
4	2.56	.XX ± .01	2.57	2.55	.02
5	7.0625	.XXXX ± .0002	7.0627	7.0623	.0004
6	3.875	.XXX ± .010	3.885	3.865	.020
7	1.0032	.XXXX ± .0001	1.0033	1.0031	.0002
8	.250-.251	(AS GIVEN)	.251	.250	.001
9	0.625	.XXX ± .010	.635	.615	.020
10	3/4	FRAC. ± 1/64	49/64	47/64	1/32

WORKSHEET 1-2

1	SMALLEST SIZE (A) COULD BE	.47
2	LARGEST SIZE (A) COULD BE	.52
3	SMALLEST SIZE (B) COULD BE	1.194
4	LARGEST SIZE (B) COULD BE	1.206
5	SMALLEST SIZE OF HOLE	.5623
6	LARGEST SIZE OF HOLE	.5627

WORKSHEET 1-3

1. Swivel support
2. A080635
3. BOW1755319
4. February, 1988
5. Full size
6. "A" size (8 1/2 × 11)
7. No revisions or changes

8. ± .005
9. .500 dia. hole, ± .005
10. "A" = 3.875 (2.500 + .75 + .625 = 3.875)
11. 4.25 (4 1/2)
12. .625 thick
13. Aluminum
14. Model no. 1124
15. .76/.74 (.75 + .01 = .76 / .75 − .01 = .74)
16. .75 dimension

WORKSHEET 2-1

PROB.	INCH FRACTION	INCH DECIMAL	METRIC (mm)
1	1/2	.500	12.7
2	1/16	.062	1.5875
3	11/16	.6875	17.46
4	2 15/16	2.9375	74.6125
5	7 7/16	7.44	188.9125
6	2	2.00	50.80
7	11/16	.6875	14.4625
8	1 7/16	1.4374	36.51
9	4 1/2	4.50	114.30
10	10 7/8	10.88	276.225

WORKSHEET 2-2

PROB	FRACTION	DECIMAL	METRIC
1	3/4	.75	19.05
2	1 5/8	1.625	41.275
3	2 13/16	2.8125	71.4375
4	3 9/16	3.5625	90.4875
5	1/2	.50	.01969
6	1 19/64	1.30	33.02
7	2 41/64	2.65	67.31
8	3 1/4	3.25	82.55
9	51/64	.7874	20
10	1 47/64	1.73228	44
11	2 29/32	2.91339	74
12	3 17/64	3.26772	83

WORKSHEET 2-3

1. 3.255
2. C795580
3. .500
4. 1.000/.998 diameter
5. X-76441
6. May, 1988
7. Ø 3.500
8. Ø 2.00 ± .01
9. 34°30′ − 35°30′
10. Full scale
11. 63.5 mm
12. Steel
13. JAO, April 1, 1988
14. None
15. .745

WORKSHEET 3-1 (Part One)

PROB.	LET	KIND OF LINE	HOW LINE IS IDENTIFIED
EXAMPLE	A	HIDDEN LINE	THIN DASHED LINE
1	B	CENTER	THIN LONG & SHORT DASHES
2	C	OBJECT	THICK - SOLID
3	D	HIDDEN	THIN - DASHED
4	E	HIDDEN	THIN - DASHED
5	F	DIMENSION	THIN - SOLID
6	G	LEADER	THIN - SOLID
7	H	DIMENSION	THIN - SOLID
8	I	EXTENSION	THIN - SOLID
9	J	HIDDEN	THIN - DASHED
10	K	EXTENSION	THIN - SOLID

WORKSHEET 3-1 (Part Two)

1. A40937143
2. Full
3. Two
4. There is no Ø 3,500 hole
5. Chrome plate
6. 1.995
7. 8½ × 11
8. Brass
9. 1.235 (1.995 − .76 = 1.235)
10. 4 × Ø .50 thru ⊽ (hole call-off)

WORKSHEET 4-1

1. 8.50 (5.25 + .75 + 2.5 = 8.50)
2. 1.25
3. 47.498 (22.098 + 25.4 = 47.498)
4. 3.25
5. 4.37
6. 1.25
7. Object line
8. Leader line
9. .06
10. 1.75
11. 1.01
12. ∅ .56
13. Object line
14. T2-A399751
15. Dimension line

WORKSHEET 4-2

1. 6.13 (5.25 + .88 = 6.13)
2. 3.24
3. 2.02
4. 2.27
5. Object line
6. .015
7. August 3, 1989
8. 2.02
9. Foreshortened leader line
10. ±.01
11. 1.50
12. Full
13. R.75
14. EN57921117
15. 1.38

WORKSHEET 4-3

1. 29°
2. 3.56 B.C.
3. 60° 30′
4. Size
5. Location
6. Object
7. The diameter of the hole pattern.
8. 1/2
9. .03 thick phenolic
10. Object line
11. Centerline
12. 2.25

13. .03
14. 149°
15. 4.75

WORKSHEET 5-1

1. Line F is an extension line.
2. .75
3. 1.62
4. 2.25
5. 1.38
6. Dimension line
7. .50
8. J
9. .75
10. Object line
11. .63
12. .25
13. Hidden line
14. .75
15. Centerline

WORKSHEET 5-2

1. 36
2. V
3. 16
4. F
5. 14 (34 − 20 = 14)
6. K
7. 11 (20 − 9 = 11)
8. A
9. 38
10. Centerline
11. P
12. D
13. 28 (M = 152 − 86 = 66 − 38 = 28)
14. .354
15. 1

WORKSHEET 5-3

1. 2.875
2. J
3. R
4. W
5. .75
6. M, U
7. K

8. .938
9. Y
10. 2.875
11. 1.125 (2.0613 − .938 = 1.125)
12. L
13. 2.063
14. Y
15. .245

WORKSHEET 5-4

1. Thru .56
2. 1.063 (2.188 + 1.125 = 1.063)
3. .75
4. J
5. 3.00
6. Y
7. .62
8. B
9. N, U
10. Depth of counterbore, .25
11. C
12. .74
13. Z
14. The .81 hole
15. D
16. Not to scale. The straight line is the not to scale symbol.
17. Reference dimension
18. 1.50
19. 63214
20. Object line

WORKSHEET 5-5

1. Unidirectional system
2. 1.25 and 2.50
3. .75
4. T
5. 2.00
6. Y
7. F
8. 10.76 (8.00 + 1.38 + 1.38 = 10.76)
9. T
10. G
11. L
12. Was 1.96, thick line = out of scale dimension
13. 3.00
14. 5
15. The 1.25 radius
16. 1.07, 1.05, 8.01
17. 1.38

18. Hole B
19. Half
20. 2.23 (1.49 + .74 = 2.23)

WORKSHEET 6-1

1. Full section
2. 2.25
3. M
4. 1.50 (2.25 − .75 = 1.50)
5. Full
6. 2.75 (3.50 − .75 = 2.75)
7. C
8. 1.75 (3.50 − .75 = 1.75)
9. N
10. .18 = Depth of counterbored hole
11. Cast iron
12. Two
13. 1.01 (1.00 + .01 = 1.01)
14. 1.12 (1.50 − .38 = 1.12)
15. P

WORKSHEET 6-2

1. Half section
2. M
3. 1.25 diameter
4. Hidden line
5. 1.88 diameter
6. .50 radius
7. .38 (2.00 − 1.62 = .38)
8. .500/.505 diameter
9. A
10. H
11. S
12. Brass
13. .31 (1.88 − 1.29 = .62 ÷ 2 = .31)
14. Cutting plane line
15. 2.00

WORKSHEET 6-3

1. Offset section
2. 2.00
3. 1.00
4. .44 radius
5. Bottom of counterbore
6. .62 diameter
7. 5.88 (4.88 + 1.00 = 5.88)

8. F
9. 1.12 (1.62 − .50 = 1.12)
10. Top of pad "A"
11. D
12. 1.12
13. "B" hole
14. 2.00
15. 1.00 (1.62 − .62 = 1.00)
16. .6 R
17. .62
18. U
19. PR 379882
20. S

WORKSHEET 6-4

1. Removed section
2. Left-end, front, right-end views
3. .72
4. .81 diameter
5. 5.25 (see "material" block)
6. .50
7. 1.51 max. (1.50 + .01 = 1.51)
8. .25 diameter
9. 1.11 min. (1.12 − .01 = 1.11)
10. F
11. R
12. P
13. M
14. 1.38
15. .75
16. .75
17. .50 large/.25 small
18. .68 deep
19. 1.502
20. 2.121 diameter (see "material" block)

WORKSHEET 6-5

1. Rotated section
2. 3.74 (3.75 − .01 = 3.74)
3. Fillet
4. 3.62
5. .12 R
6. 1.80 (2.50 − .01 = 2.49, .68 + .01 = .69, 2.49 − .69 = 1.80)
7. .3000 diameter
8. F
9. .72
10. C
11. Break line
12. 1.005 diameter

13. .75
14. Drawn out-of-scale
15. Half
16. .38
17. 13.250 (9.500 + 2.50 + 1.25 = 13.250)
18. 9.505 (9.500 + .005 = 9.505)
19. 9.375
20. K

WORKSHEET 6-6

1. Broken-out section
2. .31 + .01 = .32 upper
   .31 − .01 = .30 lower
3. O
4. Break line
5. 1.250 (1.500 − .25 = 1.250)
6. Aluminum
7. 2:1
8. .500/.501 diameter
9. F
10. H
11. 1.32 (1.500 − .18 = 1.320)
12. 1.095 max. size (.18 − .01 = .17, .25 − .01 = .24, .17 + .24 = .41, 1.500 + .005 = 1.505, 1.505 − .410 = 1.095)
13. Out-of-scale dimension
14. September 4, 1988
15. .001 tolerance (.626 − .625 = .001)

WORKSHEET 6-7

1. Assembly section
2. 6.62 − 3.88 = 2.74 × 5.48 diameter
3. Bushing A193507
4. B193508
5. B193504
6. A193503 1 required
7. Thin wall section
8. Web or MB − webs or ribs are not sectioned
9. 2
10. Shafts are not sectioned
11. Nine (9)
12. Fifteen (15)
13. D3307
14. 1/4−20 UNC-2B Hex Head Nut
15. Nut − hex hd. 1/4 − 20 unc − 2B
    Bolt − hex hd. 1/4 − 20 unc − 2A
    Washer − plane. 1/4 size

WORKSHEET 6-8

1. Full section
2. Object line
3. 1.25 diameter
4. .468
5. 3.50 diameter
6. 2.75 diameter
7. 2.59 min. size (.50 + .01 = .51, .38 + .01 = .39, .51 + .39 = .90, 3.50 − .01 = 3.49, 3.49 − .90 = 2.59)
8. Extension line (centerline)
9. .58 diameter
10. 4.25 B.C.
11. 1.49 (.50 + .38 = .88, .75 + .38 = 1.13, 1.13 + .88 = 2.01, 3.50 − 2.01 = 1.49)
12. 3.00 (3.50 − .50 = 3.00)
13. 3.12 (3.90 − .38 = 3.12)
14. .438 + .005 = .443
    .438 − .005 = .433
15. Brass

WORKSHEET 6-9

1. Full section
2. G
3. U
4. 4.50 diameter
5. 1.18 diameter
6. A or E
7. 5.00 − 4.50 = .50 .2 = .25
8. D
9. B
10. .500 diameter
11. .578 (Keyway)
12. .06R
13. .50 + .50 = 1.00
    2.50 − 1.00 = 1.50
14. 240° (360° − 120° = 240°)
15. 5.00 diameter

WORKSHEET 7-1

1. Left side, front, right side, auxiliary
2. V, G
3. .74
4. P
5. H, E
6. M
7. 25°
8. T
9. Y

10. H
11. .56
12. 1.12
13. (3.06) because the distance can be found by adding dim.
14. 1.00
15. 5.50 × 2.68 × 3.06

WORKSHEET 7-2

1. Not shown in the auxiliary view
2. Partial
3. .34
4. D, W
5. 1.50
6. .09 represents a fillet
7. 3.750 (1.06 + 1.750 + .94 = 3.750)
8. 3.015 (1.755 + .63 + .63 = 3.015)
9. .12
10. 1.88
11. 2.38
12. 1.12
13. O
14. 1.00
15. 45°
16. Z
17. AB
18. Full

WORKSHEET 7-3

1. .50
2. G, R
3. P
4. 4.00
5. .50, 3.00
6. .22
7. D
8. V
9. There is no top view.
10. 3.12 ÷ 4 = .78
11. L
12. .12
13. .12
14. .37, .39
15. .50

WORKSHEET 7-4

1. Q
2. E
3. Partial
4. C – W
5. G
6. 1.25
7. .38
8. .44
9. .12
10. .25
11. H
12. .75
13. 1.50
14. Not indicated. Need a designation for that surface in the auxiliary view.
15. 1.12

WORKSHEET 8-1

1. M
2. G – J
3. V – R
4. Fine
5. Countersunk thru hole
6. .31 thru .56 .18
7. 28
8. 4.87
9. 3/8 – 24
10. N
11. .12 fillet
12. .630
13. .50
14. .50
15. .87 (1.62 – .75 = .87)

WORKSHEET 9-1

1. R.25
2. .06
3. D
4. 1.38
5. T
6. Undercut – .06 × depth of threads
7. .938
8. J
9. 5/8 – 11 UNC – 2B thread, 1.38 deep
10. 4 (four)
11. 5 (five) threads per inch

12. 1.520 max. (2.625 + .005 = 2.630, 1.12 − .01 = 1.11, 2.630 − 1.110 = 1.520)
    1.490 min. (2.625 − .005 = 2.620, 1.12 + 01 = 1.13, 2.620 − 1.130 = 1.490)
13. Chamfer
14. 1.06
15. 5.625
16. W
17. Brass
18. S
19. 1.00 diameter × 1.88 wide
20. G3

WORKSHEET 9-2

1. Half size, full size
2. 8.76 (.62 + .62 = 1.24, 10.00 − 1.24 = 8.76)
3. 7.00 diameter
4. K
5. F
6. .75
7. B
8. .75
9. Pressboard
10. Min. radii
11. R
12. .75
13. 1.00
14. 4.26 (.87 + .87 = 1.74, 6.00 − 1.74 = 4.26)
15. A22574-5 and A22574-7

WORKSHEET 9-3

1. .75
2. .77 max. (.75 − .01 = .74, 1.50 + .01 = 1.51, 1.51 − .74 = .77 max.)
3. .50 (4.50 − 4.00 = .50)
4. .86 min. (1.62 + .01 = 1.63, 2.50 − .01 = 2.49, 2.49 − 1.63 = .86 min.)
5. .63
6. 1.03
7. 1.27 max. (1.50 − .01 = 1.49, 2.75 + .01 = 2.76, 2.76 − 1.49 = 1.27)
   1.23 min. (1.50 + .01 = 1.51, 2.75 − .01 = 2.74, 2.74 − 1.51 = 1.23)
8. .31 diameter
9. 1.38 (4.500 − 3.12 = 1.38)
10. .50
11. .25 deep
12. Countersunk thru hole
13. Bund hole
14. Threaded thru hole

15. .500/505 ↧ thru, ⌴ ⌀ . 75 ↧ 25

16. G3 Micro inches
17. .170/.172 diameter
18. .405/.410 diameter

WORKSHEET 10-1

1. Ferrous contains iron; nonferrous does not contain iron.
2. A mixture of two or more metals
3. Gray cast iron, white cast iron, malleable iron, ductile iron, wrought iron, (the carbon content makes them different)
4. Carbon steel, alloy steel
5. Nickel-chromium steel; manganese steel; molybdenum steel; vanadium steel; tungsten steel
6. Society of automotive engineers
7. Aluminum; copper; brass; bronze; zinc; magnesium; lead; tin; babbit; pewter; nickel
8. The ability to withstand deformation without breaking
9. ·High cost
10. Less than 1%
11. Breaks with little deformation
12. Controlled heating and cooling prices used to change their internal properties
13. Brinell and Rockwell. (Brinell uses a ball pressed into the surface. Rockwell is read directly.).
14. Relieves internal stresses
15. 2.0 to 2.5% carbon
16. Wrought iron
17. Low carbon steel
18. Molybdenum steel, 10% carbon
19. Brass
20. Tin-based, lead-based. Used for bearing in machines and engines.

WORKSHEET 10-2

1. 125
2. O
3. Front view, left-side view, right-side view
4. 24
5. .06
6. 63
7. 1.27 (1.25 + .01 = 1.26, 1.26 + 1.26 = 2.52, 3.78 + .01 = 3.79, 3.79 − 2.92 = 1.27)
8. Coarse
9. H
10. .12
11. Chamfer
12. diamond
13. .06

14. 20
15. 1.55 max. (3.78 + .01 = 3.79, 1.25 − .01 = 1.24, 3.79 − 1.24 = 1.55)
16. 210°
17. 2.50 (1.25 + 1.25 = 2.50)
18. S

WORKSHEET 11-1

1. .18

2. .12

3.

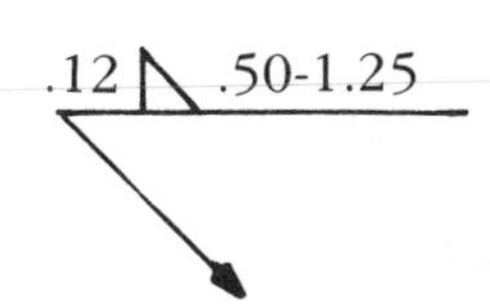

4.

5.

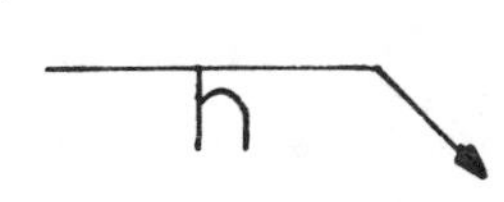

6.

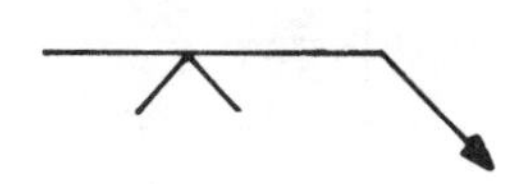

7.

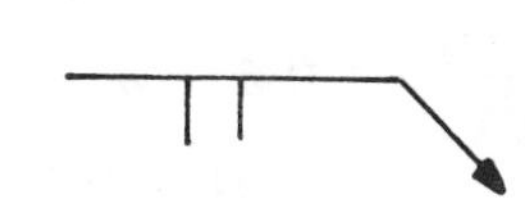

8.

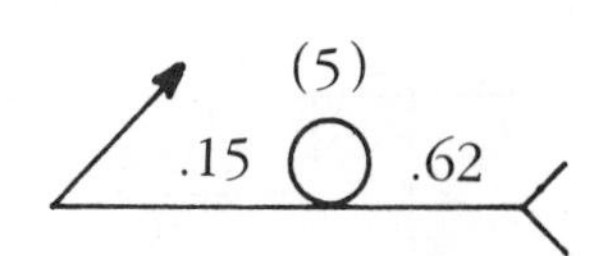

WORKSHEET 11-2

1. V
2. B
3. .18 fillet weld, .75 long, 1.50 center to center
4. U
5. Weld all around
6. J
7. G
8. F
9. D
10. M
11. .18 Fillet weld
12. B
13. Plug
14. 1010
15. The welds are on the arrow side.

WORKSHEET 12-1

1)

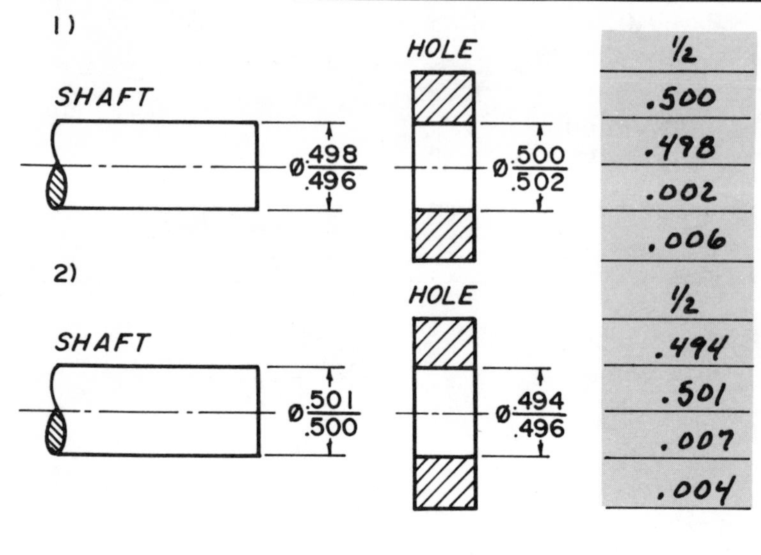

½	NOMINAL SIZE
.500	MMC OF HOLE
.498	MMC OF SHAFT
.002	ALLOWANCE
.006	CLEARANCE

2)

½	NOMINAL SIZE
.494	MMC OF HOLE
.501	MMC OF SHAFT
.007	ALLOWANCE
.004	CLEARANCE

3)

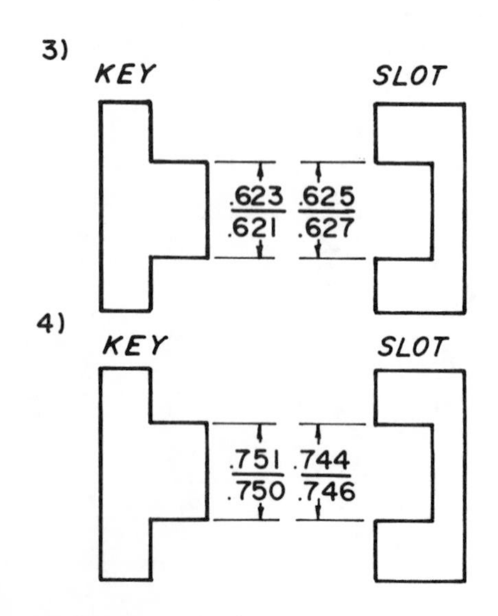

5/8	NOMINAL SIZE
.625	MMC OF SLOT
.623	MMC OF KEY
.002	ALLOWANCE
.006	CLEARANCE

4)

3/4	NOMINAL SIZE
.744	MMC OF SLOT
.751	MMC OF KEY
.007	ALLOWANCE
.004	CLEARANCE

WORKSHEET 12-2

1)

	SYMBOL	GEOMETRIC TOLERANCE
ANGULARITY	∠	ORIENTATION
TRUE POSITION	⌖	LOCATION
FLATNESS	▱	FORM
PROFILE OF A SURFACE	⌓	PROFILE
PERPENDICULARITY	⊥	ORIENTATION
CIRCULAR RUNOUT	↗	RUN OUT
STRAIGHTNESS	—	FORM
TOTAL RUNOUT	⌰	RUNOUT
PROFILE OF A LINE	⌒	PROFILE
CYLINDRICITY	⌭	FORM
CIRCULARITY	○	FORM

2)

	SYMBOL		SYMBOL
Ⓜ	MAXIMUM MATERIAL CONDITION	◎	CONCENTRICITY
-A-	DATUM	⌰	TOTAL RUNOUT
( )	REFERENCE ONLY	Ø	DIA.
R	RADIUS	Ⓢ	REGARDLESS OF FEATURE SIZE
↧	DEPTH	.25	FEATURE TOLERANCE
∨	COUNTER SINK	⌴	COUNTERBORE

3)

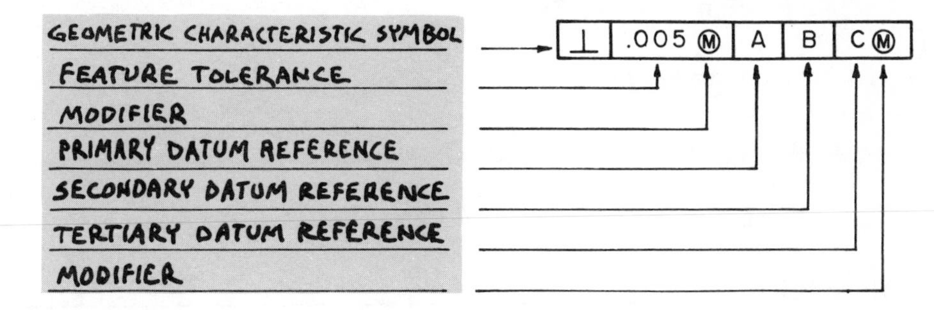

WORKSHEET 12-3

1. H
2. .005
3. 2.00
4. .41 (.62 − .21 = .41)
5. 16
6. 7
7. This surface must be parallel to surface C within .002.
8. Diameter B − (3.00)
9. 3
10. 1.50 diameter
11. 4.000
12. Datum surface (feature control)
13. 1.000 + .005 = 1.005 upper limit
    1.000 − .005 - .995 lower limit
14. .500
15. .50 (3.00 − 2.0 = 1.00 ÷ 2 = .50)

WORKSHEET 12-4

1. 1020 steel
2. .755 (.750 + .005 = .755)
3. Exact size
4. Datum surfaces (feature control)
5. .75
6. This surface must be parallel to surface C within .005, and flat within .001.
7. .20%
8. 2.500 diameter
9. Datum A
10. .015 (.010 + .005 = .015)

WORKSHEET 12-5

1. Coarse
2. .75 (full thread)
3. .0004 (.6244 − .6240 = .0004)
4. .18% − .75 long to .81 long
5. .001
6. 5
7. 63
8. .005
9. .003
10. .6244/.6240, microfinish 32
11. C1117
12. 1.002 diameter
13. 2.500
14. Typical
15. 502/.500

WORKSHEET 13-1

1. Tabulation drawing
2. Tab dimensions (variable)
3. 6.50
4. 1010 steel = .10% carbon
5. None
6. Breaks at ends, flat within .002, 1010 steel, name, finish of 125
7. 375/.378 diameter
8. No—.001 too large.
9. .002
10. .125

WORKSHEET 13-2

1. Detail drawing
2. Half section
3. 2.18
4. Was .75
5. .001
6. C.I. (cast iron)
7. .002
8. 2
9. .005
10. Fine

WORKSHEET 13-3

1. Detail drawing
2. 3
3. .003 tolerance
4. .68 full

5. 8 – size (.1640 diameter)
   36 – threads per inch
   UN – unified national
   F – fine
   2 – class 2
   B – internal threads
6. .563 diameter
7. .09 wide
8. .44
9. .001
10. .002

WORKSHEET 13-4

1. Detail drawing
2. .20%
3. .005
4. .250 deep
5. .18
6. Diamond knurl, 33 raised
7. 2
8. .43 long
9. .001
10. Yes.

WORKSHEET 13-5

1. Subassembly
2. 3
3. 4
4. 2.81
5. 2:1 (double)
6. Screw threads
7. As noted - Each detail drawing calls off material of each part.
8. No.
9. Hidden lines are not used on subassembly drawings unless necessary.
10. A13169-4

WORKSHEET 13-6

1. Detail drawing
2. 3, bottom surface (63)
3. Full section
4. 8 notches—.18 wide × .09 deep each
5. (Reference only)
6. 1.000 diameter
7. .005
8. .002
9. .02
10. Yes.

WORKSHEET 13-7

1. Assembly drawing
2. 3.26 (.38 − .25 = .13 × 2 = .26 + 3.00 = 3.26)
3. 5
4. 6
5. 1.50 movement (4.50 − 3.00 = 1.50)
6. Pan head machine screw and plain washer
7. They are used on the subassembly A16834.
8. .065 (see master parts list)
9. Five (5)
10. Two (2)

WORKSHEET 13-8

1. Press fit
2. .010 (.255 − .245 = .010)
3. .020 (.220 − .200 = .020)
4. .002 (.200 − .198 = .002)
   .007 (.203 − .196 = .007)
5. Below, .039 (.096 + .065 = .161 total height of washer/screw head, .400 − .200 = .200, .200 − .161 = .039)
6. .007 (.002 + .005 = .007)
7. So jack screw could be lowered tightly down and into jack base.
8. 2.599 distance (.563 2 = .281, .380 + .281 = .661, 3.260 − .661 = 2.599)
9. 2.15 depth (.44 + .22 = .66, 2.81 − .66 = 2.15 depth)
10. 24 T.P.I., therefore 12 turns = $^{1}/_{2}$ inch up.

Conversion Chart

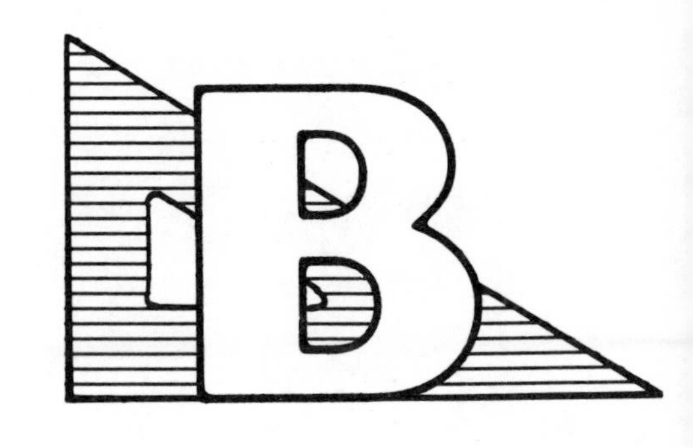

INCH/METRIC – EQUIVALENTS					
Fraction	Decimal Equivalent		Fraction	Decimal Equivalent	
	Customary (in.)	Metric (mm)		Customary (in.)	Metric (mm)
1/64	.015625	0.3969	33/64	.515625	13.0969
1/32	.03125	0.7938	17/32	.53125	13.4938
3/64	.046875	1.1906	35/64	.546875	13.8906
1/16	.0625	1.5875	9/16	.5625	14.2875
5/64	.078125	1.9844	37/64	.578125	14.6844
3/32	.09375	2.3813	19/32	.59375	15.0813
7/64	.109375	2.7781	39/64	.609375	15.4781
1/8	.1250	3.1750	5/8	.6250	15.8750
9/64	.140625	3.5719	41/64	.640625	16.2719
5/32	.15625	3.9688	21/32	.65625	16.6688
11/64	.171875	4.3656	43/64	.671875	17.0656
3/16	.1875	4.7625	11/16	.6875	17.4625
13/64	.203125	5.1594	45/64	.703125	17.8594
7/32	.21875	5.5563	23/32	.71875	18.2563
15/64	.234375	5.9531	47/64	.734375	18.6531
1/4	.250	6.3500	3/4	.750	19.0500
17/64	.265625	6.7469	49/64	.765625	19.4469
9/32	.28125	7.1438	25/32	.78125	19.8438
19/64	.296875	7.5406	51/64	.796875	20.2406
5/16	.3125	7.9375	13/16	.8125	20.6375
21/64	.328125	8.3384	53/64	.828125	21.0344
11/32	.34375	8.7313	27/32	.84375	21.4313
23/64	.359375	9.1281	55/64	.859375	21.8281
3/8	.3750	9.5250	7/8	.8750	22.2250
25/64	.390625	9.9219	57/64	.890625	22.6219
13/32	.40625	10.3188	29/32	.90625	23.0188
27/64	.421875	10.7156	59/64	.921875	23.4156
7/16	.4375	11.1125	15/16	.9375	23.8125
29/64	.453125	11.5904	61/64	.953125	24.2094
15/32	.46875	11.9063	31/32	.96875	24.6063
31/64	.484375	12.3031	63/64	.984375	25.0031
1/2	.500	12.7000	1	1.000	25.4000

Weights of Materials

Material	Avg. Lbs. per Cu. Ft.	Avg. Kg. per Cu. Meter
Aluminum	167.1	2676
Brass, cast	519	8296
Brass, rolled	527	8437
Brick, common and hard	125	2012
Bronze, copper 8, tin 1	546	8754
Cement, Portland, 376 lbs. net per bbl	110 – 115	1765 – 1836
Concrete, conglomerate, with Portland cement	150	2400
Copper, cast	542	8684
Copper, rolled	555	8896
Fibre, hard	87	1377
Fir, Douglas	31	494
Glass, window or plate	162	2577
Gravel, round	100 – 125	1586 – 2012
Iron, cast	450	7201
Iron, wrought	480	7695
Lead, commercial	710	11,367
Mahogany, Honduras, dry	35	564
Manganese	465	7448
Masonry, granite or limestone	165	2648
Nickel, rolled	541	8649
Oak, live, perfectly dry .88 to 1.02	59.3	953
Pine, white, perfectly dry	25	388
Pine, yellow, southern dry	45	706
Plastics, molded	74 – 137	1200 – 2187
Rubber, manufactured	95	1518
Slate, granulated	95	1518
Snow, freshly fallen	5 – 15	70 – 247
Spruce, dry	29	459
Steel	489.6	7837
Tin, cast	459	7342
Walnut, black, perfectly dry	38	600
Water, distilled or pure rain	62.4	988
Zinc or spelter, cast	443	7095

Index

A

acme thread form, 128
aligned dimensioning system, 31, 32
allowances, 190
alloy steels, 156, 157
alloys, characteristics of, 160
aluminum, 158
angularity, orientation tolerances, 201
annealing, 161, 162
applications block, 6
arrowheads, microfilm, 11-12
artist's name, 5
assembly drawings, 217-218
assembly section views, 95, 98
auxiliary views, 111-121

B

babbitt, 160
balloons, 97
bevel groove weld joint, 176, 177
blind holes, 53, 131-133
bolt circles 42-43
brass, 159
break lines, 21, 25-27, 112-113
Brinell hardness testing, 162
brittleness, 161
broken-out section views, 93
bronze, 159
butt joint, welding, 174, 175
buttress thread form, 129

C

call-offs
  holes, 53-57
  thread, 134, 135
cap screws, 136
carbon steel, 156, 157
carburizing, 161, 162
case hardening, 161, 162
cast iron, 155, 156
center lines, 21, 23
chamfer, threads and fasteners, 134, 135
change block, 35
checked date, 5
chromium steel, 157, 158
circles, holes and bolt circles, 42-43
circular runout, 207
circularity, form tolerances, 199
clearance fit, 191
clearance, 190-191
company name and address, 1
concave welding, 176, 177
concentricity, location tolerances, 203
conductivity, 161
contour symbols, welding, 176, 177, 182-183
contract number, 5
conversion chart, 13, 14, 15, 261
convex welding, 176, 177
copies, folding, 1, 3
copper, 159
corner joint,welding, 174, 175
corrosion resistance, 161
counterbored holes, 53, 54
countersunk holes, 53, 54
crest, threads, 126
cutting plane lines, 21, 27, 71, 72
cylindricity, form tolerances, 199

D

dadum planes, 150
dates, 5
datum feature symbol, tolerances, 194
decimal values, 13-20
decimal-inch scale, 16-17
depth of thread, 126
design layout drawings, 217
design size tolerances, 189
detail drawings, 220
diameters
  dimensioning, 33, 35
  major vs. minor, threads, 126
  pitch, threads, 126
dies, 123, 124
dimension lines, 21, 24
dimensions, 5
  aligned system, 31, 32
  basic, tolerances, 194
  diameters, 33, 35
  holes, 42-43
  kinds of, 31, 32
  locations, 31, 32
  out-of-scale, 34, 36
  radius, 33-34
  reference, 35, 37
  size, 31, 32
  spot welding, 181
  typical, 36
  unidirectional system, 31, 32
  welding, 169-171
double weld, 176
drawing number/title, 2
drawings, 217-239
drill and tap sizes, 125-126
ductile cast iron, 156
ductility, 161

E

edge joint, welding, 174, 176
edge views, 112
elasticity, 161
engineering change order (ECO), 222
engineering change request (ECR), 221
extension lines, 21, 23
  surface finishes, 141, 143, 144
external threads, 126

F

fasteners (see also threads and fasteners), 93, 96-97, 123
feature control frame, tolerances, 195
ferrous metals, 255-156
field welding, 173-174
fillet welding, 86, 87, 168, 172
fillister head screws, 136
finish requirements, 5
finish symbols, welding, 182-183
finished contour welding, 176, 177
finishing, surface finishes, 141, 143, 144
fit, 191
flange welding, 168, 179
flare bevel groove weld joint, 176, 177
flat head screws, 136
flat key, 142
flatness, form tolerances, 196
flush welding, 176, 177
folding copies, 1, 3
forge welding, 167
form tolerances, 189, 192, 196
  circularity, 199
  cylindricity, 199
  flatness, 196
  straightness, 196
fractional-inch scale, 14-16
fractions, conversions, 13
front view, 37, 47, 52, 58, 61
FSCM block, 5
full indicator movement (FIM), 207
full indicator reading (FIR), 207
full section views, 74-75, 74
fusion welding, 167

G

gib head key, 142
gray cast iron, 155, 156
groove welding, 168

H

half section views, 78-79

hardening, 161
hardness testing, 162
heat treatment, 5, 161
hidden lines, 21, 22
  full section views, 75
  half section views, 79
high carbon steel, 156, 157
holes
  blind, 53
  blind, threaded, 131, 132, 133
  call-offs, 53-57
  counterbored, 53, 54
  countersunk, 53, 54
  dimensioning, 42-43
  locating, 42-43
  section views, 102-105
  slot, 55
  spot-faced, 54-55
  threaded, 131, 132, 133
  thru, 53
  thru, threaded, 131, 132, 133

I

indents, 222
interference fit, 191
internal threads, 126
iron, 155, 156
issuance date, 5

J

J-groove weld joint, 176, 177
joints, welding, 174, 175
  grooved, 176, 177

K

keys, 142
knurling, 141, 143

L

lap joint, welding, 174, 176
lead, 160
leader lines, 21, 25
least material condition (LMC) tolerances, 190, 193, 198-199
left-hand threads, 123, 124
limits, tolerances and, 7
line of sight, one-view drawings, 37
lines, 21-29
location tolerances, 189, 192, 196
  concentricity, 203
  symmetry, 206
  true position, 203-205
lower limits, 7

M

machine screws, 136
magnesium, 160
major diameter, threads, 126
malleability, 161
malleable iron, 155, 156
manganese steel, 157, 158
master parts lists (MPL), 222
materials list, 5
materials, weights of, 263
maximum material condition (MMC) tolerances, 189-190, 193, 198
maximum size tolerances, 189
medium carbon steel, 156, 157
metallurgy, 155-165
metric drawings, 17
metric measurements, conversions, 13
metric scale, 17
metric thread, 128
microfilming, arrowheads for, 11-12
mild carbon steel, 156, 157
minimum size tolerances, 189
minor diameters, 126
modified purchased parts drawings, 220
modifiers, tolerances, 196-200
molybdenum steel, 157, 158
multiple threads, 130
multiview drawings, 47-70

N

name of artist, 5
nickel, 160
nickel steel, 158
nickel-chromium steel, 157, 158
nonferrous metals, 158-160
normalizing, 161, 162

O

object lines, 21
offset section views, 82-85
one-view drawings, 31-45
orientation tolerances, 189, 192, 196
  angularity, 201
  parallelism, 200
  perpendicularity, 201
out-of-scale dimensions, 34, 36

P

paper, 1, 2
parallelism, orientation tolerances, 200
partial auxiliary views, 112
parts lists, 6, 97, 222, 223
permanent fasteners, 123
perpendicularity, orientation tolerances, 201
pewter, 160
phantom lines, 21, 25
pictorial view, 50, 58, 61
pig iron, 155, 156
pitch, 126, 129-130
pitch diameter, 126
plasticity, 161
plug or slot welding, 168, 176, 177, 178
Pratt & Whitney key, 142
profile tolerances, 189, 192, 196, 200
projection lines, 21, 27, 47-49, 61
purchased parts drawings, 220

R

radius
  dimensioning, 33-34
  fillets and, 86, 87
  rounds and, 86, 87
reference dimensions, 35, 37
reference lines, multiple, welding, 168, 180
relief, threads and fasteners, 132, 134
removable fasteners, 123
removed section views, 86
resistance welding, 167
revisions block, 5-6, 221
revolved section views (see rotated section views)
ribs, section views, 98, 102
right-hand threads, 123, 124
right-side view, 47, 58, 61
rivets, 134, 136-137
Rockwell hardness testing, 162
roll stock, 1
root, threads, 126
rotated or revolved section views, 90-92
round head screws, 136
rounding off decimals, 13-20
rounds, 86, 87
runout tolerances, 189, 192, 196
  circular, 207
  total, 207
runouts, 86, 87, 88

S

scale, 3, 5, 14-17
schematic representation, threads and fasteners, 130
screws, 134, 136
  cap, 136
  head forms, 136
  machine, 136
  representation of, 130
  thread forms, 128-129
seam welding, 183
section lining lines, 21, 27, 71, 73
section views, 71-109
series of thread, 126
shafts, section views, 93, 96, 97
sheet number block, 3
simplified representation, threads and fasteners, 130
single threads, 130
size, drawing, 3, 31, 32
slot holes, 55
socket head screws, 136
spokes, section views, 102-105
spot welding, 168, 180-182
spot-faced holes, 54-55
square groove weld joint, 176, 177
square key, 142
square thread, 128
standards, title block, 1
steel, 156, 157
  numbering system for, 158
  properties, grade numbers, usages, 159
straightness, form tolerances, 196
strength, 161
subassembly drawings, 219
surface finishes, 141, 143
  machine characteristics of, 144
symmetry, location tolerances, 206

T

T-joint, welding, 174, 176
tabulated drawings, 144-145
taps, 123, 124
  drill sizes and, 125-126
technical information, 141-153
tempering, 161, 162
thickness of lines, 21
thin wall section views, 93-96

threads and fasteners, 123-139
three-view drawings, 58
thru holes, 53
  threaded, 131, 132, 133
tin, 160
title block, 1-12
tolerances, 6-9, 189-215
  allowances, 190
  basic dimensions, 194
  clearance, 190-191
  datum feature symbol, 194
  decimal notation of, 8
  design size, 189
  feature control frame, 195
  fit, 191, 192
  form, 189, 192, 196
  fractions in, 8
  full indicator movement (FIM), 207
  full indicator reading (FIR), 207
  geometric characteristic symbols, 196
  interference fit, 191
  least material condition (LMC), 190
  location, 189, 192, 196, 203
  maximum material condition (MMC), 189-190, 193
  minimum and maximum size, 189
  modifiers, 193, 196-200
  orientation, 189, 192, 196, 200
  profile, 189, 192, 196, 200
  required, 207
  runout, 189, 192, 196, 207
  symbols for, 192
  tolerance zones, 205, 206
  transition fit, 192
  upper and lower limits, 7
tool steel, 156, 157
top view, 52, 58
total runout, 207
toughness, 161
transition fit, 192
true position, location tolerances, 203-205
true size/shape projections, 111
tungston steel, 157, 158
two-view drawings, 47, 52

U

U-groove weld joint, 176, 177
undercut, threads and fasteners, 132, 134
unidirectional dimensioning system, 31, 32
unified national thread, 128
upper limits, 7

V

V-groove weld joint, 176, 177
vanadium steel, 157, 158
visible lines, 21, 22

W

webs, section views, 98, 102
weights of materials chart, 263
welding, 167-187
white cast iron, 155, 156
Woodruff key, 142
worksheet answers, 241-260
worm thread form, 129
wrought iron, 156

Z

zinc, 160
zoning, 2, 9-10